A PROBLEM-BASED INTRODUCTION TO
SPECIAL AND GENERAL RELATIVITY

**Including Worked Problems and Discussions of
the Ehrenfest Paradox and the Maxwell Equations**

A PROBLEM-BASED INTRODUCTION TO SPECIAL AND GENERAL RELATIVITY

Including Worked Problems and Discussions of the Ehrenfest Paradox and the Maxwell Equations

An introduction to the connections between space and time, the geometry of physical reality, and the inevitability of magnetism.

George D Gollin

University of Illinois at Urbana-Champaign, USA

World Scientific

NEW JERSEY · LONDON · SINGAPORE · BEIJING · SHANGHAI · HONG KONG · TAIPEI · CHENNAI · TOKYO

Published by

World Scientific Publishing Co. Pte. Ltd.

5 Toh Tuck Link, Singapore 596224

USA office: 27 Warren Street, Suite 401-402, Hackensack, NJ 07601

UK office: 57 Shelton Street, Covent Garden, London WC2H 9HE

Library of Congress Control Number: 2025027422

British Library Cataloguing-in-Publication Data
A catalogue record for this book is available from the British Library.

A PROBLEM-BASED INTRODUCTION TO SPECIAL AND GENERAL RELATIVITY
Including Worked Problems and Discussions of the Ehrenfest Paradox and
the Maxwell Equations

ISBN 978-981-98-1150-2 (hardcover)
ISBN 978-981-98-1239-4 (paperback)
ISBN 978-981-98-1151-9 (ebook for institutions)
ISBN 978-981-98-1152-6 (ebook for individuals)

For any available supplementary material, please visit
https://www.worldscientific.com/worldscibooks/10.1142/14266#t=suppl

Desk Editors: Nambirajan Karuppiah/Carmen Teo Bin Jie

Typeset by Stallion Press
Email: enquiries@stallionpress.com

Preface

It is Einstein's Theory of Relativity that describes how space and time are braided together into the threads from which the fabric of existence is woven and from which life can arise. A Newtonian universe would be sterile, unable to support the rich, complicated biochemistry of living things. And a flat space–time, in which the geometrical disruptions of general relativity were absent, would be empty of the binary neutron star systems that undergo gravitational collapse and seed the universe with elements too heavy to have been created by supernovas.

This book carries the entire content—including hundreds of worked problems and examinations (the exams, without solutions, are available online)—of a stand-alone course in special and general relativity that I taught a half dozen times at the University of Illinois. The course is usually taken by undergraduate physics majors during their third semester. It is, to my mind, the first time our students see why those of us who have made a career of physics did so. It opens their eyes to the hallucinatory strangeness of physical reality.

This book's first seven units—which comprise half the course—cover special relativity. I start with the fact that the speed of a light beam traveling through empty space past an observer is *always* (seen by the observer to be) exactly 299,792,458 meters per second. Careful analysis of the consequences leads to time dilation, Lorentz contraction, the loss of simultaneity between separated events, and so forth. I develop some of the mathematical tools that will be useful later, including four-vectors and Lorentz tensors, and introduce relativistic kinematics.

The next three units are an introduction to general relativity. I take a very geometrical approach, moving from Euclidean geometry to Elliptic geometry and introducing metric tensors and curvature. I introduce the Einstein equations and the Schwarzschild metric, the stress–energy tensor and Kerr metric, parallel transport, black holes, and (free-falling) motion in curved space–time. To my surprise, most of my students felt they became more comfortable with general than special relativity by the end of the course!

In the final four units, I develop the vector calculus needed to produce a covariant formulation of electrodynamics, in which the four Maxwell equations reduce to a pair of seemingly simple equations about the electromagnetic field strength tensor and its dual.

This is a wonderful subject, both to learn and to teach.

George Gollin
September 2025

About the Author

George Gollin grew up in Freeport, New York. He attended Harvard University as an undergraduate, receiving an AB in physics in 1975, and then Princeton University as a graduate student, earning a PhD (also in physics) in 1981. After a period as a Robert R. McCormick postdoctoral fellow at the University of Chicago, he returned to Princeton as an assistant professor in 1983. Gollin moved to the University of Illinois at Urbana-Champaign as an associate professor in 1989, becoming a full professor in 1996. He and his wife live in Champaign and have one daughter.

His research in experimental elementary particle physics has included participation in experiments at the Fermi National Accelerator Laboratory (near Chicago), the Cornell Electron Storage Ring (at Cornell University), and the European Organization for Nuclear Research (known by its French acronym, CERN, near Geneva, Switzerland). His scientific interests have tended to focus on questions concerning the structure and origins of the equations describing the interactions between matter and energy, as well as the fundamental nature of space and time at very small distance scales. For a few years, he collaborated in efforts to design the International Linear Collider, a very large electron-positron accelerator that would, if ever built, provide insight into the origins of particle masses and the differences between the nuclear and electromagnetic forces. In recent years, he has turned his attention towards the development of small medical devices enabled by the rapidly evolving landscape of precision sensors.

Gollin teaches both undergraduate and graduate students and works occasionally with faculty in the departments of Dance, Architecture, Library Science, and Speech Communication on interdisciplinary projects. He has received teaching awards for his leadership in restructuring the undergraduate physics curriculum at Illinois. Most recently, he designed and commissioned the Physics Department's new Master's in Instrumentation and Applied Physics and served as its founding director.

Between 2004 and 2008, Gollin assisted the "Operation Gold Seal" task force that dismantled the fraudulent "St. Regis University" and convicted its owners and staff of mail fraud and other felonies. Gollin wrote thousands of lines of analysis software for the Department of Justice and organized much of the electronic evidence seized from the defendants' computers.

Press coverage of the case led to new federal legislation. Gollin worked with Congressional staff as they developed the "Diploma Integrity Protection Act of 2007." Part of the bill was included in the Higher Education Opportunity Act, which became law in 2008.

Gollin was elected to the Board of Directors of the Council for Higher Education Accreditation in 2006, served on UNESCO's diploma mills working group in 2008, and regularly speaks on the subject of academic fraud. This work has led to appearances on CNN, CBS, Fox, and NPR as well as profiles in the Chicago Tribune and Wired magazine.

In 2009, Gollin received a John Simon Guggenheim Memorial Foundation Fellowship to support research for writing an account of the St. Regis episode and its harm to post-civil war Liberia.

Contents

Marseille, France

Unit 1

Special Relativity: Time Dilation and Lorentz Contraction

But why?

Quite some years ago, one of my professors would tell us that "Physics is supposed to make your head feel funny."[1] He was right:

---

[1]Sidney Coleman (1937–2007), "Sidney Coleman dies at 70: Luminary of theoretical physics was a member of the Harvard faculty for 43 years," *The Harvard Gazette*, November 29, 2007. (https://news.harvard.edu/gazette/story/2007/11/sidney-coleman-dies-at-70/).

those of us who've become physicists didn't do so because of a burning desire to analyze an object of mass m sliding down an inclined plane for the rest of our careers. Early on, we learned that there are much more interesting things to think about. We know that a Newtonian universe cannot support life and that a Newtonian space–time would render the Earth as hot as the surface of a star.[2] Why is it possible for carbon-based life to exist? Why isn't the Earth just a cloud of superheated plasma? And non-relativistic quantum mechanics does not resolve these difficulties: organic chemistry, without the purely relativistic Pauli Exclusion Principle,[3] just doesn't work.

So what's going on?

We are accustomed to observers sometimes disagreeing about distance intervals. If you reserve a sleeping compartment in the *City of New Orleans* and settle in for the long ride to Chicago, you'll have spent the night in a comfortable bed and awakened in the same compartment you had entered in New Orleans.

As far as you're concerned, the distance between where you went to sleep and where you woke up is at most a few feet. But an observer at rest in Union Station will conclude that the distance between your starting and stopping points was about 900 miles.

This is not a big deal; it's something we're used to.

But what we expect (in a world governed by Newtonian Physics) is that we'll agree *exactly* with the Chicago-based observer that the duration of our trip was 19 hours.

In a relativistic world, the train passenger and the Chicago resident will disagree—only slightly, but disagree nonetheless—about the duration of the journey.[4] That takes some getting used to, but it really happens, and it is why Newtonian Physics can't really be

---

[2] "Olber's Paradox: Why Is The Sky Dark at Night?," American Museum of Natural History, https://www.amnh.org/exhibitions/journey-to-the-stars/educator-resources/stars/olbers-paradox, visited September 5, 2024.

[3] "January 1925: Wolfgang Pauli announces the exclusion principle," American Physical Society Archives, https://www.aps.org/archives/publications/apsnews/200701/history.cfm, visited September 5, 2024.

[4] The disagreement is about a third of a nanosecond.

the way of the world. And it is the nature of space and time, that admits of this small disagreement, that changes everything.

It is Einstein's Theory of Relativity that describes how space and time are braided together into the threads from which the fabric of existence is woven and from which life can arise.

Here are the (observational and intellectual) underpinnings of the entire subject, in three sentences:

1. An observer in freefall (in a force-free inertial frame of reference) will always measure the speed of light in vacuum to be *exactly* 299,792,458 meters per second.
2. The presence of matter (and energy) in the universe changes the geometry of space–time: the Euclidean circumference-to-radius relationship for a circle is no longer exactly true in a matter-filled universe, and clocks in gravitational fields run slowly.
3. Things tend to roll downhill, jamming more stuff into a suitcase will make it heavier, and electromagnetism is entirely a consequence of special relativity.

The last of these will comprise the mathematics-heavy part of this book. I will develop with you the vector calculus expressions that correspond to statements about flows into/out of volumes and so forth.

I could take an approach in which I explained everything to you in terms of the transformation properties of space and time intervals under changes of reference frame. But an equivalent, and pedagogically more useful tack, is to begin with the constancy of the speed of light.

Each unit will address new material that you'll work with, through "discussion problems" whose solutions are included for you. At the end of each unit are more challenging problems and their solutions. These are the kinds of problems I would assign for homework in a course based on the material in this text; you can expect each of the problems to require anywhere from a few minutes to half an hour to solve. If you can, try working the problems on your own before reading how I would solve them.

By the end of this book, you will know why magnetic fields must exist, what a "rank-4 curvature tensor" is, that $G^{\mu\nu} = 8\pi G T^{\mu\nu}/c^4$, and that there is no such thing as a "gravitational force."

Oh, a note of thanks to one of my teachers. The approach I take in introducing you to some parts of special relativity follows closely how it was originally taught to me, many years ago, by Paul Bamberg at Harvard. Paul was (and continues to be, I am sure) a brilliant, committed teacher. I owe him a great deal.

Introduction: The speed of light is finite, and constant

You're going to love this: its consequences are the weirdest things you'll see in physics before you learn about quantum mechanics.

All the strange features of special relativity come from the fact that space and time conspire to guarantee that the speed of light is *exactly* 299,792,458 meters per second. It doesn't matter if the source of light moves with respect to the measuring apparatus: any device measuring c (the speed of light) will obtain, within experimental error, this value. Nothing else works this way. Sound travels at a fixed speed with respect to the air, for example.

The value 2.99792458×10^8 meters per second really *is* exact: it serves as a definition, along with some time standard, of the length of $1\,\mathrm{m}$. The speed of light is very close to 1 foot (about $30\,\mathrm{cm}$) per nanosecond, so I'll use that as a convenient approximation off and on.

Here's what you'll discover in this unit:

- Events that are simultaneous in one frame of reference are not necessarily simultaneous in another frame.
- The rate at which time passes in a moving frame of reference is reduced.
- Moving objects become shorter along their direction of motion.

I should explain what I mean by an inertial frame of reference. There's a pretty good paraphrase of Landau and Lifschitz's 1960 text *Mechanics* to be found online:

> In classical physics and special relativity, an inertial frame of reference... is a frame of reference not undergoing any acceleration. It is a frame in which an isolated physical object—an object with zero net force acting on it—is perceived to move

with a constant velocity or, equivalently, it is a frame of reference in which Newton's first law of motion holds.[5]

Now, there are lots of words in this description, but the idea is simple: if you find yourself in an environment in which there are no forces acting on anything—no gravity, no electric or magnetic forces, no strap-on rocket engines—then things stay put if they're initially at rest or else move with constant speed and direction if they aren't stationary.

One more thing about the strange word problems I write: I mentioned that my colleagues and I do not burn with desire to analyze inclined planes. That also goes for writing yet again another homework problem in which a boring object of mass m slides down a stultifyingly dull inclined plane. So, I use word problems as vehicles of vengeance, casting people who have behaved badly into unpleasant situations. I've changed their names, of course.

You will meet the "Blepons,"[6] who owned a hardware store in Central New Jersey. When I was a graduate student, Mrs. Blepon would follow me closely whenever I went into her store to buy things to fix stuff in my apartment. She thought I was going to shoplift!

The "Rumpdocks" were American criminals who sold fake university degrees, including MDs. They bought their way into Liberia before the end of that country's decades-long civil war and claimed that most of their fake schools were legitimate African universities. At a time when life expectancy was 39 years and a seventh of the country's children died in infancy, they controlled significant portions of Liberia's diplomatic corps, Ministry of Education, and Ministry of Justice. After I published a report of their activities, they came after me, and their supporters went after my wife and daughter. I helped the Department of Justice send them to prison.[7] So, they too end up in my word problems.

---

[5]Landau, L.D. and Lifshitz, E.M., *Mechanics,* Pergamon Press, pp. 4–6 (1960). https://en.wikipedia.org/wiki/Inertial_frame_of_reference.

[6]A supermarket in Champaign would announce "double coupons" savings, but the announcement's placement in four of the store's picture windows when read left to right was DOU COU BLE PONS.

[7]David Wolman, "Fraud U: Toppling a Bogus-Diploma Empire," *Wired,* December 21, 2009. https://www.wired.com/2009/12/ff-fake-physics/.

Now, it's time to get started. Let's look at this matter of the constancy of the speed of light.

Exercise 1.1: Constant c; simultaneity

(a) Let's do this one from an entirely Newtonian perspective.

Mr. Blepon, sociopathic co-proprietor of Blepon's Deli and Hardware Emporium, is capable of throwing a snowball at 50 miles per hour. While riding as a passenger in their second-hand delivery truck traveling 40 miles per hour, he hurls snowballs at stationary pedestrians ahead of and behind the moving truck.

Calculate the speeds of the snowballs according to the two unfortunate pedestrians shown in the diagram. Assume for the moment that Newtonian physics is accurate.

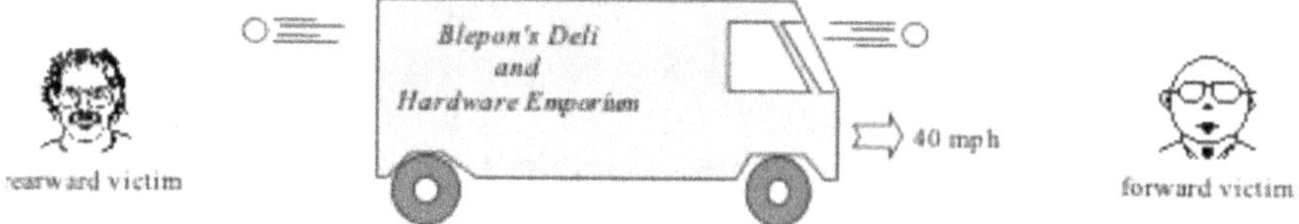

Solution

This one's easy. The velocities of the truck and snowballs need to be added to determine the snowball speeds from the perspective of the forward and rearward victims:

$$v_{\text{forward victim}} = 50 + 40 = 90$$

$$v_{\text{rearward victim}} = 50 - 40 = 10$$

(b) The starship *Nostromo*,[8] drifting in deep space, passes a large interstellar recreational facility which is also adrift, as shown in the figure. Observers on the recreational facility see the *Nostromo* as moving to the right, with speed $0.8c$. While the ship is between the ends of the rec facility, it fires its communications lasers to send signal pulses in the forward and aft directions to receivers built into the ends of the rec facility.

---

[8] *Nostromo* is the title of a 1904 novel by Joseph Conrad. The story unfolds in an imaginary city named Sulaco.

According to technicians *on the recreation facility*, how fast are the two light pulses traveling as they pass through the receiver hardware built into each end of the rec facility? (Naturally, this is a fully relativistic situation.)

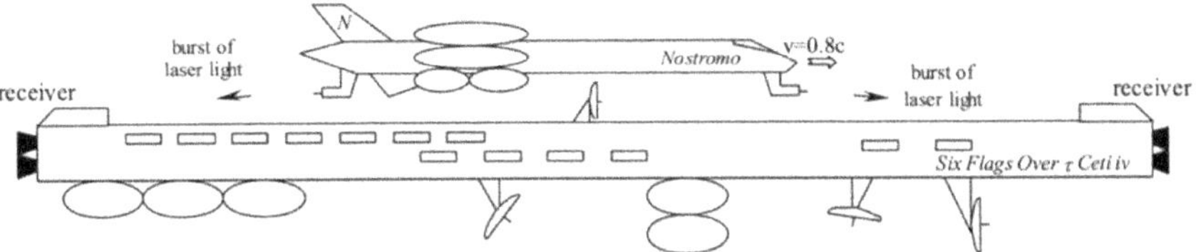

Solution

This one's even easier. The speed of light measured by any observer is always *c*. Note that (slowly moving) snowballs do not behave the same way as bursts of light!

(c) The *Sulaco*, another fast ship, glides past the same drifting interstellar recreational facility. As before, observers on the recreational facility see the ship moving to the right with speed 0.8*c*.

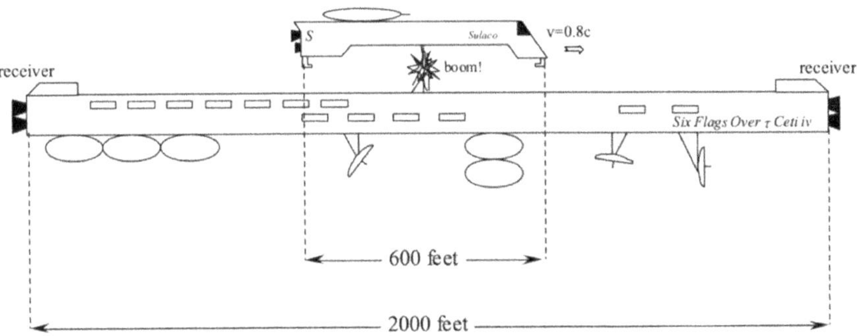

Unfortunately, *Sulaco's* X-band antenna strikes the rec facility's TV parabola, interrupting the final episode of *Pride and Prejudice*. *Sulaco's* wrecked antenna is midway along the ship's length; the rec facility's ruined parabola is also midway along the facility's length. Observers *on the recreation facility* see the *Sulaco's* length as 600 feet (about 180 m). The rec facility's length, according to the manufacturer, is 2000 feet (about 600 m).

According to technicians *on the recreation facility*, when does the flash of light from the collision reach the rec facility's right and left receivers (assume the flash occurs at $t = 0$)?

According to technicians *on the recreation facility*, when does light from the collision reach *Sulaco's aft* periscope (located at the rear of the ship)? When does light reach the forward periscope?

Solution

Let's use $c = 1$ foot per nanosecond. In the rec facility frame, both flashes travel 1000 feet to reach the rec facility receivers, so they arrive simultaneously at $t = 1000$ nanosecond.

The rec facility sees the Sulaco's forward periscope 300 feet from the point of production of the spark and moving to the right at $0.8c$. Solve for the time t when the (moving) forward periscope and light flash coincide:

$$300 + 0.8ct = ct$$

$$300 = 0.2ct$$

$$\frac{1500}{c} = t$$

$$1500 = t$$

since c is 1 foot per nanosecond. (I'm using $x = 0$ as the point of production of the spark.)

Do the same for the aft periscope, which runs into the flash:

$$-300 + 0.8ct = -ct$$

$$-300 = -1.8ct$$

$$166.67 = t$$

It is unsurprising that the aft periscope, which runs into the flash, sees the flash (according to the rec facility!) before the forward periscope.

(d) From the perspective of the *Sulaco*, the collision happens because the incompetently piloted recreational facility smashes into *Sulaco's* antenna as it lumbers past at $0.8c$, moving to the left, as shown in the following figure.

According to cosmonauts *on the Sulaco*, does the flash of light from the collision reach the *Sulaco's* forward and aft periscopes simultaneously?

According to cosmonauts *on the Sulaco*, does the flash of light from the collision reach the *rec facility's* right and left receivers simultaneously?

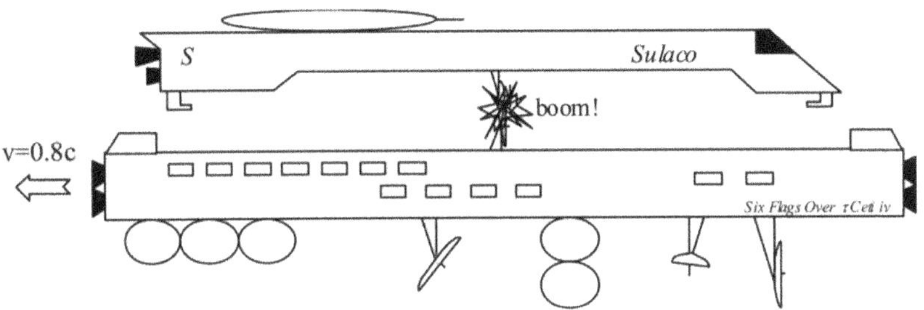

Solution

From *Sulaco's* perspective, both flashes travel half the (rest) length of *Sulaco* and must arrive at *Sulaco's* periscopes simultaneously. But since *Sulaco* sees the rec facility moving to the left, the flash (according to *Sulaco!*) takes longer to catch up to the rec facility's left receiver than it does to run into the facility's right receiver.

Discussion

Imagine we use the receipt of the light flash as a way to synchronize clocks on the rec facility: we set a pair of *stopped* clocks, one at each end of the rec facility, to $t = +1000$ nanosecond. When the light flash arrives at the left end, we start the left clock ticking. When the light flash arrives at the right end, we start the right clock ticking.

Any other sensible method the staff of the rec facility might use to synchronize their clocks—for example, placing both of them at the middle of the rec facility, starting them simultaneously, then transporting each (at identical speeds) to the left and right ends of the rec facility—will agree with the light flash method.

But how do the rec facility clocks look to *Sulaco's* crew? And how do *Sulaco's* clocks look to the rec facility staff?

In general, the crew of one spacecraft will see the moving clocks on the other spacecraft begin ticking at different times: the forward clock is running away from the synchronizing light pulse, while the aft clock is running into it and starts ticking sooner. As a result, the aft clock will be seen to read a later time than the forward clock since it began ticking sooner.

Exercise 1.2: A light clock; time dilation

Let's analyze the rate of passage of time when observers in one frame of reference look at moving clocks in a different frame of reference.

(a) Mrs. Blepon, the irritable but surprisingly fit co-proprietress of Blepon's D&HE, drifts in deep space aboard the gigantic (second-hand) starship *Vulcan Flashrocket*. Since she realizes that the speed of light is approximately one foot (30 cm) per nanosecond, Mrs. Blepon constructs a clock using two mirrors, separated by a distance of 500 million feet (about 150 million meters), and a bouncing light pulse. The pulse is of brief duration—the laser that produced it fired for only a nanosecond—and bounces between the mirrors without attenuation.

As shown in the figure, she holds the lower mirror in one hand, moving it *briefly* into place to reflect the descending light beam. After each successful reflection, she removes the mirror. Since Mrs. Blepon knows her resting pulse rate, she has constructed the device so that she will need to move the mirror into position immediately after each heartbeat. It is fortunate that she does not suffer from cardiac arrhythmia since the bomb stored below the light clock will detonate if its optical sensor is struck by the light beam.

What is Mrs. Blepon's pulse rate?

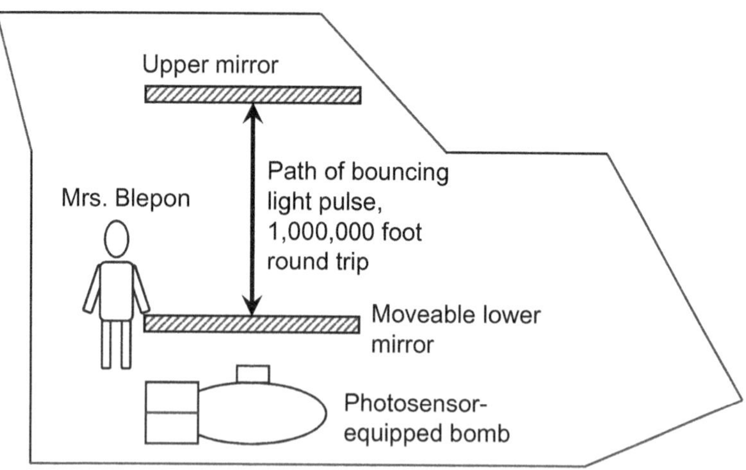

Solution

This is another easy one. The round-trip distance for the light beam is a billion feet, so it takes one second. As a result, Mrs. Blepon's pulse rate is one beat per second, or 60 beats per minute.

(b) The *Flashrocket* is passed by that hazardous interstellar recreational facility while Mrs. Blepon is occupied with her light clock. Fortunately, no collision takes place. From the perspective (the "frame of reference") of technicians on the rec facility, the *Flashrocket* is seen to glide past the facility at speed v, as shown in the following figure.

As seen from the rec facility's perspective, does Mrs. Blepon's bomb detonate?

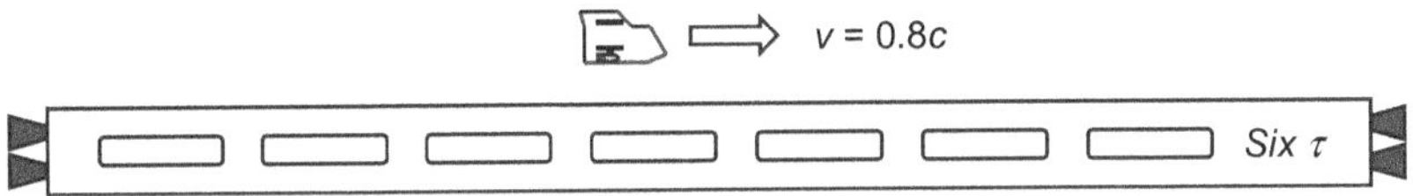

Solution

Of course not. Observers in any frame of reference will agree that the unpleasant Mrs. Blepon continues without pause to shriek insults (though possibly Doppler shifted) into her long-range communications system. This is frame-independent: if she doesn't ever blow up when observed from one frame of reference, she will remain entirely intact when observed from any other frame of reference.

(c) As seen from the recreational facility, Mrs. Blepon's light pulse travels along a diagonal path, as shown in the following. Keep in mind that the *Flashrocket* is seen to move with speed v and that the light pulse is seen to travel approximately one foot (30 cm) per nanosecond.

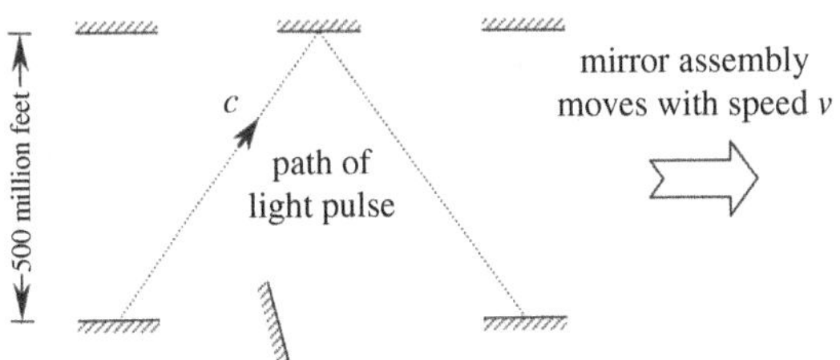

How long is the path *followed by the light pulse* during one lower-mirror-to-upper-mirror-to-lower-mirror trip? (Express your answer

in terms of v, c, and so forth.) If $v = 0.8c$, how long does it take for the light pulse to make the bottom-to-top-to-bottom trip?

Solution

The path is longer, of course: the light travels along the hypotenuses of right triangles whose vertical legs are 500 million feet. Let's call the time for the light beam to follow the path in the above diagram $\Delta t'$. The distance the beam goes as it travels from the lower mirror to the upper mirror is $c\Delta t'/2$.

The Pythagorean theorem to the rescue! We have

$$\left(c\frac{\Delta t'}{2}\right)^2 = 500{,}000^2 + \left(v\frac{\Delta t'}{2}\right)^2$$

(the last term is the distance the ship travels in time $\Delta t'/2$) so that

$$\left[\left(\frac{c}{2}\right)^2 - \left(\frac{v}{2}\right)^2\right](\Delta t')^2 = 500{,}000^2$$

$$(\Delta t')^2 = \frac{10^9}{c^2 - v^2} = \left(\frac{10^9}{c}\right)^2 \frac{1}{1 - v^2/c^2}$$

$$\Delta t' = \frac{10^9}{c} \frac{1}{\sqrt{1 - v^2/c^2}}$$

Plug in to find $\Delta t' = 5/3$ seconds. The rec facility staff see Mrs. Blepon's heart beating slower so that her pulse is 36 beats per minute of rec facility time.

(d) If we define Δt to be the time required for one bottom-to-top-to-bottom light pulse cycle *in Mrs. Blepon's frame of reference* and $\Delta t'$ to be the time required for one cycle according to *observers on the recreational facility*, what is the ratio $\Delta t'/\Delta t$? Express your result in terms of v and c.

Solution

We've already figured this out:

$$\Delta t'/\Delta t = (5/3)/1 = 1/\sqrt{1 - v^2/c^2}$$

Discussion

- To Mrs. Blepon, everything seems fine. To observers in the other frame, Mrs. Blepon is doing everything slowly.
- This MUST be true if something (light) exists which has the same velocity in all frames regardless of any source–observer relative motion.
- All systems used to measure time are affected in the same way (heartbeat, light clocks, mechanical clocks, etc.).

Exercise 1.3: Measuring the length of a moving object using a single clock

(a) Floating comfortably aboard the *Vulcan Flashrocket,* Mrs. Blepon measures the length of the *Nostromo* by timing how long it takes for the *Nostromo* to coast past her position, moving to the right at speed v.

Here's how Mrs. Blepon's length measurement looks to the crew of the *Nostromo:* they see the *Flashrocket* gliding past with speed v and, naturally, see that Mrs. Blepon's clock is ticking slowly. The *Nostromo* crew can measure the length of their own ship (let's call it L) by seeing how long it takes the *Flashrocket* to fly the full length of the *Nostromo.* (Let's call the time required Δt.)

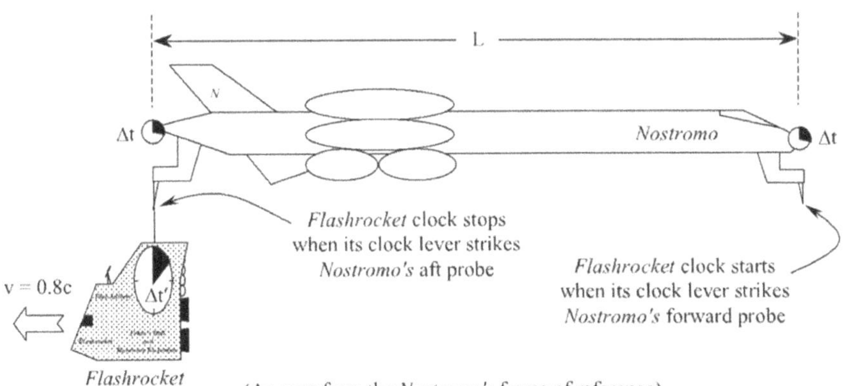

(As seen from the *Nostromo's* frame of reference)

What is L in terms of Δt and so on? (L is *Nostromo's rest length.* This is a ridiculously easy problem!)

Solution

Velocity multiplied by time is distance, of course!

$$L = v\Delta t$$

(b) In Mrs. Blepon's rest frame, the measurement looks as shown in the following figure. Her clock is started and stopped by the probes mounted near *Nostromo*'s nose and tail. Naturally, she sees that *Nostromo*'s clocks (not shown) are ticking slowly and are not synchronized. (Note that *Nostromo*'s aft clock must still read Δt, however, as it passes Mrs. Blepon.)

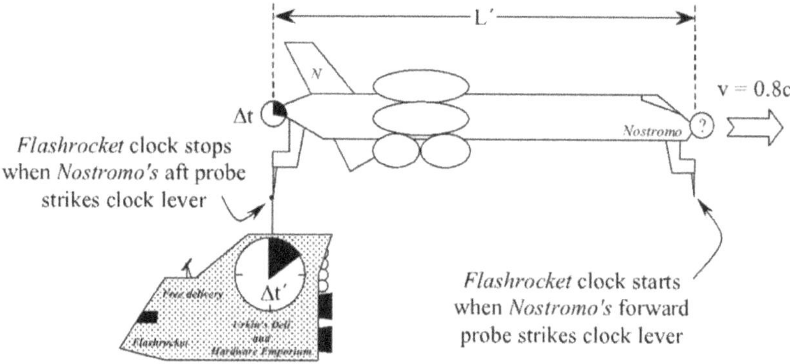

(As seen from the *Flashrocket's* frame of reference)

Define L' to be the *Nostromo* length measured by Mrs. Blepon; the time interval which passed on the *Flashrocket* clock is $\Delta t'$. In terms of v and $\Delta t'$, what is the value she obtains for L'? (Clearly, we have another difficult problem here!)

Solution

$$L' = v\Delta t'$$

(c) What is the ratio L'/L, expressed only in terms of v and c? (Lorentz contraction!) Keep in mind that $\Delta t' < \Delta t$.

Solution

$$\frac{L'}{L} = \frac{v\Delta t'}{v\Delta t} = \sqrt{1 - v^2/c^2}$$

Discussion

All methods of performing a length measurement (tape measure, timing how long it takes a moving object to coast past a fixed point, etc.) will yield consistent results in one frame of reference.

We have found that the moving *Nostromo* is shorter in Mrs. Blepon's frame than in *Nostromo's* rest frame.

Exercise 1.4: Is there any "transverse" contraction?

Lengths parallel to the direction of motion become shorter as $v \to c$. What happens to lengths perpendicular to the direction of motion?

Consider a long stick that carries a pair of paint balls, spaced along the edge of the stick that is perpendicular to its velocity. The stick crashes into a stationary ruler, as shown in the figure.

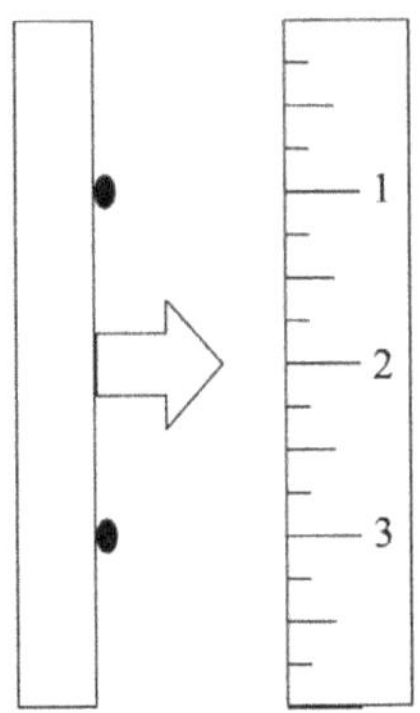

paint balls on moving stick

How far apart are the paint marks which are left on the ruler?

Solution

It's just the way it appears in the figure: the marks are 2 units apart.

Imagine that a second observer views the same process (the very same ruler–stick collision shown in the above diagram) but from a frame in which the stick is at rest and the ruler is in motion.

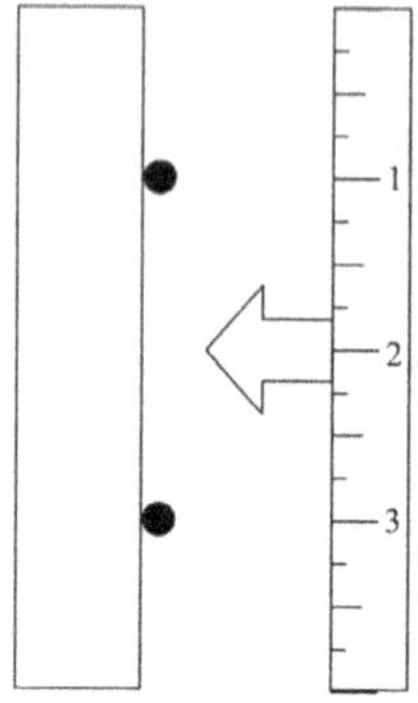

paint balls on stationary stick

How far apart will this observer think the paint marks are? What conclusion can you draw concerning contraction perpendicular to the direction of motion from this system?

Solution

Same answer: since the marks end up at the 1 and 3 marks on the ruler, that'll be what all observers in all frames will see. There's no contraction perpendicular to the direction of the relative velocity of the two frames.

Summary and wrap-up discussion

- Time is messed up:
  1. Moving clocks take longer per tick by a factor of $1/\sqrt{1 - (v^2/c^2)}$.
  2. Simultaneity doesn't work: clocks "at the front" read earlier times than clocks "at the back"

- All time references are affected identically: light clocks, heartbeat, etc.
- Conclusions are valid if *anything* (light, neutrinos, gravitons, etc.) is found to move with constant velocity, independent of source–observer motion.
- Some things don't change when one switches frames of reference (e.g. the bomb did/didn't explode).

- Length is messed up: moving objects are shorter by a factor of $\sqrt{1 - (v^2/c^2)}$ but only along their direction of motion.

Supplemental reading and problems

Einstein's Dreams, Alan Lightman. Warner Books (paperback), 179 pages, 1994. Comments from Amazon.com: "The book takes flight when Einstein takes to his bed and we share his dreams, 30 little fables about places where time behaves quite differently. In one world, time is circular; in another a man is occasionally plucked from the present and deposited in the past: "He is agonized. For if he makes the slightest alteration in anything, he may destroy the future ... he is forced to witness events without being part of them ... an inert gas, a ghost ... an exile of time."[9]

Einstein's Bridge, John Cramer. Avon Books (paperback), 1998. A high-energy physics experiment at the SSC lab in Texas opens a channel into another universe. There are good things and bad things lurking out there, and both find their way through the portal. I suppose this really is about a quantum version of General Relativity rather than Special Relativity, but it was really fun to read.

Problem 1: K_S meson decay

Neutral K mesons ("kaons") are unstable particles composed of a quark and an antiquark. They can be produced copiously in energetic collisions between stable particles. The neutral kaon mass is roughly half the proton mass; short-lived neutral kaons usually decay into pairs of (lighter) pi mesons through the decays $K_S^0 \to \pi^+\pi^-$ and $K_S^0 \to \pi^0\pi^0$. The average K_S^0 lifetime for a kaon *at rest* is 0.8935×10^{-10} seconds. Naturally, a kaon moving at relativistic speed will tend to live longer due to time dilation.

A beam of fast K_S is produced at a national laboratory. The average distance traveled by a kaon before decaying in flight is found to be approximately $9.2\,\mathrm{cm}$. How fast are the kaons moving? (They are *not* traveling faster than c!!) You should assume all kaons are

---

[9]https://www.amazon.com/Einsteins-Dreams-Alan-Lightman/dp/04 46670111.

moving at the same speed v and use the value $c = 3 \times 10^{10}$ cm/sec in your calculations.

Solution

Let's call D the distance a kaon travels in the lab frame and Δt the time until it decays, also in the lab frame:

$$\Delta t = \frac{\Delta t'}{\sqrt{1 - v^2/c^2}} \quad \text{with } \Delta t' = 0.8935 \times 10^{-10}$$

$$D = v\Delta t = v \cdot \frac{0.8935 \times 10^{-10}}{\sqrt{1 - v^2/c^2}} = 9.2$$

$$9.2^2 = v^2 \cdot \frac{(0.8935 \times 10^{-10})^2}{1 - v^2/c^2}$$

$$9.2^2 - 9.2^2 v^2/c^2 = v^2 \cdot (0.8935 \times 10^{-10})^2$$

$$9.2^2 = v^2 \cdot \left\{ (0.8935 \times 10^{-10})^2 + \frac{9.2^2}{c^2} \right\}$$

$$\frac{9.2^2}{(0.8935 \times 10^{-10})^2 + \frac{9.2^2}{c^2}} = v^2$$

$$\frac{1}{(\frac{0.8935 \times 10^{-10}}{9.2})^2 + \frac{1}{9 \times 10^{20}}} = v^2$$

$$\frac{10^{20}}{(\frac{0.8935}{9.2})^2 + \frac{1}{9}} = v^2$$

$$2.88 \times 10^{10} = v$$

So, that's about 96% of the speed of light. Pretty fast!

In this problem, I had wanted you to use in your calculation the fact that fast-moving kaons live longer in the lab frame than they do in their rest frame.

Problem 2: The space telescope sees clearly

Two photons from a pair of faint, distant objects simultaneously enter the open end of the Hubble Space Telescope's barrel. One of the objects is moving at $0.5c$ directly *toward* the telescope, while the other is moving at $0.5c$ directly *away from* the telescope. Assuming that the distance from the telescope's open end to its primary mirror is 20 feet (6 m), calculate how long it takes for each of the two photons to arrive at the primary mirror after entering the open end of the telescope barrel as determined by observers in the rest frame of the telescope.

Solution

The photons arrive simultaneously at the primary mirror: all photons travel at the same speed. Using 1 foot/nanosecond as the speed of light, they arrive 20 nanosecond after entering the barrel of the telescope.

Problem 3: The minoans synchronize their watches

The identical sister ships *Linear A* and *Linear B* pass each other in deep space, as shown from the perspective of *Linear A's* crew in the following figure. In its own rest frame, a ship is 2000 feet (600 m) long; the crew of *Linear A* sees the other ship glide by with speed $0.8c$. As the midpoint of *Linear B* passes the midpoint of *Linear A*, parabolic antennas on the two ships collide. The brief flash of light from the collision travels outwards from the point of contact.

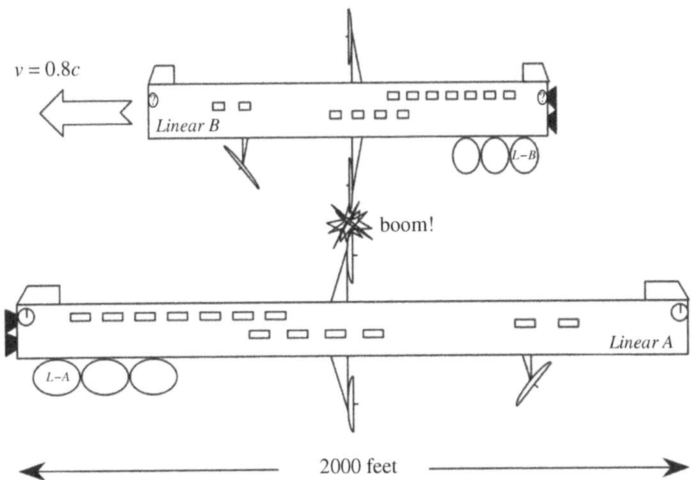

(a) According to the crew of *Linear A*, how long is the ship *Linear B*?

Solution

The moving ship is Lorentz contracted and is $2000\sqrt{1 - 0.8^2} = 1200$ feet long.

(b) According to the crew of *Linear B*, how long is the ship *Linear A*?

Solution

Same answer. The moving ship is Lorentz contracted and is $2000\sqrt{1 - 0.8^2} = 1200$ feet long.

(c) Both ships carry (initially stopped) clocks at their forward and aft ends. Crew members stationed next to each of the clocks start them ticking as soon as they see the light from the explosion. After all the clocks have finally begun ticking, how do the clocks on *Linear B* look to the crew on *Linear A*? If you conclude that the *Linear A* crew believes them to be out-of-synch, which *Linear B* clock reads the later time according to *Linear A* observers?

Solution

As we discussed above, the clock at the front of a moving ship is seen to be running away from the synchronizing light pulse. It starts ticking after the aft clock, so it reads an earlier time than the aft clock.

Problem 4: A math warmup

(a) An object moves slowly in a circle with constant (non-relativistic) speed so that its position at time t is

$$(x, y, z) = R(\cos \omega t, \sin \omega t, 0)$$

(1) Calculate the object's velocity and acceleration as functions of time.

Solution

Take derivatives!

$$(x, y, z) = R(\cos \omega t, \sin \omega t, 0)$$
$$(v_x, v_y, v_z) = \omega R(-\sin \omega t, \cos \omega t, 0)$$
$$(a_x, a_y, a_z) = -\omega^2 R(\cos \omega t, \sin \omega t, 0)$$

(2) Prove that the object's velocity is always perpendicular to its acceleration.

Solution

Recall that the dot product between perpendicular vectors is always zero:

$$\vec{v} \cdot \vec{a} = v_x a_x + v_y a_y + v_z a_z$$

$$= \omega^3 R^2 [\sin \omega t \cdot \cos \omega t - \cos \omega t \cdot \sin \omega t] = 0$$

(b) Perform the following matrix multiplications:

(1) $\begin{bmatrix} 2 & 1 \\ 3 & 5 \end{bmatrix} \begin{bmatrix} 3 & 4 \\ 2 & 1 \end{bmatrix}$ Solution: $\begin{bmatrix} 2\cdot 3 + 1\cdot 2 & 2\cdot 4 + 1\cdot 1 \\ 3\cdot 3 + 5\cdot 2 & 3\cdot 4 + 5\cdot 1 \end{bmatrix} = \begin{bmatrix} 8 & 9 \\ 19 & 17 \end{bmatrix}$

(2) $\begin{bmatrix} 2 & 1 \\ 3 & 5 \end{bmatrix} \begin{bmatrix} 5 \\ 7 \end{bmatrix}$ Solution: $\begin{bmatrix} 2\cdot 5 + 1\cdot 7 \\ 3\cdot 5 + 5\cdot 7 \end{bmatrix} = \begin{bmatrix} 17 \\ 50 \end{bmatrix}$

(3) $\begin{bmatrix} 6 & 2 \end{bmatrix} \begin{bmatrix} 2 & 1 \\ 3 & 5 \end{bmatrix}$ Solution: $\begin{bmatrix} 6\cdot 2 + 2\cdot 3 & 6\cdot 1 + 2\cdot 5 \end{bmatrix} = \begin{bmatrix} 18 & 16 \end{bmatrix}$

(4) $\begin{bmatrix} 2 & 4 \end{bmatrix} \begin{bmatrix} 2 & 1 \\ 3 & 5 \end{bmatrix} \begin{bmatrix} 5 \\ 7 \end{bmatrix}$ Solution: $\begin{bmatrix} 2 & 4 \end{bmatrix} \begin{bmatrix} 2 & 1 \\ 3 & 5 \end{bmatrix} \begin{bmatrix} 5 \\ 7 \end{bmatrix} = \{ \begin{bmatrix} 4 + 12 & 2 + 20 \end{bmatrix} \} \begin{bmatrix} 5 \\ 7 \end{bmatrix}$

$= \begin{bmatrix} 16 & 22 \end{bmatrix} \begin{bmatrix} 5 \\ 7 \end{bmatrix} = 80 + 154 = 234$

(5) $\begin{bmatrix} 2 & 4 \end{bmatrix} \begin{bmatrix} 5 \\ 7 \end{bmatrix}$ Solution: $\begin{bmatrix} 2 & 4 \end{bmatrix} \begin{bmatrix} 5 \\ 7 \end{bmatrix} = 2\cdot 5 + 4\cdot 7 = 38$

(6) $\begin{bmatrix} 5 \\ 7 \end{bmatrix} \begin{bmatrix} 2 & 4 \end{bmatrix}$ Solution: $\begin{bmatrix} 5 \\ 7 \end{bmatrix} \begin{bmatrix} 2 & 4 \end{bmatrix} = \begin{bmatrix} 5\cdot 2 & 5\cdot 4 \\ 7\cdot 2 & 7\cdot 4 \end{bmatrix} = \begin{bmatrix} 10 & 20 \\ 14 & 28 \end{bmatrix}$

Note that matrix multiplication is not commutative!

Cassis, France

Unit 2

Special Relativity—Non-Simultaneity:
The Lorentz Transformations

Introduction: Reference frames and measurements

The only quantitative features of special relativity we've seen so far are the changes in the lengths of moving objects and the duration of time intervals measured by moving clocks. If we're clever,

23

we can use this to draw quantitative conclusions regarding the non-simultaneity of events which are simultaneous when viewed from a different reference frame.

In the previous unit, we discussed briefly what we mean by our frame of reference. The idea is simple: it's just a coordinate system in which all our clocks and rulers are (and remain) at rest, and all our clocks are synchronized. A measurement of a spatial interval between two events amounts to determining the distance between the two clocks placed next to the events. The time interval is just the difference in the readings on the faces of the two clocks. For example, the space–time interval between waking up and entering my office involves a distance of about two miles and a time interval that depends on how sleepy I am.

Synchronizing clocks in a reference frame is easy: we preset all of them so that a clock which is R feet from the origin reads $+R$ nanoseconds. Each clock (that is initially paused) starts running when it senses the light from a strobe, placed at the origin, which fires at time zero.

Keep in mind that most measurements involve both position and time: measuring the length of a moving object as it glides past our (stationary) ruler requires us to see where the two ends of the object are, in comparison to the scale printed on our ruler, *at the same time.* Timing how long a moving object needs to coast past our stopwatch requires that the stopwatch *remain in the same place* during the timed interval.

We'll continue to use the approximation $c = 1$ foot per nanosecond for the time being.

Here's what you'll find/work on in this unit:

- a quantitative expression for the non-simultaneity of events viewed from one frame when the events occur simultaneously in another frame;
- putting it all together: deriving the Lorentz transformations.

Exercise 2.1: Quantitative description of non-simultaneity

We know from the last unit that clocks at the front of a *moving* object read earlier times than clocks at the back of the object. Let's work this up quantitatively.

Here's the setup: a flash is produced at the tail of the *Nostromo* as its rear clock reads zero. (The flash illuminates the clock and exposes the film that records the time on the face of the clock.) The light travels the length of the ship, arriving at a clock-equipped sensor at its nose.

We can describe the production and reception of the light as a pair of "events," where an event is something that happens at one point in space and lasts only an instant. The first event is the production of the spark near *Nostromo's* aft clock. The second event is the arrival at the forward sensor of light from the spark.

The figure shows these two events in the rest frame of *Nostromo*.

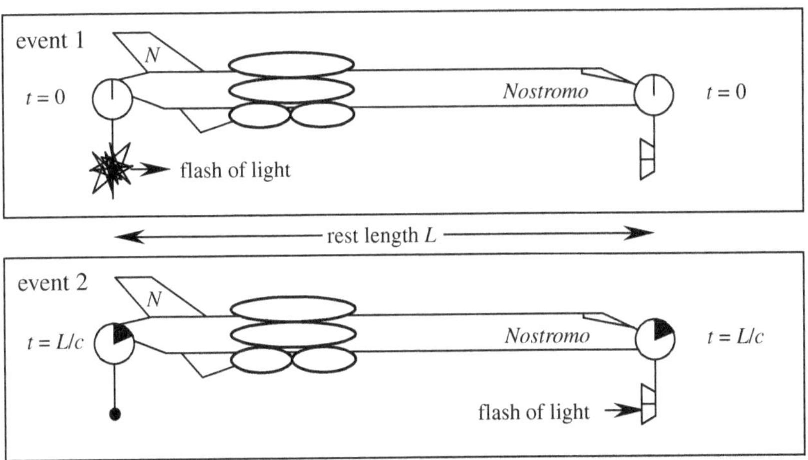

Since *Nostromo's rest length* is L, and all its clocks are synchronized in its *rest frame*, we know that the light will arrive at the forward clock when that clock reads L/c. There are well-defined spatial and temporal intervals between the two events in this reference frame: $\Delta x = L$ and $\Delta t = L/c$.

From the perspective of observers who see *Nostromo* moving to the right, the ship's length is Lorentz contracted to $L\sqrt{1 - (v/c)^2}$. *Nostromo's* forward clock shows an earlier time than its aft clock. (Let's call this time difference δ.)

The two events look like this:

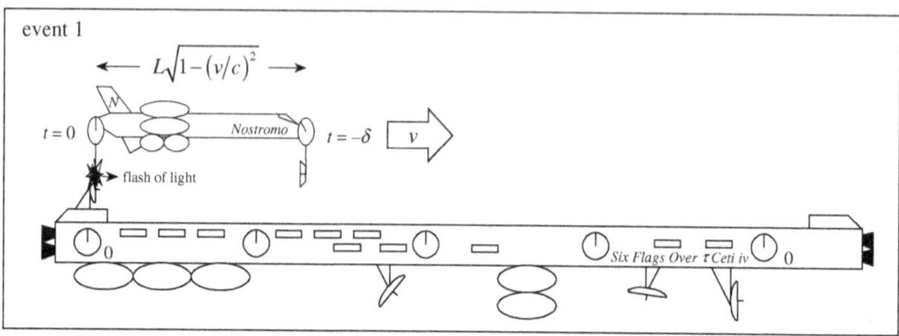

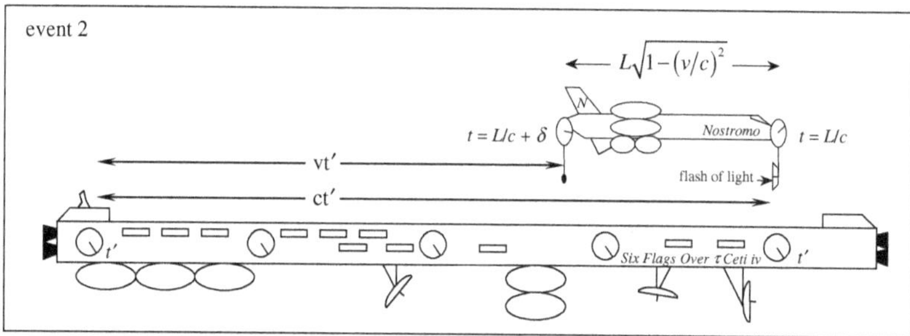

Event 1: Spark is produced at the aft end of *Nostromo*

Nostromo's tail clock reads 0; the nose clock $-\delta$. (We already know that δ is a positive number because the forward clock reads an earlier time than the aft clock.) Since we are viewing events from the rest frame of the observers on the rec facility, all clocks on the recreation facility read zero.

Event 2: Light from the spark reaches the forward end of *Nostromo*

From the photograph taken in the *Nostromo* rest frame, we know that *Nostromo's* forward clock will read L/c when the light arrives. Keep in mind that observers in all frames are going to agree that the reading in the photograph of *Nostromo's* forward clock shows the actual face of the clock when the light arrives.

The tail clock always reads δ later than the nose clock according to rec facility observers. And to these observers, the rec facility clocks all read t' when the light arrives. Naturally, in time t', the light from the spark has traveled a distance ct' to catch up with the nose of *Nostromo*, while the *Nostromo* itself has moved a distance vt'.

There is a well-defined spatial and temporal interval between the two events in this reference frame too: $\Delta x' = ct'$ and $\Delta t' = t'$. Note that I'm using coordinates with primes to describe variables in the rec facility frame.

(a) In the rec facility rest frame, the light beam travels a distance $\Delta x' = c\Delta t'$, which is the same as the sum of the Lorentz-contracted *Nostromo* length L' and the distance traveled by the *Nostromo* while the light flash travels to *Nostromo's* nose.

Write an equation that says the same thing as the previous sentence, and then use this to solve for $\Delta t'$ in terms of L, v, and c.

Solution

The primed frame is the rec facility rest frame. So,

$$\Delta x' = c\Delta t' = L' + v\Delta t' = L\sqrt{1 - \frac{v^2}{c^2}} + v\Delta t'$$

$$\therefore c\Delta t' = L\sqrt{1 - \frac{v^2}{c^2}} + v\Delta t'$$

or

$$\Delta t' = \frac{L\sqrt{1 - \frac{v^2}{c^2}}}{c - v} = \frac{L}{c}\frac{\sqrt{1 - \frac{v^2}{c^2}}}{1 - v/c}$$

Recall that $a^2 - b^2 = (a - b)(a + b)$ so that

$$\Delta t' = \frac{L}{c}\frac{\sqrt{(1 - v/c)(1 + v/c)}}{1 - v/c} = \frac{L}{c}\sqrt{\frac{1 + v/c}{1 - v/c}}$$

(b) When event 2 occurs in the rec facility rest frame, *Nostromo's* forward clock must read L/c and its aft clock must still be ahead of the forward clock by an amount δ. As a result, the reading on Nostromo's slowly ticking aft clock has advanced from 0 to $L/c + \delta$ during the time it took for rec facility clocks to advance (in the rec facility rest frame) from 0 to $\Delta t'$.

Since the moving *Nostromo* clocks are ticking slowly in comparison to the rec facility's clocks, we must have $L/c + \delta = \Delta t' \sqrt{1 - v^2/c^2}$. Use this fact to derive an expression for δ in terms of L, v, and c. Please simplify your expression as much as possible.

Solution

We'll use

$$\Delta t' = \frac{L}{c} \sqrt{\frac{1 + v/c}{1 - v/c}}$$

where

$$\frac{L}{c} + \delta = \Delta t' \sqrt{1 - \frac{v^2}{c^2}} = \frac{L}{c} \sqrt{\frac{1 + v/c}{1 - v/c}} \sqrt{1 - \frac{v^2}{c^2}} \quad \text{(from before)}$$

$$= \frac{L}{c} \sqrt{\frac{(1 + v/c)(1 - v/c)(1 + v/c)}{1 - v/c}}$$

$$= \frac{L}{c}(1 + v/c)$$

$$\therefore \frac{L}{c} + \delta = \frac{L}{c} + \frac{vL}{c^2} \quad \text{so that}$$

$$\delta = \frac{vL}{c^2}$$

The moving ship's aft clock is ahead of the forward clock by vL/c^2 where L is the *rest length* of the moving ship.

Discussion

How do the rec facility clocks look to the *Nostromo* crew? There's a nice symmetry here: clocks toward the front of the (moving) rec facility read behind in similar fashion.

Exercise 2.2: Deriving the Lorentz transformations

Observers in different frames of reference which are in relative motion will tend to disagree about lengths, distances, time intervals, and the synchronization of clocks. We now have enough information to write the equations relating the intervals in the primed frame to those in the unprimed frame.

Here's a summary of what we know so far:

1. A single, moving clock ticks slowly so that a time interval Δt between two events measured *by this one clock* will be shorter than the time interval $\Delta t'$ measured by observers who see that this clock is in motion: $\Delta t = \Delta t' \sqrt{1 - v^2/c^2}$.
2. The length L' of a moving object is shorter than the length L of the object as measured in the object's rest frame: $L\sqrt{1 - v^2/c^2} = L'$.
3. Clocks which are separated by a distance Δx *in their rest frame* and are synchronized *in their rest frame* are seen to be out of sync in a frame in which the clocks are moving with speed v by an amount $-v\Delta x/c^2$. (The negative sign means that the forward clock reads an earlier time than the aft clock.)

Imagine that a pair of events is seen by observers in two different frames whose coordinate systems we'll call O and O'. Observers in O (the "unprimed" frame) measure the space–time intervals between the events to be Δx, Δy, Δz, Δt, while observers in O' measure the interval to be $\Delta x'$, $\Delta y'$, $\Delta z'$, $\Delta t'$.

(I'm defining $\Delta x \equiv x_2 - x_1$, $\Delta t \equiv t_2 - t_1$, and so forth.)

Let's assume that the origins of O and O' coincide when clocks *at the origins* of both frames read zero. Frame O' (according to observers in O) is *moving with velocity v* along the x-axis of O, as shown in the figure. (Note that the sign of v matters.)

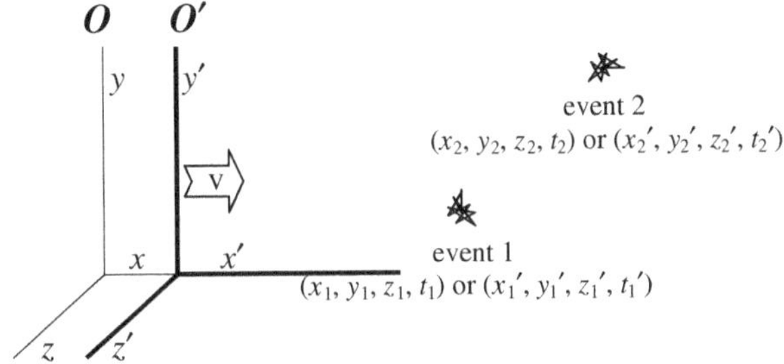

The *Lorentz transformations* allow us to calculate the space–time interval between the two events measured by observers in one frame if we know the interval between the events measured by observers in the other frame and also know the relative velocities of the two frames. They allow us to write $\Delta x'$ as a function of Δx, Δt, v, and c and also $\Delta t'$ as a (different) function of Δx, Δt, v, and c.

(a) Let's redraw the figure used in the last exercise in order to make the choice of primed and unprimed coordinates agree with the discussion in this exercise. Here's the redrawn figure:

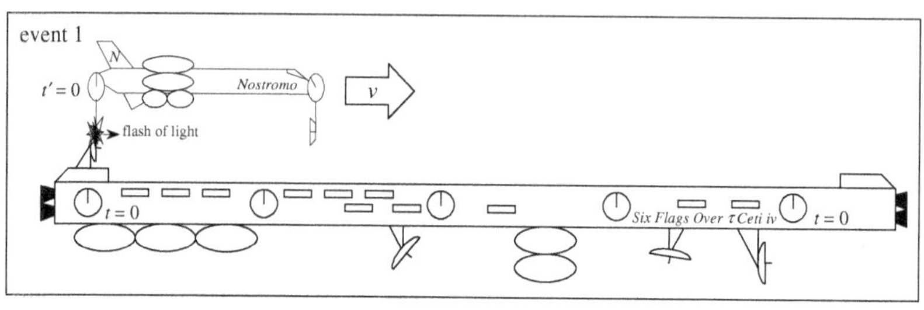

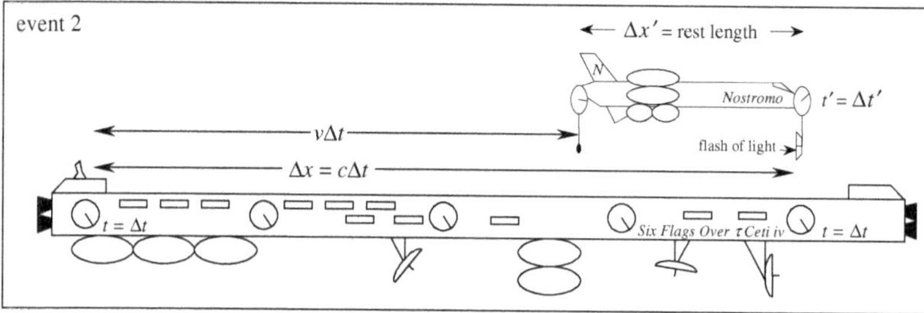

In *Nostromo's* rest frame, the light beam travels the ship's rest length, which (after renaming our variables) is $\Delta x'$.

In the rec facility frame, the light beam travels a distance which is the same as the sum of the Lorentz-contracted *Nostromo* length and the distance traveled by the *Nostromo*. The equation which says the same thing as this is $\Delta x = \Delta x' \sqrt{1 - v^2/c^2} + v\Delta t$.

Solve this equation for $\Delta x'$ as a function of c, v, Δx, and Δt.

Solution

A little bit of algebra:

$$\Delta x = \Delta x' \sqrt{1 - v^2/c^2} + v\Delta t$$

$$\Delta x' = \frac{\Delta x - v\Delta t}{\sqrt{1 - v^2/c^2}}$$

(b) Since the two events are separated by $\Delta x'$ and $\Delta t'$ in the *Nostromo* rest frame, we know that $\Delta t' = \Delta x'/c$. We also know that $\Delta t = \Delta x/c$ and $\Delta t/c = \Delta x/c^2$.

Use these facts to rewrite your answer to part (a) as an equation for $\Delta t'$, and simplify it as much as possible to express $\Delta t'$ as a function of c, v, Δx, and Δt.

Solution

More algebra: begin by dividing the above expression by c and then use our expression for $\Delta x'$:

$$\frac{\Delta x'}{c} = \frac{\frac{\Delta x}{c} - \frac{v\Delta t}{c}}{\sqrt{1 - v^2/c^2}}$$

and

$$\Delta t' = \frac{\Delta t - \frac{v\Delta x}{c^2}}{\sqrt{1 - v^2/c^2}}$$

Discussion

You have derived the Lorentz transformations! The frames O and O' correspond to the rest frames of the rec facility and the *Nostromo*, respectively; the space–time separation of *any* pair of events measured by observers in the unprimed frame can be used to derive the space–time interval between the events that would be measured by an observer in the primed frame. To write them more compactly define $\gamma \equiv \sqrt{1 - v^2/c^2}$ and $\beta \equiv v/c$ so that

$$\Delta x' = \gamma(\Delta x - \beta c\Delta t) \quad \Delta x = \gamma(\Delta x' + \beta c\Delta t')$$

$$\Delta y' = \Delta y$$

$$\Delta z' = \Delta z$$

$$\Delta t' = \gamma(\Delta t - \beta \Delta x/c) \quad \Delta t = \gamma(\Delta t' + \beta \Delta x'/c)$$

These are the Lorentz transformations. If we know the intervals in one frame, we can calculate what they'll be in the other frame. Keep in mind that *positive* v means that O' is moving to the right, as viewed from O.

Note that we can extract all of the quantities we've calculated so far using the Lorentz transformations as long as we choose the right pair of events. Keep in mind that almost all measurements require you to know something about space *and* time. For example, a time interval measured using a single clock is done so that the spatial separation between the "starting" event and the "stopping" event used to define the time interval is zero.

A small, but important point: v refers to the velocity of the primed frame's coordinate origin. It's *not necessarily* the velocity of the moving rocket in the setup to a problem. The very first thing you'll want to do is to draw the coordinate axes for the two frames attached to whatever (possibly moving) objects I've specified in the problem.

Summary and wrap-up discussion

- Time AND space are messed up.

  1. *The rate of passage of time is slowed in a moving frame:* $\Delta t = \Delta t' \sqrt{1 - (v^2/c^2)}$.
  2. *Simultaneity doesn't work:* clocks "at the front" read earlier by $v\Delta x/c^2$ than clocks "at the back," where Δx is the separation between the clocks *as measured in their own rest frame.*
  3. *Moving objects are shorter:* $L' = L\sqrt{1 - (v^2/c^2)}$, where L is the *rest length* of the object.

- The Lorentz transformations can be used to calculate the intervals Δx, Δy, Δz, Δt between events observed from one reference frame using the intervals $\Delta x'$, $\Delta y'$, $\Delta z'$, $\Delta t'$ between the same events observed from another reference frame.
- $v \leq c$ (you'll derive this in one of the homework-style problems).

Optional reading, just for fun

- *The Feynman Lectures on Physics* (Volume I), Chapter 15.
- *The Forever War*, Joe Haldeman. Avon books (paperback), 1974.

This science fiction novel won both the Hugo and Nebula awards. Time dilation plays a role in the messes the characters make of their lives.

Problem 1: Can anything go faster than c?

In this problem, you'll derive the relativistic velocity addition formula.

To replenish *Nostromo's* perilously low supply of Jell-O, Mrs. Blepon's fast shuttle is launched on a resupply mission. As seen from *Nostromo*, the shuttle streaks away with constant velocity u in the positive x' direction. At position x_1' and time t_1', the shuttle strikes a

dust mote, vaporizing it. Somewhat later, the shuttle hits a microm-
eteoroid.

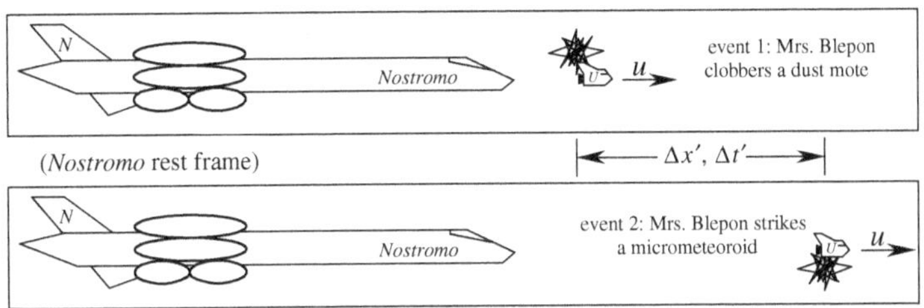

The light flashes from both events are seen by *Nostromo's* cook,
who calculates (after correcting for the light's propagation time
to the ship) that the space–time interval between the events was
$(\Delta x', \Delta t') = (x'_2 - x'_1, t'_2 - t'_1)$. Not surprisingly, the cook notes that
$u = \Delta x'/\Delta t'$.

The same events are viewed by Mrs. Blepon's personal trainer on
board the interstellar recreation facility. He sees *Nostromo* traveling
with speed v in the positive x-direction and determines that the
space–time interval between the two events was $(\Delta x, \Delta t) = (x_2 -
x_1, t_2 - t_1)$. He concludes that the shuttlecraft was moving in the
positive x-direction with speed $\Delta x/\Delta t$.

(a) By making use of the Lorentz transformations, calculate $\Delta x/\Delta t$
in terms of u (the shuttle's velocity as seen from *Nostromo*), v
(*Nostromo's* velocity, as seen from the rec facility), and c.

(b) How does your expression for $\Delta x/\Delta t$ behave in the limit that
both u and v are close to c?

Solution

We'll attach the origin of the primed coordinate system to *Nostromo*.
As a result, the velocity v to be used in the Lorentz transformations
is a positive quantity. For part (a), let's calculate the ratio $\Delta x/\Delta t$
by plugging in the Lorentz-transformed intervals we obtain from the
primed intervals in *Nostromo's* frame. Here we go:

$$\frac{\Delta x}{\Delta t} = \frac{\gamma(\Delta x' + \beta c \Delta t')}{\gamma(\Delta t' + \beta \Delta x'/c)} = \frac{\left(\frac{\Delta x'}{\Delta t'} + \frac{\beta c \Delta t'}{\Delta t'}\right)}{\left(\frac{\Delta t'}{\Delta t'} + \frac{\beta \Delta x'/c}{\Delta t'}\right)}$$

$$= \frac{(u + \beta c)}{\left(\frac{\Delta t'}{\Delta t'} + \beta u/c\right)} = \frac{(u + v)}{(1 + vu/c^2)}$$

since $\beta c = v$. The non-relativistic velocity addition formula would comprise the numerator alone, but the full, correct relativistic expression has that denominator too.

Note that in the limit that u and v approach c, we get

$$\frac{(u+v)}{(1+vu/c^2)} \rightarrow \frac{(c+c)}{(1+c^2/c^2)} = \frac{2c}{2} = c$$

In the limit that only one speed—let's say u—approaches c, we find

$$\frac{(u+v)}{(1+vu/c^2)} \rightarrow \frac{(c+v)}{(1+vc/c^2)} = c\frac{(1+v/c)}{(1+v/c)} = c$$

No matter how hard we try, we can't cause something to go faster than the speed of light.

Problem 2: The train–tunnel "paradox"

Here's how this one works. We have a tunnel, 800 feet long in its rest frame, with doors on each end which can be used to seal the tunnel. The train is 1,000 feet long in its own rest frame, as shown in the following illustration:

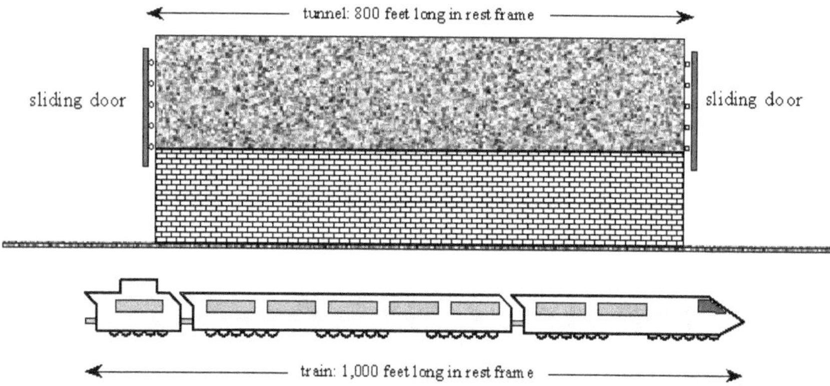

Tunnel rest frame

The train travels at a speed of 0.8 c so that its Lorentz-contracted length allows it to fit entirely inside the tunnel. When (synchronized in their rest frame) tunnel clocks by both doors read zero, just as the train is neatly centered inside the tunnel, the doors slam shut, trapping the entire train inside the tunnel. The front of the train crashes through the right-side tunnel door 125 nanoseconds later, but the (closed) left-side door was able to close without interfering with the train.

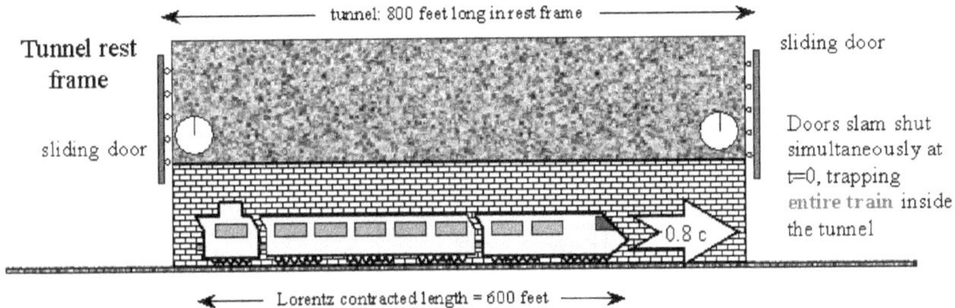

Train rest frame

From the train's rest frame, things look rather different: the Lorentz-contracted tunnel rushes towards it as shown in the following figure:

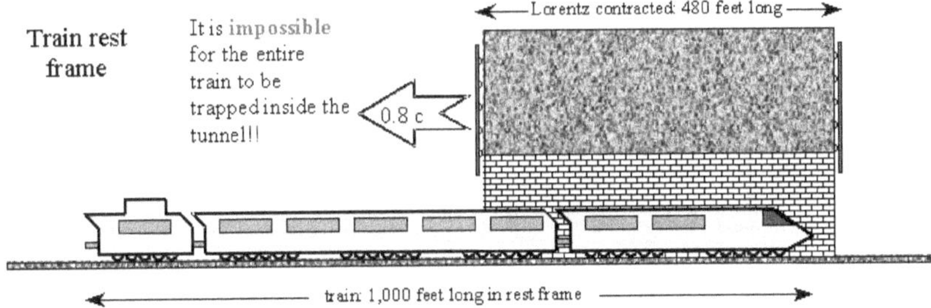

The tunnel is much too short to be able to trap the train entirely inside itself, yet both doors will drop, and neither will touch the train (although the front of the train will crash through the already closed right-side door).

How can this be possible?

Solution

The key is to bear in mind that the tunnel clocks are not synchronized when viewed from the train rest frame. As a result, each door drops when it is able to do so without striking the train. Let's fix the origin of the primed coordinate system to the center of the train, while the unprimed origin is attached to the center of the tunnel. And let's set the clocks so that a train clock at the center of the train (not shown) reads 0 just as it passes by the tunnel's center clock (also not shown).

According to the tunnel operators, the doors are separated by 800 feet and close simultaneously at time 0. As a result, we have

$\Delta x = 800$ feet, $\Delta t = 0$ nanoseconds where the two events are the closings of the tunnel doors. Let's first calculate how far apart the door closings are in the train frame. Use the Lorentz transformations, with $v = 0.8\,c$ so that $\beta = 0.8$ and $\gamma = 5/3$:

$$\Delta x' = \gamma(\Delta x - \beta c\Delta t) = \frac{5}{3}(800 - 0) = \frac{4000}{3} \text{ feet}$$

$$\Delta t' = \gamma(\Delta t - \beta\Delta x/c) = \frac{5}{3}(0 - 0.8 \cdot 800) = -\frac{3200}{3} \text{ nanoseconds}$$

So far so good: the train crew thinks the door closings are further apart in space than the length of their train. But what we'd really like to know is where the doors are, in the primed frame, when they slam shut. So, let's use the passing of the train's and tunnel's center clocks as the first event and the closing of a door as a second event.

For the tunnel's left door: $x_2 = -400$, $t_2 = 0$. Lorentz transform now, with $\Delta x = -400$ feet, $\Delta t = 0$ nanoseconds. Keep in mind that the back of the train is, in its frame, at -500 feet:

$$\Delta x' = \gamma(\Delta x - \beta c\Delta t) = \frac{5}{3}(-400 - 0) = \frac{-2000}{3} \text{ feet}$$

$$\Delta t' = \gamma(\Delta t - \beta\Delta x/c) = \frac{5}{3}(0 - 0.8 \cdot (-400)) = \frac{1600}{3} \text{ nanoseconds}$$

Ah, so the tunnel's left door has moved past the back of the train by the time it closes!

Now, do the same calculation for the tunnel's right door:

$$\Delta x' = \gamma(\Delta x - \beta c\Delta t) = \frac{5}{3}(400 - 0) = \frac{2000}{3} \text{ feet}$$

$$\Delta t' = \gamma(\Delta t - \beta\Delta x/c) = \frac{5}{3}(0 - 0.8 \cdot (400)) = \frac{-1600}{3} \text{ nanoseconds}$$

We find that the tunnel's right door closes early when it hasn't yet reached the front of the train, which is at $x' = 500$ feet.

Problem 3: Fast food

The owners of *Le Four de Naïs*[1] have grown tired of running out of baguettes before running out of customers. To increase their rate of

---

[1] Naïs is, in Greek mythology, the mother of Chiron, noblest of the centaurs. "Le Four de Naïs" would be a perfectly serviceable name for a Provençal bakery.

bread production, they have acquired an Asher-Danzig FlashOven, which is guaranteed to bake a loaf of bread in less than a picosecond. Unfortunately, after installing the oven, they discover that it is only one foot long, while their baguettes are two feet long. The oven is open at both ends, as illustrated in the following:

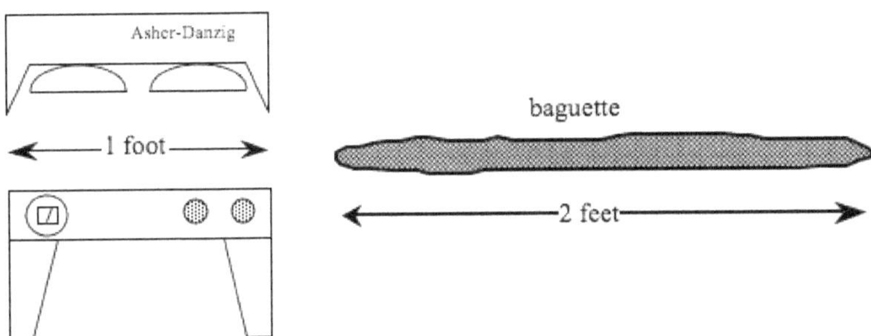

The *Four de Naïs* staff decide to solve their problem by buying a used St. Regulus catapult. They prepare bread dough and catapult-launch ready-to-be-baked loaves at relativistic speed into the oven. For an instant when a loaf is *entirely inside the oven*, the oven flashes ON, baking the loaf in a picosecond. (Somehow, the loaves are not damaged by the action of the catapult.)

(a) How fast must the bread be traveling to be entirely contained inside the oven?

(b) By mistake, the bakers mix their watches into one of the loaves of bread. One watch is at the tip of the loaf which enters the oven first, while the other is at the end of the loaf which enters the oven last. Somehow, the St. Regulus catapult causes the watches to become synchronized *in the rest frame of the relativistic loaf* as it leaves the catapult. While passing through the oven, the watch mechanisms are destroyed by the heat flash. Calculate the difference in the times on the faces of the broken watches.

Solution

(a) We need the bread to Lorentz contract to half its original length. So, $L' = L\sqrt{1 - (v^2/c^2)}$ requires

$$\sqrt{1 - v^2/c^2} = \frac{1}{2}$$

$$1 - v^2/c^2 = \frac{1}{4}$$

$$\frac{\sqrt{3}}{2}c = v$$

(b) The difference in the readings on the faces of the two watches, as seen by the oven's illuminating flash lamps, is

$$\Delta t = \frac{vL}{c^2} = \frac{\sqrt{3}}{2} \cdot 2 = \sqrt{3} \text{ nsec}$$

Recall that L is the *rest length* of the bread.

Problem 4: An introduction to partial derivatives

Here's a reminder about the definition of an ordinary derivative, and an introduction to partial derivatives.

The Blepons are hiking in eastern France's Jura mountains. They have been told by residents of a nearby village that the altitude above sea level is well described by the function $h(x, y) = 10^{-7}x^3$, where x and y are the distances due east and due north (in meters) of the center of town. A plot of the elevation near the town would therefore look something like this:

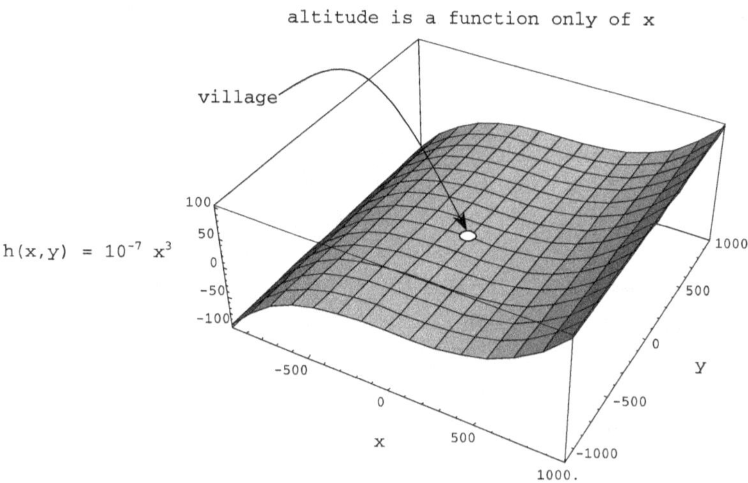

(a) The Blepons hike due east from the point (x, y) one morning, stopping for lunch at the point $(x + a, y)$. Write an algebraic expression for their change in altitude $\Delta h = h(x+a, y) - h(x, y)$ and for the average slope of their path, $m = \Delta h/a$. Please simplify your result as much as possible, expressing it as a function of x and a.

(b) Evaluate the average slope in the limit that $a \to 0$, expressing your answer as a function of x. (Since the definition of the derivative of a function of one variable is $df(x)/dx = \lim_{\Delta x \to 0}[f(x + \Delta x) - f(x)]/\Delta x$, your answer shouldn't come as a surprise to you!)

The next day the Blepons are hiking in a region where the altitude is well described by the function $h(x, y) = 10^{-10}x^3 y$, where x and y are the distances due east and due north of the center of a different town. An elevation plot would therefore look something like the following:

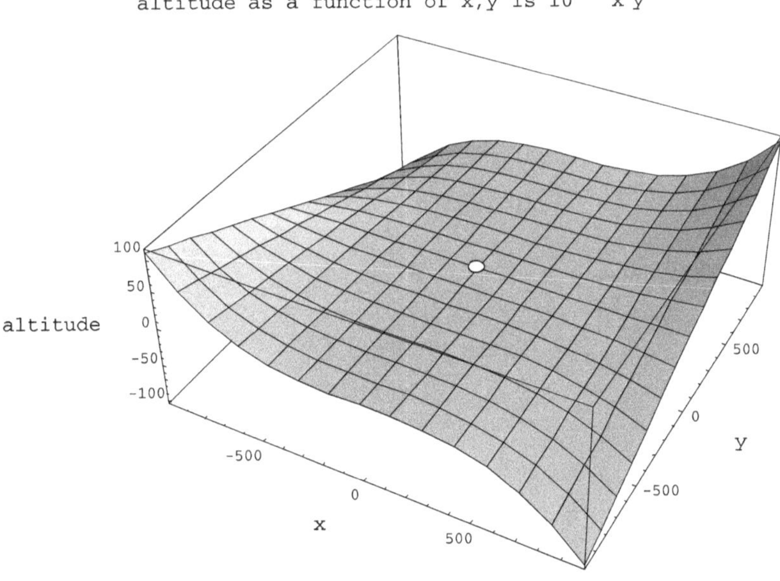

(c) The Blepons hike due east from the point (x, y) to the point $(x + a, y)$. Write an algebraic expression for their change in altitude $\Delta h = h(x+a, y) - h(x, y)$ and for the average slope of their path, $m = \Delta h/a$. Simplify your result, expressing it as a function of x, y, and a.

(d) Evaluate the average slope in the limit that $a \to 0$ as a function of x and y.

You have just evaluated a partial derivative! The definition of a partial derivative of a function of several variables is quite similar to the definition of the derivative of a function of one variable:

$$\frac{\partial f(x,y)}{\partial x} = \lim_{\substack{\Delta x \to 0 \\ y \text{ constant}}} \frac{f(x + \Delta x, y) - f(x,y)}{\Delta x} \quad \text{and}$$

$$\frac{\partial f(x,y)}{\partial y} = \lim_{\substack{\Delta y \to 0 \\ x \text{ constant}}} \frac{f(x, y + \Delta y) - f(x,y)}{\Delta y}$$

The machinery is easy to use: just treat all "the other" variables as if they were constants and take the derivative of the function in the usual way.

Solution

(a)

$$h(x,y) = 10^{-7}x^3, \quad h(x + a, y) = 10^{-7}(x + a)^3$$
$$\Delta h = 10^{-7}[x^3 + 3x^2 a + 3xa^2 + a^3 - x^3]$$
$$= 10^{-7}[3x^2 a + 3xa^2 + a^3]$$

Average slope is

$$\frac{\Delta h}{a} = 10^{-7}[3x^2 + 3xa + a^2]$$

(b)

$$\lim_{a \to 0} \frac{\Delta h}{a} = 10^{-7} \cdot 3x^2 = 3 \times 10^{-7}x^2$$

(c) Now, we have

$$\Delta h = 10^{-10}[(x + a)^3 y - x^3 y]$$
$$= 10^{-10}y[3x^2 a + 3xa^2 + a^3]$$

so the average slope is

$$\frac{\Delta h}{a} = 10^{-10} y [3x^2 + 3xa + a^2]$$

(d)
$$\lim_{a \to 0} \frac{\Delta h}{a} = 10^{-10} y \cdot 3x^2 = 3 \times 10^{-7} x^2 y$$

France

Unit 3

The Origin of the Magnetic Field as a Consequence of Special Relativity

In this unit, we review matters from before pertaining to the synchronization of clocks and also the use of the Lorentz transformations, then move on to something new: how Coulomb's law and special relativity necessitate the existence of magnetic fields and forces.

A quick review of portions of the previous unit

Clock synchronization

You may be a little fuzzy about how clocks that are synchronized in their rest frame can actually be out of synch when viewed from a different frame. Here are a couple of pictures that might help, as long as you keep in mind that observers in any frame of reference will always agree that the speed of light is constant.

Both diagrams show the creation of that synchronizing light flash, with the first drawn from the *Sulaco's* reference frame and the second from *Nostromo's* frame.

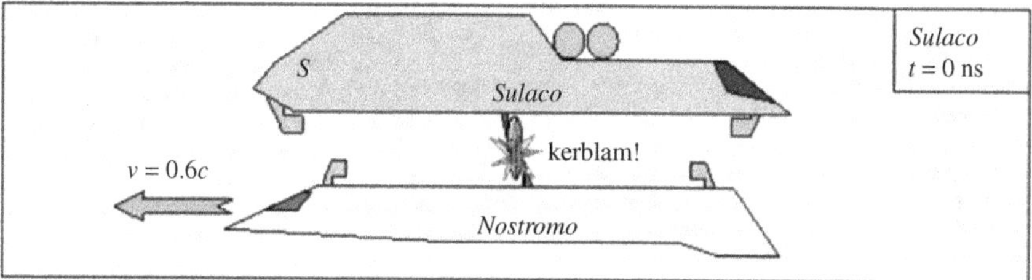

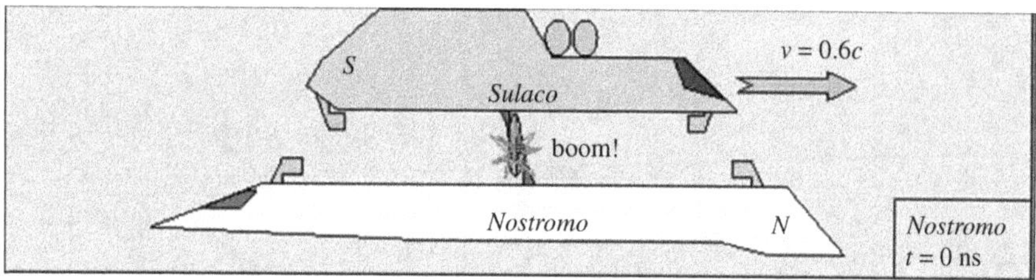

In both reference frames, the forward periscope on the moving spaceship is running away from the flash and consequently is illuminated after the aft periscope, which runs into the flash. Due to this, the aft clock on the moving ship begins ticking sooner and will read a later time than the forward clock.

Using the Lorentz transformations

Here's a concrete example of how to set up the Lorentz transformations to determine the space-time interval between two events as determined by observers in one frame when we know the intervals determined by observers in another frame. The diagrams ought to help you sort this out:

Step 1: Draw the systems you'll be working with, placing axes and origins O and O' on the drawing. (We are at rest in frame O.) Take note that the velocity v represents motion of the O' **origin** as seen by observers in O. In the instant case, it's $+0.8c$ in the x-direction.

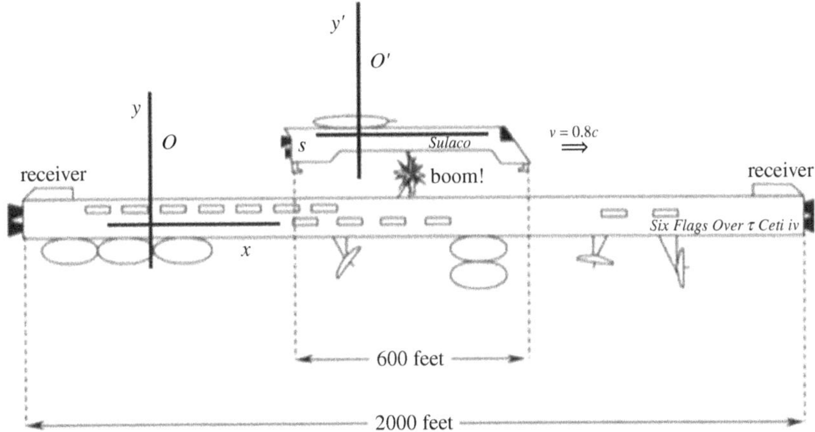

Step 2: Identify the two events whose separation you plan to transform. Label their coordinates on your diagram and define $\Delta x = x_2 - x_1$, $\Delta t = t_2 - t_1$. (In this example, the events are the arrivals of the light flashes at the rec facility's left and right periscopes. Since they happen simultaneously in frame O, I can draw both events in a single diagram.)

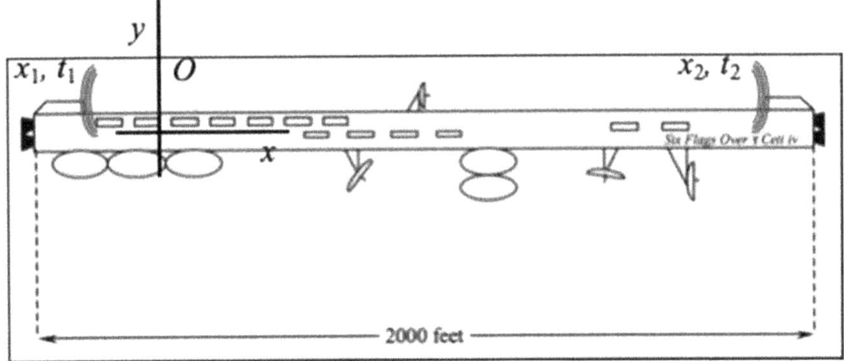

Step 3: Calculate the interval between the two events according to observers in the other frame: $\Delta x' = \gamma(\Delta x - v\Delta t), \Delta t' = \gamma(\Delta t - v\Delta x/c^2)$. Since the events are not simultaneous in the other frame, I need to show them in separate diagrams. Here are the two diagrams:

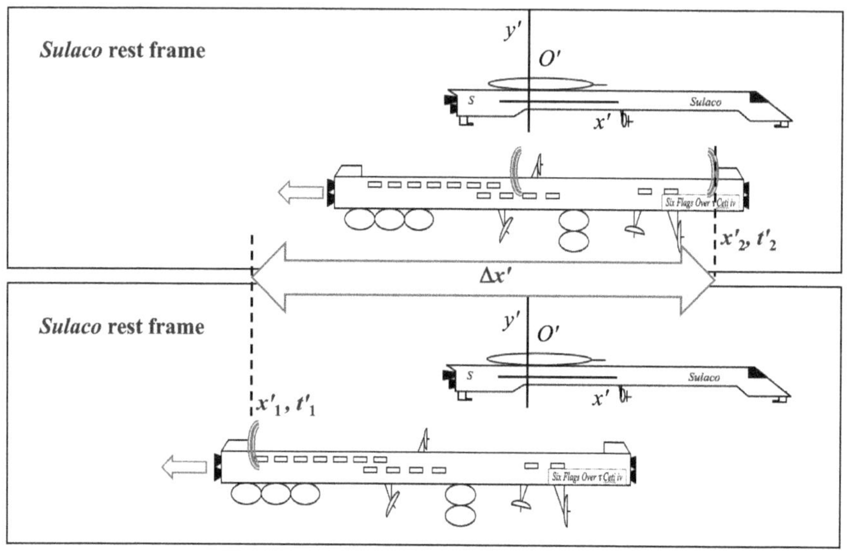

Be careful: $v, \beta = v/c$, and γ refer to the motion of the primed frame's coordinate origin. Draw the axes onto your diagrams.

Note that to shift to a frame in which a stationary object in the original, unprimed frame is seen (by observers in the new, primed frame) to move in the $+x'$-direction, the primed frame origin will need to move in the $-x$-direction, so β will be **negative**. (But in the case we're considering, observers in the primed frame see the unprimed frame's stuff moving to the left, not to the right.)

Something new: Magnetic fields are just electric fields seen from a different perspective

And now for something new.

You probably learned in an introductory course that the force exerted on a charge q moving with velocity $\vec{v}$ in a magnetic field $\vec{B}$ is

$$\vec{F} = q\vec{v} \times \vec{B}$$

This is a very odd force law! The force depends on how fast the charge is moving and is also perpendicular to both the charge's velocity and the local magnetic field.

One small point: this is the "Lorentz force" in SI ("Systéme International") units, in which we use kilograms, meters, Coulombs, amperes, Teslas, and seconds. But a more natural convention for electrodynamics is to use the CGS system, in which the corresponding units are grams, centimeters, ESUs (electrostatic units), and so forth. The CGS version of the Lorentz force replaces v with v/c. We'll stick to SI, but keep in mind that many of your professors might have learned electrodynamics in CGS.

Magnetic fields and magnetic forces exist **entirely** because of Coulomb's law and special relativity. Though observers in one frame may determine that a current-carrying wire is electrically neutral, observers in another frame might conclude that the wire *does* carry a net electric charge.

We'll derive this today, for the particular case of motion parallel (or antiparallel) to the current.

Here's what you'll work on today:

- analyzing the charge density in a wire loop that suddenly finds itself carrying a current, according to observers in two different frames of reference;
- describing how electric fields from a static charge transform when changing to a frame in which the charge is in motion.

Exercise 3.1: Whap! They're moving

Consider a rectangular loop that is made from a thin filament of wire as drawn in the following figure. Initially, no current flows in the wire; there are as many (positive) metal ions in the wire's crystal lattice as there are conduction electrons. The spacing between adjacent electrons is a while the spacing between positive metal ions is b. Naturally, since the wire holds no net charge, $a = b$.

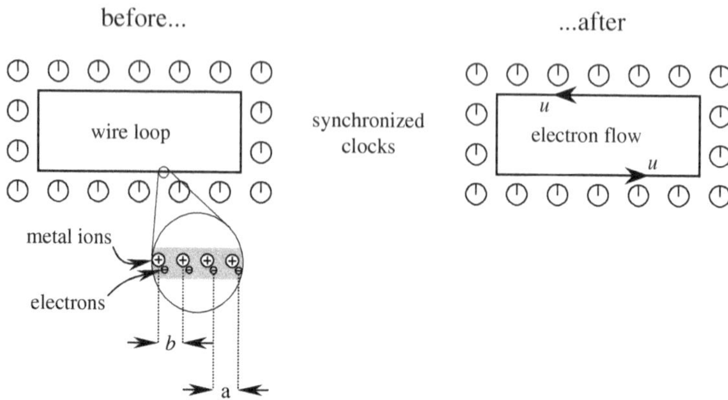

When all (synchronized) clocks in the lab frame read zero, all the electrons are suddenly accelerated from rest to a speed u so that the flow of electrons in the wire loop is in a counter-clockwise sense. This corresponds to a circulating current that flows in a ***clockwise*** sense since we describe a current as if it comprises moving ***positive*** charges:

(a) According to observers in the lab frame (the rest frame of the still-stationary metal ions), what is the spacing between adjacent (but now-moving) electrons?

For discussion: what happened to Lorentz contraction?

Solution

The spacing is still a. Think about the process of bringing all electrons up to speed: since they all begin moving at exactly the same time ***in the wire rest frame***, there's no opportunity for one electron to move closer to or farther from any other electron.

So, what happened to Lorentz contraction? To answer that, you should shift to the rest frame of the now-moving electrons.

In this frame, metal ions in the two horizontal lengths of wire are seen moving with speed u. They are, of course, closer together than in the other frame due to Lorentz contraction. And in this frame, the electrons are seen as having come to rest ***in this frame*** at different times since each comes to rest just as the clock next to it reads zero, and these clocks are ***not*** synchronized.

So, the electron-electron spacing has ***changed*** and is no longer a.

(b) An observer in a different frame of reference sees the wire loop glide past with speed v, moving to the left as shown in the following figure. When the observer's clock reads zero, clocks close to the right-most edge of the (moving) wire loop are also seen to read zero.

Sketch the readings on the other (moving) clocks' faces at the instant the right-side clocks read zero.

Solution

Recall from the previous unit that clocks separated by a distance L in their rest frame are out of synch by vL/c^2, with the forward clock reading an earlier time than the aft clock. With separation b in the wire loop rest frame, successive clocks, moving to the left, read $-vb/c^2, -2vb/c^2, -3vb/c^2, -4vb/c^2, -5vb/c^2, -6vb/c^2$.

(c) Keep in mind that an electron is accelerated when the lab-frame clock next to it reads zero. From the perspective of the observer who sees the wire loop moving to the left, the densities of electrons and metal ions are identical *before* any of the electrons begin circulating in the wire loop.

Describe qualitatively the relative densities of electrons and metal ions in each of the four sides of the loop *after* all the electrons have been set in motion. It may help to think in terms of which electrons begin moving when. Do this from the perspective of an observer who sees the wire loop moving to the left.

Solution

The separations of positive metal ions in the top and bottom portions of the loop are Lorentz contracted, so their density (number of ions per unit length of wire) increases. However, the spacing between ions in the left and right vertical legs of the loop is unchanged since the inter-ion spacing is still b.

Now, for the electrons. Before the current begins to flow, the electron spacings are the same as the ion spacings. Since electrons are set in motion relative to the ions when a clock by an electron reads 0, electrons begin flowing **into** the right side of the **top** of the loop before electrons begin flowing **out of** the left side of the top of the loop. So, the electron density on the top section of the loop increases steadily until electrons begin flowing out of the left side. From then on, for every electron flowing into the top right of the horizontal portion of the loop, an electron flows out at the left. As a result, the increased electron density at the top of the loop stabilizes and does not change further.

The bottom of the loop begins losing electrons out of the right end before it begins receiving electrons at its left end, so it becomes somewhat depleted of electrons: the electron density *decreases*.

The electron densities in the vertical portions of the loop do not change.

As a result, in a frame in which the loop is moving to the left, the top of the loop appears to hold net negative charge while the bottom appears to hold net positive charge.

(d) What would happen to a test charge, stationary in the frame of the observer who sees the wire loop moving to the left, if the test charge were placed just below (and close to) the bottom side of the loop? How would this charge behave when seen from the rest frame of the wire loop?

Solution

In the frame of the test charge, the bottom of the loop has net positive charge, so the stationary test charge is repelled from it. Note that the test charge, pushed away from the loop, will be seen in **any** frame to accelerate away from the bottom of the loop.

In the rest frame of the loop, the test charge is initially seen as moving to the right, and its velocity is deflected downwards, away from the current-carrying loop.

Think about the magnetic field associated with the current flowing to the left in the bottom part of the loop: in the plane of the paper, the B field is perpendicular to the plane of the paper, and, below the loop, points toward the reader. (Keep in mind that electrons moving to the right correspond to a current flowing to the left.) From the right-hand rule, the Lorentz force $F = qv \times B$ points down, toward the bottom of the page: v is to the right and B is out of the page.

In one frame (the charge's rest frame), the origin of the force is ascribed to the net charge on the loop's horizontal segments. In the other frame, we invent something we call the "magnetic field" and employ a strange-looking force law to describe the behavior of the moving test charge.

Exercise 3.2: Quantitatively, now...

Let's consider the bottom leg of the wire loop now, after the electrons have started moving with speed u.

You should have already concluded that the electron spacing, viewed from the lab frame (the rest frame of the wire), does not change after the electrons are accelerated. (How could it possibly change? In this frame, all electrons started moving simultaneously, so there was never a time interval during which one electron was moving but its neighbor was not.)

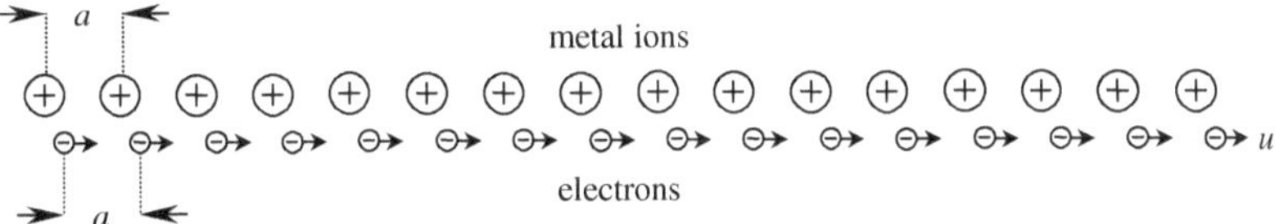

(a) Calculate (in terms of a and u) the spacing between electrons in the electrons' *rest frame*. (Even though $u \ll c$ for realistic currents, you'll need to consider the effects of special relativity.)

Solution

In the wire's rest frame, the electron spacing is/has Lorentz-contracted to a along the top and bottom of the loop. As a result, in the electrons' rest frame, the larger inter-electron spacing is $\dfrac{a}{\sqrt{1-\beta^2}}$ along the top and bottom of the loop. If you prefer, you can approximate this using a binomial expansion if you're familiar with this as

$$a\left(1 + \frac{1}{2}\frac{u^2}{c^2}\right)$$

(b) Do not use a calculator for this problem. The "binomial expansion" for $(1+\varepsilon)^n$ is

$$(1+\varepsilon)^n = 1 + \frac{n}{1!}\varepsilon + \frac{n(n-1)}{2!}\varepsilon^2 + \cdots$$

This works even when n isn't an integer and is useful for approximations. Some examples for $\varepsilon \ll 1$: $\sqrt{1+\varepsilon} \approx 1+\varepsilon/2$; $1/(1+\varepsilon) \approx (1-\varepsilon)$;

$$1/\sqrt{1+\varepsilon} = (1+\varepsilon)^{-1/2} \approx 1 - \varepsilon/2$$

In the following exercises, you'll only need to calculate things to *second-order* accuracy in any of the velocities present in your equations: terms proportional to $v^3, u^3, uv^2 \ldots$ can be ignored.

Use the binomial expansion to calculate a version of the relativistic velocity addition formula which is accurate to second order in the velocities. What I mean by this is that you should turn the expression

$$\frac{u+v}{1+\frac{uv}{c^2}}$$

into something in the form $u + v + a_1 u^2/c + a_2 uv/c + a_3 v^2/c$. (The coefficients a_1, a_2, a_3 might be zero. You should drop all terms proportional to $u^3, u^2 v, uv^2$, and v^3.)

Solution

Here we go:

$$\frac{u+v}{1+\frac{uv}{c^2}} = (u+v)\left(1+\frac{uv}{c^2}\right)^{-1} \approx (u+v)\left(1-\frac{uv}{c^2}\right)$$

$$= (u+v) - \frac{u^2 v}{c^2} - \frac{uv^2}{c^2} \Rightarrow (u+v)$$

so there **are** no terms such as u^2/c or v^2/c! To this order, we can use the non-relativistic velocity addition formula.

(c) In another frame of reference, the wire is seen to move to the left at speed $v \ll c$. (By this, I mean that the metal lattice is moving with speed v.) How fast are the electrons moving in this frame? (Use the second-order expression you just derived.)

Solution

Simple: non-relativistic velocity addition, so speed is $|u-v|$ or $|u+v|$.

(d) Calculate the spacing between metal ions in the frame in which the wire is moving. Do the same for the spacing between electrons in this frame.

Solution

The metal ion spacing is Lorentz contracted, so it becomes $a\sqrt{1 - v^2/c^2}$. But for the electrons, we first need to get into their rest frame, then Lorentz contract. In their rest frame, the electron separation is greater, namely $a/\sqrt{1 - u^2/c^2}$. This contracts to

$$\frac{a}{\sqrt{1 - u^2/c^2}}\sqrt{1 - (u - v)^2/c^2}$$

You **must** shift to the electrons' rest frame as your first step!

(e) Use the binomial expansion to generate an approximate expression for the spacing between adjacent metal ions in the frame with the wire moving. Include all terms up to second order in the ratios of small velocities (u and v). Do the same thing for the electrons.

Solution

It's just algebra. Watch this: for metal ions,

$$a\sqrt{1 - v^2/c^2} \approx a - \frac{av^2}{2c^2}$$

For electrons,

$$\left(a/\sqrt{1 - u^2/c^2}\right)\sqrt{1 - (u - v)^2/c^2}$$

$$\approx a\left(1 + \frac{u^2}{2c^2}\right)\left(1 - \frac{1}{2}\left(\frac{u - v}{c}\right)^2\right)$$

$$= a\left(1 + \frac{u^2}{2c^2}\right)\left(1 - \frac{1}{2}\left(\frac{u^2}{c^2} + \frac{v^2}{c^2} - 2\frac{uv}{c^2}\right)\right)$$

$$= a\left(1 - \frac{u^2}{2c^2} - \frac{v^2}{2c^2} + \frac{uv}{c^2} + \frac{u^2}{2c^2} - \frac{u^4}{4c^4} - \frac{u^2v^2}{4c^4} + \frac{u^3v}{2c^4}\right)$$

$$\approx a\left(1 - \frac{v^2}{2c^2} + \frac{uv}{c^2}\right)$$

(f) Calculate the apparent charge density in the wire in terms of the fundamental charge (e), a, u, v, and c. Verify that your expression yields zero if $v = 0$ or $u = 0$.

Solution

The apparent charge density is the net charge per meter. For positive metal ions, there's an ion every $a - \frac{av^2}{2c^2}$ meters, so the charge density carried by ions is

$$e \left/ \left(a - \frac{av^2}{2c^2} \right) \right. \approx \frac{e}{a} \left(1 + \frac{v^2}{2c^2} \right)$$

(here e is the charge on a proton), while that carried by electrons is

$$-e \left/ a \left(1 - \frac{v^2}{2c^2} + \frac{uv}{c^2} \right) \right. \approx \frac{-e}{a} \left(1 + \frac{v^2}{2c^2} - \frac{uv}{c^2} \right)$$

As a result, the net charge density is $+\frac{e}{a}\frac{uv}{c^2}$ Coulombs per meter. This is obviously zero when either u or v (or both) is zero.

Exercise 3.3: Real numbers

Let's calculate what happens in a real wire.

Imagine that we have a 1000 ampere current flowing in a heavy copper cable with a cross-sectional area of $1\,\text{cm}^2$. The density of copper is about 9 grams per cm^3, while the mass of a single copper atom is approximately 10^{-22} grams. As a result, every cubic centimeter of copper cable holds 9×10^{22} copper atoms. Each copper atom gives up two conduction electrons so that there are 1.8×10^{23} mobile electrons per cm^3 of cable. The charge on each electron is 1.6×10^{-19} Coulombs.

(a) How fast are the electrons moving, on average?

Solution

$1000\,\text{amps}$ is 1000 Coulombs per second going past a point along the wire, so current will consist of $1000/1.6 \times 10^{-19}$ electrons per second or about 6.25×10^{21} electrons per second. Since there are about 1.8×10^{23} electrons per cubic centimeter, we'll need all the electrons in $6.25 \times 10^{21}/1.8 \times 10^{23} = 0.0347$ cubic centimeters of wire to move past a point every second. So, the speed is very slow: it's only $0.0347\,\text{cm/sec}$ since the cross-sectional area of the wire is $1\,\text{cm}^2$!

(b) How large is the electric field generated by the (charged!!) wire at a distance of 10 centimeters according to a test charge which observes the wire moving to the left with speed $100\,\text{m/s}$? (Use Gauss' law!)

Solution

We found that the effective charge density in the wire is $\frac{euv}{ac^2}$ Coulombs/meter and that there are about 1.8×10^{23} electrons per cubic centimeter. Every centimeter of wire holds 1.8×10^{23} electrons, so $a = 1\,\text{cm}/1.8 \times 10^{23} = 5.55 \times 10^{-26}\,\text{m}$. (Note the switch from centimeters to meters.) As a result, the (linear) charge density in Coulombs per meter is

$$\frac{euv}{ac^2} = \frac{(1.6 \times 10^{-19})(3.47 \times 10^{-4}) \times 100}{(5.55 \times 10^{-26})(9 \times 10^{16})}$$

$$= 1.11 \times 10^{-12} \text{ Coulombs/meter.}$$

Gauss's law, for a cylindrical enclosing surface of length L and radius r:

$$E \times 2\pi r L = \lambda L/\varepsilon_0$$

$$E = \frac{1}{r}\frac{\lambda}{2\pi\varepsilon_0} = \frac{1}{0.1}1.11 \times 10^{-12} \times 1.8 \times 10^{10} = 0.2\,\text{N/C}$$

(c) If the test charge is an electron (with mass 9.1×10^{-31} kg), how large is the acceleration it experiences due to the passage of the current-carrying wire?

Solution

Since $a = F/m, = eE/m$, plug in to find that the acceleration is huge: $3.5 \times 10^{10}\,\text{m/sec}^2$. Note that this is several billion times greater than the Earth's gravitational acceleration!

(d) In the wire loop's rest frame, the charge moves with an initial velocity of $100\,\text{m/s}$ parallel to the wire. Its acceleration is nearly the same in this frame. What is the radius of curvature of its path at the instant that it is moving parallel to the wire? Recall that $a = v^2/r$.

Solution

Solve for r and plug in to find $r = 2.84 \times 10^{-7}$, which is really small.

Summary and wrap-up discussion

- Magnetic force is just a consequence of relativity: observers in different frames do not agree about charge densities!

Optional reading: *The Feynman Lectures on Physics (volume II)*, Chapter 26.
The Feynman Lectures on Physics (volume II), Chapter 42.

Problem 1: More on Lorentz transformations

(Some of these problems require you to know how to multiply matrices and to remember that the *transpose* of a matrix is what you get when you swap its rows and columns.)

Imagine observers in the frame $\boldsymbol{O}$ see another frame $\boldsymbol{O'}$ moving in the positive x-direction with velocity v so that the origins of the two coordinate systems overlap at $t = t' = 0$. We can use the Lorentz transformations to relate the space-time coordinates of an event in one frame with those in another frame. Defining $\beta \equiv v/c$ and $\gamma \equiv 1/\sqrt{1 - \beta^2}$, the Lorentz transformations become

$$x' = \gamma(x - \beta ct), \quad y' = y, \quad z' = z, \quad t' = \gamma(t - \beta x/c)$$
$$x = \gamma(x' + \beta ct'), \quad y = y', \quad z = z', \quad t = \gamma(t' + \beta x'/c).$$

Keep in mind that v is the velocity of the primed frame's coordinate axes as seen from the unprimed frame.

We can multiply the time equation by c and add some parentheses to rewrite these as follows:

$$x' = \gamma(x - \beta(ct)), \quad y' = y, \quad z' = z, \quad (ct') = \gamma((ct) - \beta x)$$
$$x = \gamma(x' + \beta(ct')), \quad y = y', \quad z = z', \quad (ct) = \gamma((ct') + \beta x')$$

We can then write the Lorentz transformations in matrix form this way:

$$\begin{bmatrix} ct' \\ x' \\ y' \\ z' \end{bmatrix} = \begin{bmatrix} \gamma & -\gamma\beta & 0 & 0 \\ -\gamma\beta & \gamma & 0 & 0 \\ 0 & 0 & 1 & 0 \\ 0 & 0 & 0 & 1 \end{bmatrix} \begin{bmatrix} ct \\ x \\ y \\ z \end{bmatrix}$$

and

$$\begin{bmatrix} ct \\ x \\ y \\ z \end{bmatrix} = \begin{bmatrix} \gamma & +\gamma\beta & 0 & 0 \\ +\gamma\beta & \gamma & 0 & 0 \\ 0 & 0 & 1 & 0 \\ 0 & 0 & 0 & 1 \end{bmatrix} \begin{bmatrix} ct' \\ x' \\ y' \\ z' \end{bmatrix}$$

More compactly, defining

$$\underset{\sim}{x'} \equiv \begin{bmatrix} ct' \\ x' \\ y' \\ z' \end{bmatrix}, \quad \underset{\sim}{x} \equiv \begin{bmatrix} ct \\ x \\ y \\ z \end{bmatrix}, \quad \text{and} \quad \underset{\approx}{\Lambda} \equiv \begin{bmatrix} \gamma & -\gamma\beta & 0 & 0 \\ -\gamma\beta & \gamma & 0 & 0 \\ 0 & 0 & 1 & 0 \\ 0 & 0 & 0 & 1 \end{bmatrix}$$

allows us to write the Lorentz transformations as

$$\underset{\sim}{x'} = \underset{\approx}{\Lambda}\underset{\sim}{x}, \quad \underset{\sim}{x} = \underset{\approx}{\Lambda}^{-1}\underset{\sim}{x'} \quad \text{where} \quad \underset{\approx}{\Lambda}\underset{\approx}{\Lambda}^{-1} = \underset{\approx}{\Lambda}^{-1}\underset{\approx}{\Lambda} = \underset{\approx}{I} \equiv \begin{bmatrix} 1 & 0 & 0 & 0 \\ 0 & 1 & 0 & 0 \\ 0 & 0 & 1 & 0 \\ 0 & 0 & 0 & 1 \end{bmatrix}$$

Any quantity that transforms this way under changes of reference frame is called a four-vector.

Another four-vector can be built from a particle's energy E and momentum $\vec{p}$:

$$\underset{\sim}{p} \equiv \begin{bmatrix} E/c \\ p_x \\ p_y \\ p_z \end{bmatrix}$$

since it transforms like this: $\underset{\sim}{p'} = \underset{\approx}{\Lambda}\underset{\sim}{p}$. (We discuss this more in later units.)

Naturally, this is just shorthand for the four equations:

$$p'_x = \gamma(p_x - \beta E/c), \quad p'_y = p_y, \quad p'_z = p_z, \quad E'/c = \gamma(E/c - \beta p_x).$$

(a) For $\underset{\approx}{\Lambda}$ defined as

$$\begin{bmatrix} \gamma & -\gamma\beta & 0 & 0 \\ -\gamma\beta & \gamma & 0 & 0 \\ 0 & 0 & 1 & 0 \\ 0 & 0 & 0 & 1 \end{bmatrix}$$

prove that its inverse $\underset{\approx}{\Lambda}^{-1}$ is

$$\begin{bmatrix} \gamma & +\gamma\beta & 0 & 0 \\ +\gamma\beta & \gamma & 0 & 0 \\ 0 & 0 & 1 & 0 \\ 0 & 0 & 0 & 1 \end{bmatrix}$$

Not surprisingly, the inverse of $\underset{\approx}{\Lambda}$ is the transformation with the velocity reversed in sign.

Solution

All we need to do is show that the product of the two matrices is the identity matrix. Here, it is

$$\begin{bmatrix} \gamma & -\gamma\beta & 0 & 0 \\ -\gamma\beta & \gamma & 0 & 0 \\ 0 & 0 & 1 & 0 \\ 0 & 0 & 0 & 1 \end{bmatrix} \times \begin{bmatrix} \gamma & +\gamma\beta & 0 & 0 \\ +\gamma\beta & \gamma & 0 & 0 \\ 0 & 0 & 1 & 0 \\ 0 & 0 & 0 & 1 \end{bmatrix}$$

$$= \begin{bmatrix} \gamma^2 - \gamma^2\beta^2 + 0 + 0 & \gamma^2\beta - \gamma^2\beta + 0 + 0 & 0+0+0+0 & 0+0+0+0 \\ -\gamma^2\beta + \gamma^2\beta + 0 + 0 & -\gamma^2\beta^2 + \gamma^2 + 0 + 0 & 0+0+0+0 & 0+0+0+0 \\ 0+0+0+0 & 0+0+0+0 & 0+0+1+0 & 0+0+0+0 \\ 0+0+0+0 & 0+0+0+0 & 0+0+0+0 & 0+0+0+1 \end{bmatrix}$$

$$= \begin{bmatrix} \gamma^2\left(1-\beta^2\right) & 0 & 0 & 0 \\ 0 & \gamma^2\left(1-\beta^2\right) & 0 & 0 \\ 0 & 0 & 1 & 0 \\ 0 & 0 & 0 & 1 \end{bmatrix} = \begin{bmatrix} 1 & 0 & 0 & 0 \\ 0 & 1 & 0 & 0 \\ 0 & 0 & 1 & 0 \\ 0 & 0 & 0 & 1 \end{bmatrix}$$

If you're fuzzy about that last step, recall that $\gamma^2 = 1/(1-\beta^2)$.

(b) A proton *at rest* has energy $E_{\text{proton}} = m_p c^2 \approx 938\,\text{MeV}$. (Its vector momentum, not surprisingly, is zero.) The protons which circulated inside the Fermilab Tevatron had energies close to $1{,}000\,\text{GeV}$. ($1\,\text{GeV} = 1{,}000\,\text{MeV}$.) What value of γ did a Tevatron proton have?

Solution

Lorentz transform the four-momentum of a proton at rest (the "unprimed frame") to the Fermilab (primed) frame, or at least (for now) the energy component. From a few pages ago,

$$E'/c = \gamma(E/c - \beta p_x) = \gamma E/c$$

$$\gamma = E'/E = 1000/0.938 = 1066.1$$

(c) Somehow, a Tevatron proton captures an electron, becoming a fast-moving hydrogen atom without changing its speed.

An electron at rest has $E_{\text{electron}} = m_e c^2 \approx 0.511\,\text{MeV}$. What is the electron's energy in the rest frame of Fermilab's sedentary buffalo herd?

Solution

The electron is moving at the same speed, approximately, as the proton. As a result, just multiply the rest energy of the electron by the Lorentz γ for the proton you just calculated:

$$E'_{\text{electron}} = \gamma m_e c^2 \approx 0.511\,\text{MeV} \times 1066.1 = 544.78\,\text{MeV}$$

Problem 2: Field transformation laws

As you've seen, the electric and magnetic fields are intimately connected by relativity. We can construct the "electromagnetic field strength tensor" out of the fields' components like this:

$$\underset{\approx}{F} \equiv \begin{bmatrix} 0 & -E_x/c & -E_y/c & -E_z/c \\ E_x/c & 0 & -B_z & B_y \\ E_y/c & B_z & 0 & -B_x \\ E_z/c & -B_y & B_x & 0 \end{bmatrix}$$

What makes this a "second rank tensor" (a four-vector is a "first rank tensor") is that the electric and magnetic fields transform when one changes frames this way:

$$\underset{\approx}{F'} = \begin{bmatrix} 0 & -E'_x/c & -E'_y/c & -E'_z/c \\ E'_x/c & 0 & -B'_z & B'_y \\ E'_y/c & B'_z & 0 & -B'_x \\ E'_z/c & -B'_y & B'_x & 0 \end{bmatrix} = \underset{\approx}{\Lambda}\,\underset{\approx}{F}\,\underset{\approx}{\Lambda}^T = \underset{\approx}{\Lambda}\,\underset{\approx}{F}\,\underset{\approx}{\Lambda}$$

If an observer in one frame of reference measures $\vec{E}$ and $\vec{B}$ at the point (x, y, z) at time t, that observer can predict what an observer in a different frame would measure for $\vec{E}'$ and $\vec{B}'$ should their field-measuring devices glide past each other just as the measurements are being performed:

(a) The electric and magnetic fields of a 1 Coulomb charge Q are measured by a pair of gizmos, as shown in the figure. From the perspective of observers in frame O, the charge is at rest at the origin and one of the field-measuring devices is also at rest, with position $(x, y, z) = (0, 1, 0)$. Observers in O see the second field measuring gizmo moving at high speed, with velocity $v = \beta c \hat{x}$.

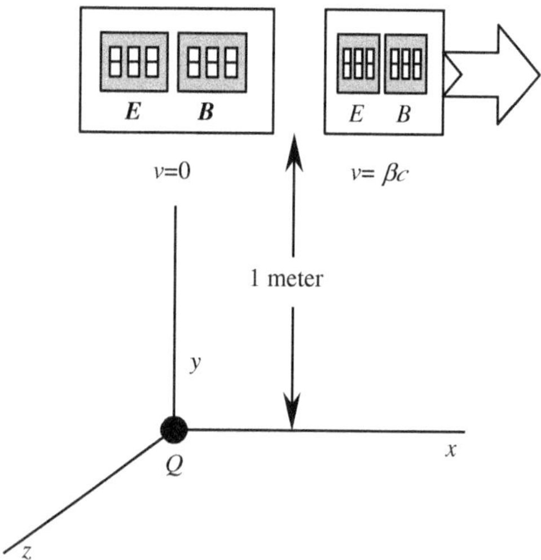

At time $t = 0$, this gizmo passed very close to the stationary device at $(x, y, z) = (0, 1, 0)$. The readings on the dials of the moving device were photographed by observers in O as it was illuminated by the lights from the stationary gizmo's panel lights:

- Calculate the components of the electric and magnetic field vectors measured by the *stationary device* at the time the photograph was taken.
- Calculate the components of the electric and magnetic field vectors measured by the *moving device.*
- According to observers in the rest frame of the moving device, how far away is the charge Q?

Solution

First bullet point: In the unprimed frame, the magnitude of the electric field measured 1 m up the y-axis is just $E = kQ/r^2 = k$ for our one Coulomb charge. The field is entirely in the $+y$-direction. There is no magnetic field since the charge is not moving.

Second bullet point: Transform to the frame of the moving device (the "primed frame") to figure out the electric and magnetic field it sees, plugging in for the components of the field strength tensor:

$$
F' = \begin{bmatrix} 0 & -E'_x/c & -E'_y/c & -E'_z/c \\ E'_x/c & 0 & -B'_z & B'_y \\ E'_y/c & B'_z & 0 & -B'_x \\ E'_z/c & -B'_y & B'_x & 0 \end{bmatrix} = \Lambda \begin{bmatrix} 0 & 0 & -k/c & 0 \\ 0 & 0 & 0 & 0 \\ k/c & 0 & 0 & 0 \\ 0 & 0 & 0 & 0 \end{bmatrix} \Lambda
$$

Do the left multiplication first:

$$
\Lambda \begin{bmatrix} 0 & 0 & -k/c & 0 \\ 0 & 0 & 0 & 0 \\ k/c & 0 & 0 & 0 \\ 0 & 0 & 0 & 0 \end{bmatrix} = \begin{bmatrix} \gamma & -\gamma\beta & 0 & 0 \\ -\gamma\beta & \gamma & 0 & 0 \\ 0 & 0 & 1 & 0 \\ 0 & 0 & 0 & 1 \end{bmatrix} \begin{bmatrix} 0 & 0 & -k/c & 0 \\ 0 & 0 & 0 & 0 \\ k/c & 0 & 0 & 0 \\ 0 & 0 & 0 & 0 \end{bmatrix}
$$

$$
= \begin{bmatrix} 0 & 0 & -k\gamma/c & 0 \\ 0 & 0 & k\gamma\beta/c & 0 \\ k/c & 0 & 0 & 0 \\ 0 & 0 & 0 & 0 \end{bmatrix}
$$

Now, do the right multiplication:

$$
\begin{bmatrix} 0 & 0 & -k\gamma/c & 0 \\ 0 & 0 & k\gamma\beta/c & 0 \\ k/c & 0 & 0 & 0 \\ 0 & 0 & 0 & 0 \end{bmatrix} \Lambda = \begin{bmatrix} 0 & 0 & -k\gamma/c & 0 \\ 0 & 0 & k\gamma\beta/c & 0 \\ k/c & 0 & 0 & 0 \\ 0 & 0 & 0 & 0 \end{bmatrix} \begin{bmatrix} \gamma & -\gamma\beta & 0 & 0 \\ -\gamma\beta & \gamma & 0 & 0 \\ 0 & 0 & 1 & 0 \\ 0 & 0 & 0 & 1 \end{bmatrix}
$$

$$
= \begin{bmatrix} 0 & 0 & -k\gamma/c & 0 \\ 0 & 0 & k\gamma\beta/c & 0 \\ k\gamma/c & -k\gamma\beta/c & 0 & 0 \\ 0 & 0 & 0 & 0 \end{bmatrix}
$$

As a result, we have

$$E'_x = 0$$

$$E'_y = \gamma k = \gamma E_y$$

$$E'_z = 0$$

$$B'_x = 0$$

$$B'_y = 0$$

$$B'_z = -\gamma \beta k / c = -\gamma \beta E_y / c$$

Third bullet point: There's no Lorentz contraction along the axes perpendicular to the velocity of the primed frame, so the charge zooms by $1\,\mathrm{m}$ from the sensor as observed from the primed frame.

(b) The same gizmos are now placed on the x-axis with the stationary device at position $(x, y, z) = (1, 0, 0)$ and the moving device traveling *along the x-axis* with velocity $v = \beta c \hat{x}$. As the moving device brushes past the stationary device, photographs are taken. (The charge is still at the origin.)

- Calculate the components of the electric and magnetic field vectors measured by the *stationary device* at the time the photograph was taken.
- Calculate the components of the electric and magnetic field vectors measured by the *moving device*.
- According to observers in the rest frame of the moving device, how far away is the charge Q?

Solution

First bullet point: In the unprimed frame, the electric field is along the positive x-axis, with strength k; the magnetic field is still zero.

Second bullet point: Just do the Lorentz transformation on the electromagnetic field strength tensor in the unprimed frame to determine its components in the primed frame. In an equation,

$$\underset{\approx}{F'} = \begin{bmatrix} 0 & -E'_x/c & -E'_y/c & -E'_z/c \\ E'_x/c & 0 & -B'_z & B'_y \\ E'_y/c & B'_z & 0 & -B'_x \\ E'_z/c & -B'_y & B'_x & 0 \end{bmatrix} = \underset{\approx}{\Lambda} \begin{bmatrix} 0 & -k/c & 0 & 0 \\ k/c & 0 & 0 & 0 \\ 0 & 0 & 0 & 0 \\ 0 & 0 & 0 & 0 \end{bmatrix} \underset{\approx}{\Lambda}$$

Do the left- and right-multiplications in the usual way, just like we did in the previous problem (you don't really need me to show all the steps this time!) to find

$$\underset{\approx}{F'} = \begin{bmatrix} 0 & -E'_x/c & -E'_y/c & -E'_z/c \\ E'_x/c & 0 & -B'_z & B'_y \\ E'_y/c & B'_z & 0 & -B'_x \\ E'_z/c & -B'_y & B'_x & 0 \end{bmatrix}$$

$$= \begin{bmatrix} 0 & -\gamma^2(1-\beta^2)k/c & 0 & 0 \\ \gamma^2(1-\beta^2)k/c & 0 & 0 & 0 \\ 0 & 0 & 0 & 0 \\ 0 & 0 & 0 & 0 \end{bmatrix}$$

$$= \begin{bmatrix} 0 & -k/c & 0 & 0 \\ k/c & 0 & 0 & 0 \\ 0 & 0 & 0 & 0 \\ 0 & 0 & 0 & 0 \end{bmatrix}$$

As a result, we find that the electric field is the same in both frames, and the magnetic field is zero in both frames.

Third bullet point: Observers in the primed frame think that the separation between the charge and the measuring device in the unprimed frame has been Lorentz contracted to $1/\gamma$ meters.

Problem 3: Interstellar unofficial

A group of irresponsible cosmonauts flies off towards a large interstellar cloud of ethanol. In the party-hearty cosmonauts' rest frame, all the clocks aboard their string of 101 spaceships are synchronized.

Their spaceships travel in a single file, equally spaced by 1000 feet (as measured in their rest frame), as shown in the following diagram. Cosmonaut Alexander Swift-Beagle is in the lead ship.

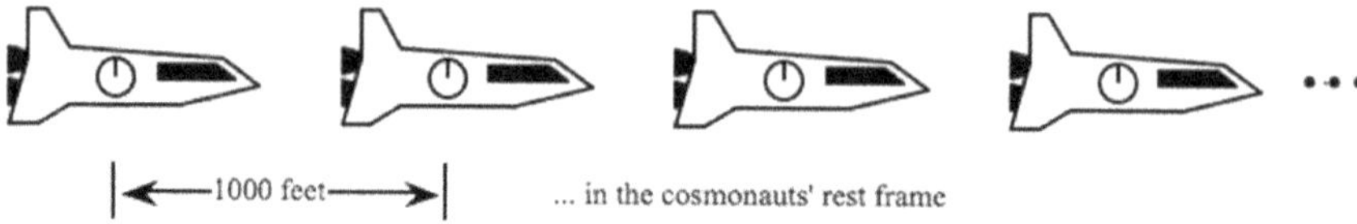

An agent from the Galactic treasury's Bureau of Alcohol, Tobacco, and Firearms is at rest with respect to the ethanol cloud and sees the cosmonauts heading his way with speed $4/5c$. He measures the cloud to be 80,000 feet thick, as shown in the following figure.

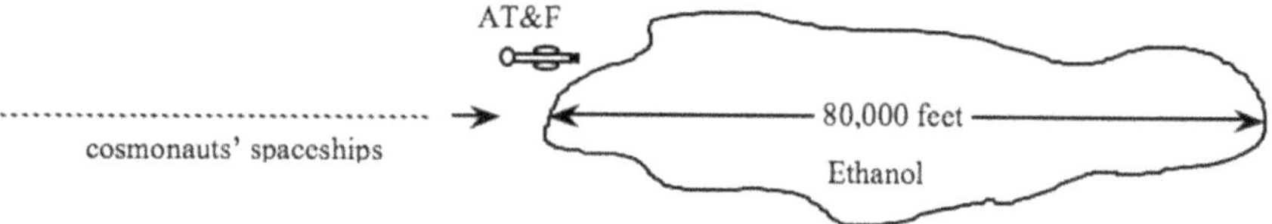

The line of cosmonauts flies through the cloud without slowing down. As Swift-Beagle enters the cloud, his clock reads zero, as does the Alcohol, Tobacco, and Firearms agent's clock.

(a) According to an observer in the ATF agent's rest frame, when does Swift-Beagle leave the ethanol cloud?

Solution

Swift-Beagle is traveling at $0.8c$, so he needs $80{,}000\,\text{ft}/0.8\,\text{ft}/$ nsec or 100,000 nsec to exit the cloud.

(b) According to the ATF agent, when does the last cosmonaut *enter* the ethanol cloud?

Solution

Note that there are 100 gaps between spaceships, not 101. These gaps are all Lorentz contracted from 1,000 ft to 600 ft, so the line of ships is only 60,000 feet long. It requires $60{,}000\,\text{ft}/0.8\,\text{ft}/\text{nsec} = 75{,}000\,\text{nsec}$ for the last cosmonaut to enter the cloud.

(c) What does Swift-Beagle's clock read when he leaves the ethanol cloud?

Solution

There are several ways to think about this. Swift-Beagle sees the "moving" cloud Lorentz contracted to 48,000 ft in length and traveling at $0.8c$, so it takes $48,000\,\text{ft}/0.8\,\text{ft}/\text{nsec} = 60,000\,\text{nsec}$ to move past him, in his rest frame.

(d) What does the last cosmonaut's clock read when he/she enters the ethanol cloud?

Solution

Easy: this cosmonaut thinks his/her ship is 100,000 feet away from the lead ship when all spaceship clocks read zero. The cloud is approaching at $0.8c$, so the cloud takes $100,000\,\text{ft}/0.8\,\text{ft}/\text{nsec} = 125,000$ to reach the last cosmonaut.

Problem 4: A little more calculus practice

An object of mass m moves through a gooey fluid so that it slows down due to viscous drag. At time $t = 0$, the object has initial velocity v_0; the drag force is velocity-dependent, so the net force on the object is

$$F = m\frac{dv}{dt} = -bv$$

with b a positive constant. (Ignore relativistic effects for the time being.)

This is a differential equation, of course, and to a great extent, we can treat the derivative as if it were a fraction. This lets us rewrite the above force equation as

$$\frac{dv}{v} = -\frac{b}{m}dt$$

which we can integrate:

$$\int \frac{dv}{v} = -\int \frac{b}{m}dt$$

From this, determine the object's velocity as a function of time, $v(t)$. Note that you will use the initial condition $v(0) = v_0$ to determine the value of the otherwise undetermined constant of integration that arises when you do the (indefinite) integral.

Solution

$$\int \frac{dv}{v} = -\int \frac{b}{m} dt$$

$$\ln(v(t)) = -bt/m + \ln(v(0))$$

$$e^{\ln(v(t))} = e^{-bt/m + \ln(v(0))}$$

$$v(t) = v(0)e^{-bt/m}$$

Roussillon, France

Unit 4

Developing the Mathematical Tools of Relativity—Scalars, Four-Vectors, Lorentz Tensors, the Metric Tensor, Covariant Notation: "The Ehrenfest Paradox"

In this unit, we streamline the mathematical notation we use to write equations describing changes of frame and relativistic transformations. But—entirely for fun—we begin with a description

67

of the "Ehrenfest paradox," in which a spoked wheel (whose center is kept fixed) is suddenly made to rotate so that its rim is moving relativistically.

Just for fun: The Ehrenfest Paradox

What happens if we suddenly set a spoked wheel to spinning so rapidly that its rim is moving relativistically? Does the wheel contract? (The wheel's center stays fixed in our frame.) If so, what happens to the spokes, whose motion is transverse to their lengths?

To answer these questions, we need to consider how we go about making the wheel spin.

I suspect that most physicists get this one wrong and flail heroically, but futilely, in their attempts to analyze the system. That's been my observation, discussing this with my fellow professors at Illinois. For sport, you could ask your own professors!

Here's a diagram:

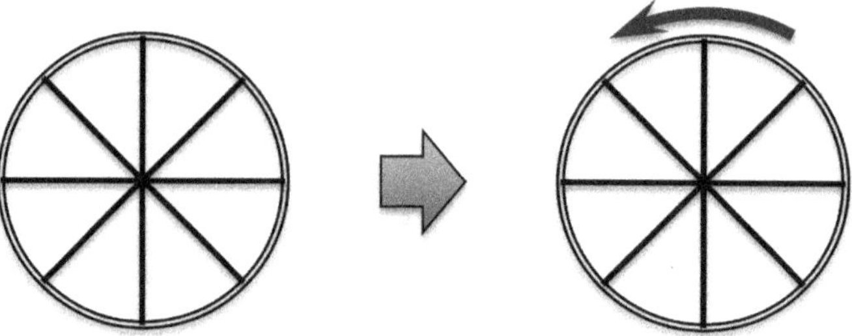

There is no single *inertial (non-rotating) frame* in which we can see the stationary-in-the-classroom wheel to be rotating rapidly. We can only make it spin by doing something to induce rapid rotations rather than just jumping into a different frame of reference.

Let's analyze how we go about setting it spinning.

Place the non-rotating wheel on a mat of synchronized clocks, each attached to a super-fast mechanical driver that will (just as its clock reads zero) kick the part of the wheel at its location into motion. Perhaps this process takes only 10^{-24} seconds so that nothing can move more than 3×10^{10} cm/sec $\times 10^{-24}$ sec $= 3 \times 10^{-14}$ cm during this process. (Recall that the diameter of a proton is roughly 1.7×10^{-13} cm.)

The wheel cannot possibly change shape because the relative spacing between parts of the wheel (when viewed from the lab frame) cannot change as it spins up! Each piece of stuff is given the necessary velocity to be executing a rotation around the center with angular speed ω, and each piece is given this kick at exactly the same instant. Nothing has a chance to move toward (or away from) a piece of the wheel that is still stationary.

See the following figure to help visualize what I mean.

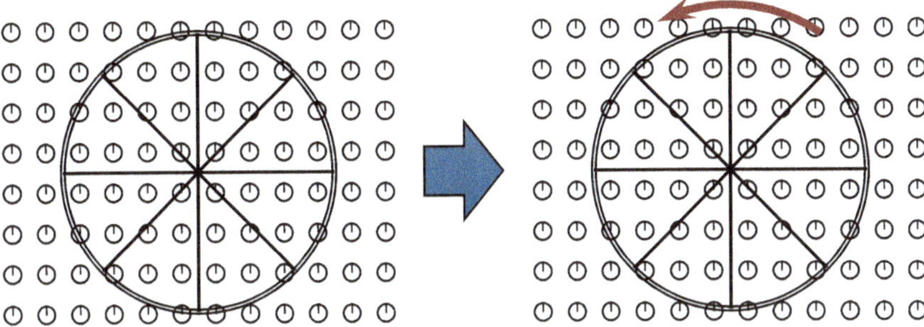

Viewed from the lab frame, there is no change in the shape of the wheel. There appear to be no Lorentz contractions. Curious!

Here's a diagram of a small part of a (very large) wheel, with stationary clocks installed on the wheel itself, as well as free-floating clocks already in motion to the right at $0.8c$. In the initial frame, the spacing between the clocks in motion is Lorentz contracted to $30\,\text{m}$, matching the $30\,\text{m}$ separation between clocks on the wheel. (Naturally, the moving clocks are considerably farther apart in their rest frame!) The stationary accelerating devices are 30 meters apart and below the wheel; they fire simultaneously.

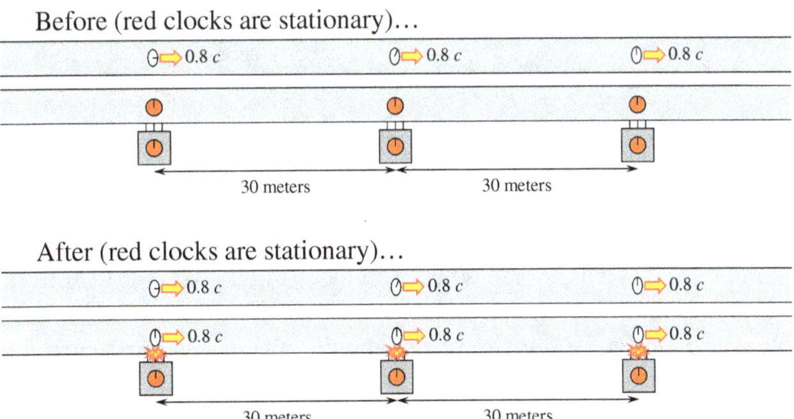

Now, switch to the frame of the upper set of moving clocks, keeping in mind that clocks synchronized in their rest frame will not be synchronized in a different frame.

In the following diagrams, the times at which the wheel rim clocks come to rest are different from each other! Take note of the spacings between clocks and the readings on each of the clocks.

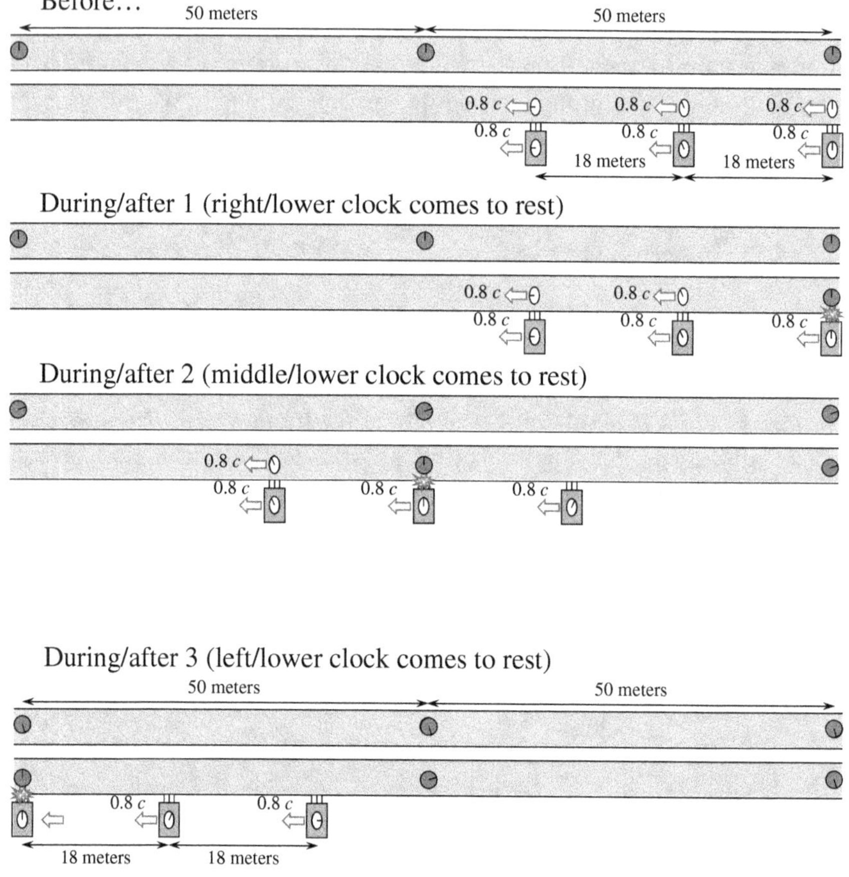

Since, in this frame, the wheel rim clocks come to rest at different times, their spacing *increases.*

If you'd prefer to see the wheel shown with some curvature, here are more diagrams, first in the original frame

Before...

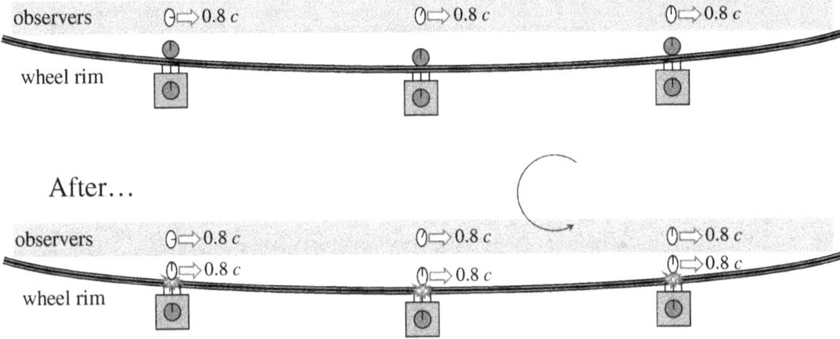

and then in a frame moving to the right.

Before...

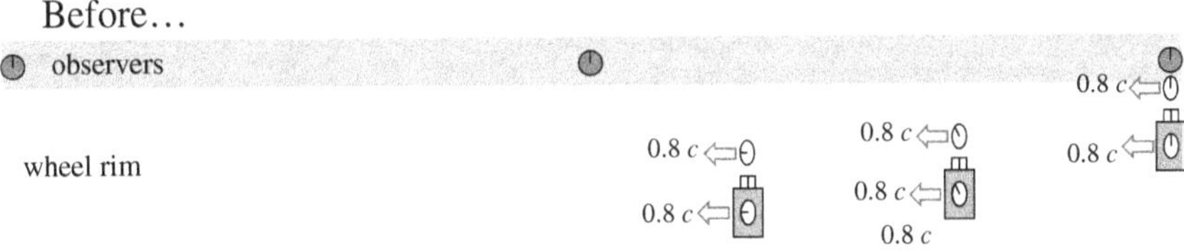

During/after 1... same as before: part of the rim comes to rest.

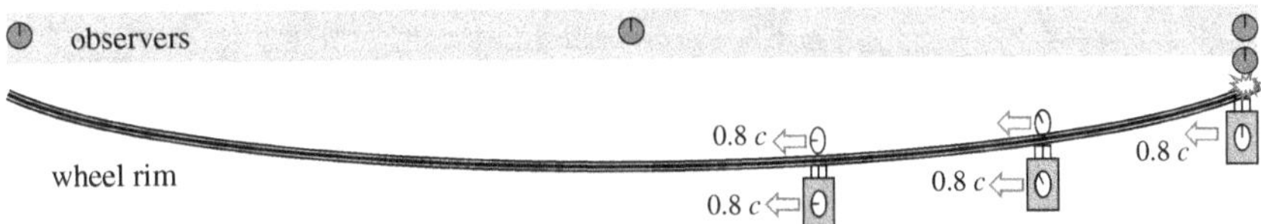

The spacing between atoms in the lower part of the rim *increases*.

Now, place a group of observers just above the rim, at the top of the wheel and moving to the right. (They are in the same frame as the line of moving observers near the bottom of the wheel.)

In the following diagram, we are in the rest frame of the wheel before it begins to rotate, and we stay in this frame after the wheel is rotating.

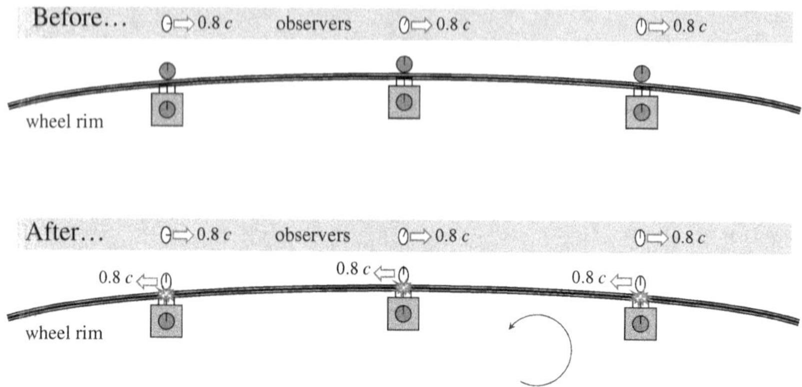

But now look at how this presents, from a frame that is moving to the right with $0.8c$ before the wheel starts rotating.

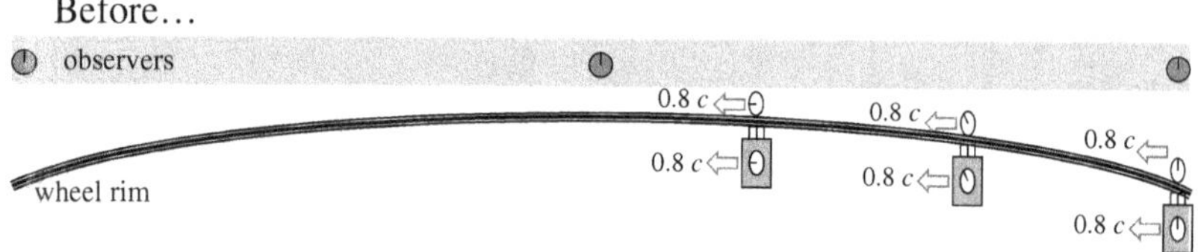

During/after 1... the right-most clock begins moving ***even faster*** than the center and left clocks. The distance between the right clock and its left-neighbors begins to decrease.

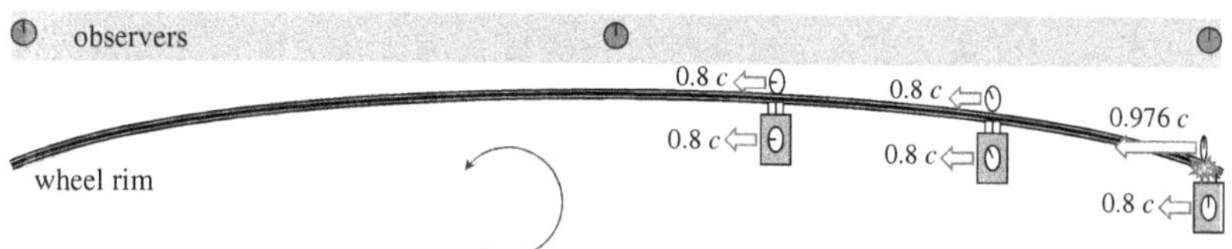

The spacing between atoms in the upper part of the rim *decreases*.

So, that's what's going on: in this frame, the inter-clock spacing at the top of the wheel decreases, while the spacing at the bottom of the wheel increases. Once all parts of the wheel are participating in its rotation, this varying clock-to-clock spacing persists since for every clock entering the top portion of the rim from the right, another exits the top portion of the rim from the left.

Clocks on the right side of the wheel begin moving earliest, feeding more clocks into the top of the wheel before clocks flow out of the top into the left side.

In words, in the original frame, in which the center of the wheel is at rest, the wheel remains round, and the spacing between clocks on its rim does not change after it is spinning.

But in the frame in which the wheel's center is always moving to the left, the wheel is an oval. Its horizontal "diameter" is Lorentz contracted by a factor of $\sqrt{1 - \beta^2} = 0.6$, while its vertical "diameter" is unchanged. Before it begins rotating, the spacing between clocks is Lorentz contracted, with the amount of contraction depending on how much of the clock-to-clock displacement is parallel to the wheel center's velocity. After it is rotating, the clock-to-clock spacing is no longer the same at the top and bottom of the spinning wheel.

OK, back to work.

Introduction: Streamlining our notation

During the last unit, you worked with a matrix equation version of the Lorentz transformations. We're going to develop a more compact version of that today, with slightly different notation. There isn't any new physics in this.

I had defined a four-vector as a four-component object built from the time and space coordinates of an event this way:

$$\underset{\sim}{x} \equiv \begin{bmatrix} ct \\ x \\ y \\ z \end{bmatrix}$$

Under the Lorentz transformation to a different frame of reference moving along the x-axis of the original frame, the four-vector changed this way:

$$\begin{bmatrix} ct' \\ x' \\ y' \\ z' \end{bmatrix} = \begin{bmatrix} \gamma & -\gamma\beta & 0 & 0 \\ -\gamma\beta & \gamma & 0 & 0 \\ 0 & 0 & 1 & 0 \\ 0 & 0 & 0 & 1 \end{bmatrix} \begin{bmatrix} ct \\ x \\ y \\ z \end{bmatrix} \qquad \text{or} \quad \underset{\sim}{x}' = \underset{\approx}{\Lambda}\,\underset{\sim}{x}$$

This matrix equation is just a convenient way to write the four equations that give us the space–time coordinates in the new frame. (We assume in almost all cases that the origins of the two frames coincide at $t = t' = 0$.)

Let's refer to specific components of these four-vectors using a raised index, rather than a subscript, with the time component assigned the index 0, and the space components the indices $1, 2, 3$. I refer to components of the Lorentz transformation matrix with a raised first index for the row and a lowered second index for the column.

Due to the way matrix multiplication works, the equation for the μth term of x' is this:

$$x'^{\mu} = \Lambda^{\mu}_{0}x^{0} + \Lambda^{\mu}_{1}x^{1} + \Lambda^{\mu}_{2}x^{2} + \Lambda^{\mu}_{3}x^{3} = \sum_{v=0}^{v=3} \Lambda^{\mu}_{v}x^{v}$$

Terms such as Λ^{μ}_{0} and x^{0} are just numbers, so it doesn't matter whether we write their product as $\Lambda^{\mu}_{0}x^{0}$ or $x^{0}\Lambda^{\mu}_{0}$.

In formal terms, a four-vector defined this way (and whose components are labeled with raised indices) is a *contravariant* four-vector. It is common for physicists to be a little sloppy in how we use this notation, writing, for example:

$$x^{\mu} \equiv \begin{vmatrix} ct \\ x \\ y \\ z \end{vmatrix}$$

Here, the μ superscript just serves as a flag telling us that x^{μ} is a four-vector rather than needing to be assigned a particular value chosen from $0, 1, 2, 3$. And the three-barred equal sign means "is defined to be."

There's no new physics in any of this: it's just another way of expressing the Lorentz transformations:

$$x' = \gamma(x - \beta ct), \quad y' = y, \quad z' = z, \quad t' = \gamma(t - \beta x/c)$$
$$x = \gamma(x' + \beta ct'), \quad y = y', \quad z = z', \quad t = \gamma(t' + \beta x'/c)$$

Sometimes (especially in more advanced courses), it will be convenient to use a version of a four-vector in which the space components (but not the time component) carry negative signs. We label the components of this kind of four-vector with lowered—rather than raised—indices. In other words, the *covariant* space–time four-vector is defined this way:

$$x_\mu \equiv \begin{bmatrix} ct \\ -x \\ -y \\ -z \end{bmatrix}$$

The Lorentz transformations still work fine, but we need to take into account the negative sign that we've put in front of $x_1 \equiv -x$, $x_2 \equiv -y$, $x_3 \equiv -z$. For example, if we express $x' = \gamma(x - \beta ct)$ in terms of covariant quantities, the equation becomes $-x_1' = \gamma(-x_1 - \beta x_0)$ or $x_1' = \gamma(x_1 + \beta x_0)$.

The equation $t' = \gamma(t - \beta x/c)$ becomes $x_0'/c = \gamma(x_0/c + \beta x_1/c)$ or $x_0' = \gamma(x_0 + \beta x_1)$.

As a result, the Lorentz transformation matrix for covariant four-vectors has positive signs in front of the $\gamma\beta$ terms. We write the transformation matrix with its first index lowered and its second raised: Λ_v^μ.

It takes some getting used to, but once you're comfortable with it, you'll find it to be an improvement in notation.

Here is a summary:

Kind of four-vector	Lorentz transformation rule	Lorentz transformation matrix
Contravariant, e.g. $x^\mu \equiv (ct, x, y, z)$	$x'^\mu = \sum_v \Lambda_v^\mu x^v$	$\Lambda_v^\mu = \begin{bmatrix} \gamma & -\gamma\beta & 0 & 0 \\ -\gamma\beta & \gamma & 0 & 0 \\ 0 & 0 & 1 & 0 \\ 0 & 0 & 0 & 1 \end{bmatrix}$
Covariant, e.g. $x_\mu \equiv (ct, -x, -y, -z)$	$x_\mu' = \sum_v \Lambda_\mu^v x_v$	$\Lambda_\mu^v = \begin{bmatrix} \gamma & \gamma\beta & 0 & 0 \\ \gamma\beta & \gamma & 0 & 0 \\ 0 & 0 & 1 & 0 \\ 0 & 0 & 0 & 1 \end{bmatrix}$

Two more useful objects at our disposal are the contravariant and covariant metric tensors $g^{\mu v}$ and $g_{\mu v}$. They are defined this way:

$$g^{\mu v} = g_{\mu v} \equiv \begin{bmatrix} 1 & 0 & 0 & 0 \\ 0 & -1 & 0 & 0 \\ 0 & 0 & -1 & 0 \\ 0 & 0 & 0 & -1 \end{bmatrix}$$

Here's what we take on in this unit:

- working with co(contra)variant four-vectors and Lorentz transformations;
- exploring the Lorentz transformation properties of scalars and four-vectors.

Exercise 4.1: Turning covariants into contravariants

The metric tensor can be used to flip a covariant four-vector into a contravariant four-vector and vice versa.

(a) Prove this: $x_\mu = \sum_\nu g_{\mu\nu} x^\nu$.

Solution

Just multiply it out! If in doubt, write it out component by component:

$$\sum_\nu g_{\mu\nu} x^\nu = g_{\mu 0} x^0 + g_{\mu 1} x^1 + g_{\mu 2} x^2 + g_{\mu 3} x^3$$

$$= \begin{cases} x^0 = ct = x_0 & \text{when } \mu = 0 \\ -x^1 = -x = x_1 & \text{when } \mu = 1 \\ -x^2 = -y = x_2 & \text{when } \mu = 2 \\ -x^3 = -z = x_3 & \text{when } \mu = 3 \end{cases}$$

since $g_{\mu\nu} = 0$ when $\mu \neq \nu$.

(b) Prove this: $x^\mu = \sum_\nu g^{\mu\nu} x_\nu$.

Solution

$$\sum_\nu g^{\mu\nu} x_\nu = g^{\mu 0} x_0 + g^{\mu 1} x_1 + g^{\mu 2} x_2 + g^{\mu 3} x_3$$

$$= \begin{cases} x_0 = ct = x^0 & \text{when } \mu = 0 \\ -x_1 = +x = x^1 & \text{when } \mu = 1 \\ -x_2 = +y = x^2 & \text{when } \mu = 2 \\ -x_3 = z = x^3 & \text{when } \mu = 3 \end{cases}$$

(c) Prove this: $g^\mu_\nu \equiv \sum_\sigma g^{\mu\sigma} g_{\sigma\nu} = \begin{bmatrix} 1 & 0 & 0 & 0 \\ 0 & 1 & 0 & 0 \\ 0 & 0 & 1 & 0 \\ 0 & 0 & 0 & 1 \end{bmatrix}$

Solution

Once again, just multiply it out. Note that this sum is just matrix multiplication of the two tensors:

$$g^{\mu\nu} = g_{\mu\nu} \equiv \begin{bmatrix} 1 & 0 & 0 & 0 \\ 0 & -1 & 0 & 0 \\ 0 & 0 & -1 & 0 \\ 0 & 0 & 0 & -1 \end{bmatrix}$$

$$\sum_\sigma g^{\mu\sigma} g_{\sigma\nu} = g^{\mu 0} g_{0\nu} + g^{\mu 1} g_{1\nu} + g^{\mu 2} g_{2\nu} + g^{\mu 3} g_{3\nu}$$

$$= \begin{cases} 1 & \text{when } \mu = \nu = 0 \\ 1 & \text{when } \mu = \nu = 1 \\ 1 & \text{when } \mu = \nu = 2 \\ 1 & \text{when } \mu = \nu = 3 \\ 0 & \text{when } \mu \neq \nu \end{cases}$$

Exercise 4.2: Lorentz scalars

Even though the Lorentz transformations mix space and time intervals when one changes frames of reference, there are some things that are invariant under Lorentz transformations. These *Lorentz scalars* have the same value in all frames of reference. One example is the speed of light in vacuum, which is the same in all frames. Another example—which you will prove momentarily—is the "proper time" between two events. (An event is anything that happens at a single point in space, at a single instant in time.)

Imagine that the first event, viewed from the unprimed frame, occurs at the origin at time $t = 0$, while the second takes place at the space–time point (ct, x, y, z).

Let's assume that the primed frame is moving along the x-axis of the unprimed frame with speed βc and that, as the origins of the two frames coincide, clocks at the origins read $t = 0$ and $t' = 0$.

In the unprimed frame, the square of the proper time between the two events is defined to be

$$(\Delta\tau)^2 \equiv (t_2 - t_1)^2 - \frac{1}{c^2}\left((x_2 - x_1)^2 + (y_2 - y_1)^2 + (z_2 - z_1)^2\right)$$

$$= t^2 - \frac{1}{c^2}(x^2 + y^2 + z^2)$$

while in the primed frame, it is defined as $(\Delta\tau')^2 \equiv t'^2 - \frac{1}{c^2}(x'^2 + y'^2 + z'^2)$. (Note that I'll usually leave off the parentheses around $\Delta\tau$ to write its square as $\Delta\tau^2$.) Don't be alarmed that this can be negative: the interval between two events with large spatial (but small temporal) separation is an example.

By expressing the primed variables in terms of the unprimed variables ($x' = \gamma(x - \beta ct)$, for example), prove that $\Delta\tau'^2 = \Delta\tau^2$.

Solution

Here we go! Let's write $\Delta\tau'^2$ in terms of the primed variables, then replace them with the Lorentz-transformed unprimed variables:

$$\left(\Delta\tau'\right)^2 = t'^2 - \frac{1}{c^2}\left(x'^2 + y'^2 + z'^2\right)$$

$$= [\gamma(t - \beta x/c)]^2 - \frac{1}{c^2}[\gamma(x - \beta ct)]^2 - \frac{1}{c^2}\left[y^2 + z^2\right]$$

$$= \gamma^2\left[t^2 - 2t\beta x/c + \beta^2 x^2/c^2\right] - \frac{\gamma^2}{c^2}$$

$$\times\left[x^2 - 2x\beta ct + \beta^2 c^2 t^2\right] - \frac{1}{c^2}\left[y^2 + z^2\right]$$

$$= \left(\gamma^2\beta^2 x^2/c^2 - \frac{\gamma^2}{c^2}x^2\right) + \left(-\gamma^2 2t\beta x/c + \frac{\gamma^2}{c^2}2x\beta ct\right)$$

$$+ \left(\gamma^2 t^2 - \frac{\gamma^2}{c^2}\beta^2 c^2 t^2\right) - \frac{1}{c^2}\left[y^2 + z^2\right]$$

$$= \frac{\gamma^2}{c^2}x^2\left(\beta^2 - 1\right) + \gamma^2 t^2\left(1 - \beta^2\right) - \frac{1}{c^2}\left[y^2 + z^2\right]$$

Since $\gamma^2 \left(1 - \beta^2\right) = 1$, we have the above is equal to

$$= t^2 - \frac{1}{c^2} \left[x^2 + y^2 + z^2\right]$$
$$= \Delta\tau^2$$

As a result, the proper time interval is a *Lorentz scalar.*

Exercise 4.3: Light beam

Consider two events. The first, which occurs as the origins of our two frames coincide, is the detonation of an explosive which emits a flash of light. The second event is the arrival of the light flash at a camera at (x, y, z) just as its shutter opens at time t, recording the image of the distant explosion.

Prove that the proper time interval between these two events satisfies $\Delta\tau^2 = 0$.

Solution

Easy: since the light travels at speed c, we have

$$\sqrt{x^2 + y^2 + z^2} = ct$$
$$t^2 - \frac{x^2 + y^2 + z^2}{c^2} = \Delta\tau^2 = 0$$

Exercise 4.4: Causality

Clocks on Earth are synchronized with clocks on Mrs. Rumpdock's starship after it lands on Tau Ceti IV in search of indigenous alien life to defraud. On March 16, Mrs. Rumpdock's dimwitted daughter sends Mrs. Rumpdock birthday greetings from the family's spaceport near Spokane, Washington. The Tau Ceti system is 12 light years from Earth.

Ten years later, Mrs. Rumpdock is awakened by an alert from her starship's onboard systems.

(a) Identify the two events as the transmission of the message from Spokane and the awakening of Mrs. Rumpdock, calculate $\Delta\tau^2$ between the two events. (Convenient for you: recall that the speed of light is one lightyear per year.)

Solution

$$\Delta x = 12\text{LY}, \Delta t = 10\text{Y}$$

$$\text{so } \Delta\tau^2 = (10\text{Y})^2 - \left(\frac{12\text{LY}}{c}\right)^2$$

$$= (100 - 144)\text{Y}^2 = -44\text{Y}^2$$

$$\text{since } c = 1 \text{ lightyear per year.}$$

(b) Could reception of the signal from Earth have triggered the alert from the ship that woke Mrs. Rumpdock? Explain.

Solution

No. You can see from the velocity addition formula that nothing can go faster than light, so the signal from Earth could not have traveled 12 lightyears in only 10 years.

We refer to the interval between events with $\Delta\tau^2 < 0$ as "spacelike" while the separation of those with $\Delta\tau^2 > 0$ as "timelike."

Exercise 4.5: Using the new notation

Show that $\sum_\mu x^\mu x_\mu$ is a Lorentz invariant: that it has the same value in all frames of reference.

Solution

Just plug and chug:

$$\sum_\mu x^\mu x_\mu = (ct)^2 - x^2 - y^2 - z^2 = c^2\tau^2$$

Both c^2 and τ^2 are Lorentz invariants, so their product is too. Note that another way to have written the sum would have been this:

$$\sum_\mu x^\mu x_\mu = \sum_\mu x^\mu \left[\sum_\nu g_{\mu\nu}x^\nu\right] = \sum_\mu \sum_\nu x^\mu x^\nu g_{\mu\nu}$$

The double sum has 16 terms in all, but only those with $\mu = v$ are non-zero.

Exercise 4.6: Proper time and elapsed time on a moving clock

We note that the folks in the primed frame have a clock at their origin, which happens to read zero as it zooms past our origin in the unprimed frame. The origin of the primed frame moves along our x-axis with speed βc.

We see that the moving clock is ticking slowly since it is moving with respect to our origin. When all of our clocks read a time t, the moving clock crashes into a rock and stops running. In our (unprimed) frame, the stationary rock is on the x-axis at the point $(D, 0, 0)$.

(a) From our (unprimed) perspective, how long does it take for the clock to travel from our origin to the location of the rock?

Solution

Time is distance divided by velocity, so $t = D/\beta c$.

(b) What is the reading on the face of the broken clock?

Solution

Since it ticks slowly, we just need to apply that slow-down factor:

$$t' = t\sqrt{1 - \beta^2} = \frac{D\sqrt{1 - \beta^2}}{\beta c} = \frac{D}{\gamma \beta c}$$

(c) Taking the coincidence of the origins of the two frames as the first event and the collision with the rock as the second event, what is the proper time τ between those two events?

Solution

The easiest way is to calculate it in the frame of the clock that collides with a rock: as far as it is concerned, it is sitting at its origin and gets clobbered by a rock. So, $\tau = \frac{D}{\gamma \beta c}$.

(d) How does the proper time τ compare with the reading t' on the face of the broken clock?

Solution

They are the same, of course.

A little more on this

Any four-component object that transforms under changes of frame in the same way as the contravariant space–time four-vector x^μ is automatically a contravariant four-vector.

We can define new four-vectors from existing four-vectors by, for example, multiplying (or dividing) them by Lorentz scalars. You know of two Lorentz scalars so far: the speed of light (which is the same in all frames of reference) and the proper time interval between two events.

Imagine that an object zooms past the origin (that's the first event) and collides with something at time t and at the point (x, y, z). If the object was traveling in a straight line at constant speed, it would be natural for us to identify its conventional velocity as the ratio of its change in position and the elapsed time between events: $\vec{v} \equiv (x/t, y/t, z/t)$.

We can define an object's *four-velocity* in a similar fashion:

$$u^\nu \equiv \frac{x^\nu}{\tau} = \frac{(ct, x, y, z)}{\tau}$$

Note the use of proper time in the denominator since we need to use a Lorentz scalar.

You found that the time that passes on the clock attached to the moving object is just τ and that this clock ticks slowly by a factor of γ. As a result,

$$t = \gamma\tau$$

which lets us conclude that

$$u^\nu = \frac{(ct, x, y, z)}{t/\gamma} = \gamma\left(c, \frac{x}{t}, \frac{y}{t}, \frac{z}{t}\right) = \gamma(c, \vec{v})$$

Since the four-velocity is a four-vector, under Lorentz transformations, it will obey

$$u'^\mu = \sum \Lambda^\mu_\nu u^\nu$$

Summary and wrap-up discussion

We've discussed a number of mathematical objects that behave well under Lorentz transformations. There's no new physics here, but it is so very convenient to use the new machinery—four-vectors, scalars, tensors—that we will make extensive use of them for the rest of the course. And note that all four-vectors transform the same way under boosts: the way the various components mix together is the same for a space–time interval as for a four-momentum, and for any other four-component object we identify as a four-vector.

Problem 1: Three-vectors, rotation matrices, and the metric tensor

Let's restrict ourselves to vectors in the x, y plane for the moment. Imagine that we have a vector $\vec{A}$ with components (A_x, A_y) as shown in the figure.

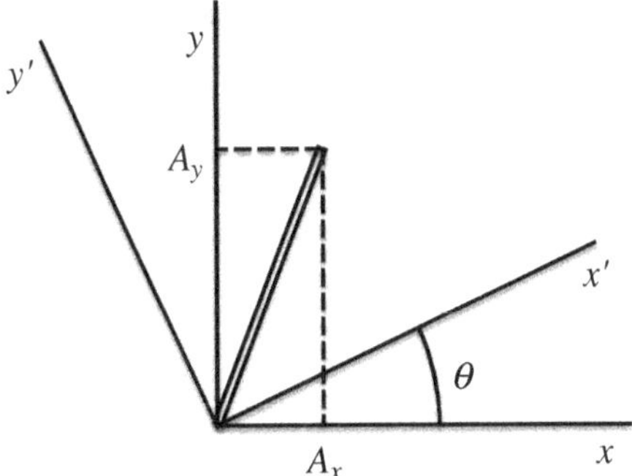

The squared length of the vector is $L^2 = A_x^2 + Ay^2$, of course: there are no terms proportional to $A_x A_y$ that enter into our calculation of length. If we wanted to express the length of a vector in an N-dimensional space, we'd go about the calculation in the same way:

$$L^2 = \sum_{i=1}^{i=N} A_i^2.$$

Again, there are no terms involving products such as $A_i A_j$ with $i \neq j$.

Observers using the primed axes x', y' will determine the components of the same vector to be (A'_x, A'_y). It is obvious that its length will be unchanged: $L'^2 = A'^2_x + Ay'^2 = L^2 = A_x^2 + Ay^2$. Note that the angle θ shown in the diagram is positive.

With some thought, you'll realize that $A'_x = A_x \cos\theta + A_y \sin\theta$ and $A'_y = -A_x \sin\theta + A_y \cos\theta$. We can write this as a matrix equation:

$$\begin{bmatrix} A'_x \\ A'_y \end{bmatrix} = \begin{bmatrix} \cos\theta & \sin\theta \\ -\sin\theta & \cos\theta \end{bmatrix} \begin{bmatrix} A_x \\ A_y \end{bmatrix}$$

(a) With this definition of the transformation of a vector under rotations, prove that the length of a vector is unchanged by a rotation of coordinate axes. Do this by showing that $A'^2_x + Ay'^2 = A^2_x + Ay^2$.

Solution

Grind it out and make use of the famous trig identity:

$$A'^2_x + A'^2_y = A^2_x \cos^2\theta + A^2_y \sin^2\theta + 2A_x A_y \cos\theta \sin\theta + A^2_x \sin^2\theta$$
$$+ A^2_y \cos^2\theta - 2A_x A_y \cos\theta \sin\theta$$
$$= A^2_x \left(\cos^2\theta + \sin^2\theta\right) + A^2_y \left(\sin^2\theta + \cos^2\theta\right) = A^2_x + A^2_y$$

What makes something a vector is the fact that it transforms under rotations according to the rule described above. If it doesn't transform that way, it isn't a vector.

(b) Imagine that we are using a coordinate system whose origin is near UIUC's Physics Department and for which the x- and y-axes lie along Green Street and Goodwin Avenue, respectively. Consider a three-component object defined this way:

$G = $ (temperature in Champaign, your age in seconds, distance from the earth to the moon).

Does G transform like a vector under rotations? Why, or why not?

Solution

It clearly does **not** transform like a vector under rotations. None of the components, because of how we've defined them, change when we rotate our coordinate axes.

The fact that the (invariant under rotations) length of a vector is the sum of the squares of the vector's components says something fundamental about the nature of Euclidean geometry in Cartesian coordinate systems.

We can rewrite the expression for the length this way:

$$L^2 = \sum_{i=1}^{i=3} A_i^2 = \sum_{i=1}^{i=3}\sum_{j=1}^{j=3} A_i A_j \eta^{ij}$$

where

$$\eta^{ij} \equiv \begin{bmatrix} 1 & 0 & 0 \\ 0 & 1 & 0 \\ 0 & 0 & 1 \end{bmatrix}$$

This way, we've put the nature of Euclidean geometry into our definition of the Euclidean metric tensor η^{ij} rather than into the vector A itself.

We're doing something akin to that with the metric tensor used in Special Relativity: the invariant under Lorentz transformations associated with four-vectors is calculated using a metric tensor with one positive, three negative, and twelve zero components. To remind you, we have

$$(c\Delta\tau)^2 = \sum_{\mu} x^\mu x_\mu = \sum_{\mu}\sum_{v} x^\mu x^v g_{\mu v} \quad \text{for} \quad g_{\mu v} \equiv \begin{bmatrix} 1 & 0 & 0 & 0 \\ 0 & -1 & 0 & 0 \\ 0 & 0 & -1 & 0 \\ 0 & 0 & 0 & -1 \end{bmatrix}$$

Problem 2: High perihelion

The starship *Asher-Danzig Kohlrabi* fires a shot at the interstellar garbage scow *Richard Howler's Rutabaga* in deep space. Both ships are observed to be traveling at relativistic speeds parallel to the x-axis by an Immigration and Customs Enforcement officer prowling the shipping lanes in search of space aliens. As measured in the ICE officer's frame, *Kohlrabi's* shot is fired at $x = -600{,}000$ feet and $t = 0$ nanoseconds. A short time later, the officer observes an explosion on board *Rutabaga* at the position $x = +400{,}000$ feet and time $t = 800{,}000$ nanoseconds. (Recall that $5, 12, 13$ is a Pythagorean triple.) See the following diagram.

Feel free to use a calculator on this one if you'd like but just for evaluating the numerical expressions, not for doing symbolic manipulations. And let's use the approximation that the speed of light is one foot per nanosecond.

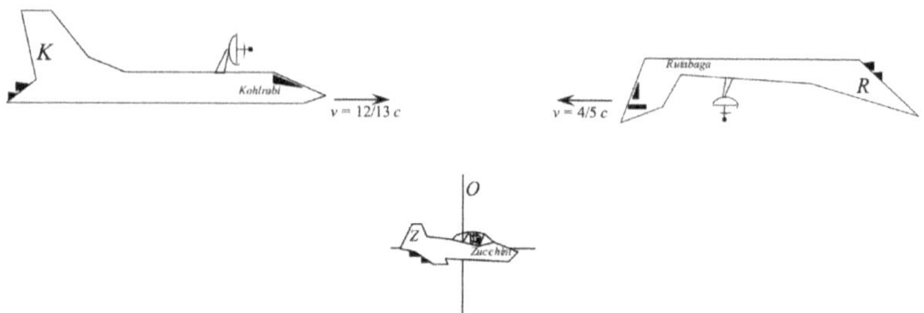

(a) What are the space and time intervals between *Kohlrabi's* shot (event 1) and the explosion on *Rutabaga* (event 2) from the perspective of the commander of the *Asher-Danzig Kohlrabi?*

Solution

Let the unprimed frame be the ICE agent's frame and the primed frame be the *Kohlrabi's*. We have the velocity of the primed frame origin to be $12/13c$. Recall the Pythagorean triple $5^2 + 12^2 = 13^2$. In the unprimed frame, we have

$$x_1 = -600{,}000, \quad t_1 = 0, \quad x_2 = 400{,}000, \quad t_2 = 800{,}000$$

$$\Delta x = 10^6, \quad \Delta t = 8 \times 10^5$$

Lorentz transform:

$$\Delta x' = \gamma(\Delta x - \beta c \Delta t), \quad \Delta t' = \gamma(\Delta t - \beta \Delta x/c)$$

$$\gamma = \frac{1}{\sqrt{1 - \beta^2}} = \frac{1}{\sqrt{1 - (\frac{12}{13})^2}} = \frac{13}{5}$$

Plug in to find $\Delta x' = 680{,}000$ feet, $\Delta t' = -320{,}000$ nanoseconds. Note that in this frame, the explosion happens *before* the shot is fired.

(b) What are the space and time intervals between the shot and the explosion from the perspective of the *Rutabaga?* Assume that the *Rutabaga* uses a coordinate system with the x''-axis pointing to the *right* in the figure so that the bow of the *Rutabaga* is pointing in the negative x''-direction.

Solution

The *Rutabaga* frame's origin has $\beta = -4/5, \gamma = 5/3$. So, we can Lorentz transform from the unprimed frame:

$$\Delta x_{Rutabaga} = \frac{5}{3}\left(10^6 - \left(-\frac{4}{5}\right) 8 \times 10^5\right) = 2.733 \times 10^6 \text{ feet}$$

$$\Delta t_{Rutabaga} = \frac{5}{3}\left(8 \times 10^5 - \left(-\frac{4}{5}\right) 10^6\right) = 2.667 \times 10^6 \text{ nsec}$$

(c) The ICE officer investigates. *Rutabaga* claims that the shot from *Kohlrabi* struck the *Rutabaga's* food locker, causing the explosion. *Kohlrabi* disagrees, stating that its shot could not possibly have caused the explosion: the projectile carried no warhead and did not come anywhere near the *Rutabaga*. Which *one* of the two accounts of the incident *must* be *incorrect*? Be sure to explain your reasoning (backed up by equations as necessary) clearly and coherently.

Solution

In all three frames, the spatial separation between the events is greater than the speed of light times their temporal separation, so the shot could NOT have caused the explosion. *Rutabaga's* description is incorrect.

Problem 3: Some things never change

The components of the Lorentz transformation matrix Λ^{ν}_{μ} used to effect "boosts" along the x-direction are defined in the in-class material. Recall that covariant four-vectors transform under boosts this way:

$$A'_{\mu} = \sum_{v=0}^{v=3} \Lambda^{v}_{\mu} A_v \qquad A'^{\mu} = \sum_{v=0}^{v=3} \Lambda^{\mu}_{v} A^v$$

Also, recall that we can obtain the components of the *contravariant* version of A this way: $A^{\mu} = \sum_{v=0}^{v=3} g^{\mu v} A_v$ and that the quantity $\sum_{\mu=0}^{\mu=3} \sum_{v=0}^{v=3} g^{\mu v} A_{\mu} A_v = \sum_{v=0}^{v=3} A^v A_v$ is a Lorentz invariant. (Don't be intimidated by any of this: all we're doing is employing matrix multiplication to write what is nothing more than a bunch of Lorentz transformations.)

Also, recall that the components of a particle's four-velocity u^{μ} are $(c, 0, 0, 0)$ in the particle's rest frame.

One can define a four-vector called the "four-spin" to describe particles with intrinsic angular momentum (such as electrons) so that $s^\mu \equiv (0, s_x, s_y, s_z)$ *in the particle's rest frame. Here, s_x, s_y, s_z are the three components of the particle's angular momentum in its rest frame.*

(a) An electron with spin four-vector $s^\mu = (0, 0.5, 0, 0)$ in its rest frame is viewed from another frame in which the electron is moving along the x-axis with velocity $0.8c$. Calculate the components of the electron's four-velocity u'^μ and four-spin s'^μ in this frame. Keep in mind that transforming to a frame in which the electron is seen to be moving to the right means that the origin of this new frame is seen to move *to the left* from the original (electron rest) frame, so β is negative.

Solution

Do the Lorentz transformation:

$$s'^\mu = \begin{bmatrix} 5/3 & +4/3 & 0 & 0 \\ +4/3 & 5/3 & 0 & 0 \\ 0 & 0 & 1 & 0 \\ 0 & 0 & 0 & 1 \end{bmatrix} \begin{bmatrix} 0 \\ 0.5 \\ 0 \\ 0 \end{bmatrix} = \begin{bmatrix} 2/3 \\ 5/6 \\ 0 \\ 0 \end{bmatrix}$$

$$u'^\mu = \begin{bmatrix} 5/3 & +4/3 & 0 & 0 \\ +4/3 & 5/3 & 0 & 0 \\ 0 & 0 & 1 & 0 \\ 0 & 0 & 0 & 1 \end{bmatrix} \begin{bmatrix} c \\ 0 \\ 0 \\ 0 \end{bmatrix} = \begin{bmatrix} 5c/3 \\ 4c/3 \\ 0 \\ 0 \end{bmatrix}$$

(b) Calculate the value of $\sum_{v=0}^{v=3} u'^v s'_v$ for an electron moving along the x-axis with velocity $0.8c$.

Solution

The easiest way is to recall that the scalar product of any two four-vectors is *always* a Lorentz invariant, so it is easiest to calculate it in the electron's rest frame:

$$\sum_\mu u'^\mu s'_\mu = \sum_\mu u^\mu s_\mu = 0 + 0 + 0 + 0 = 0$$

(c) Prove that $\sum_{v=0}^{v=3} u'^v u'_v = c^2$ for a particle moving with speed v parallel to the x-axis.

Solution

Again, the easiest way is to recall that the scalar product of any two four-vectors is *always* a Lorentz invariant, so it is easiest to calculate it in the electron's rest frame:

$$\sum_\mu u'^\mu u'_\mu = \sum_\mu u^\mu s_\mu = c^2 + 0 + 0 + 0 = c^2$$

Problem 4: Even more calculus practice

A particle of mass m, initially at the origin, experiences a conservative radial force:

$$\vec{F} = -\alpha m r e^{-\beta r^2}\,\hat{r}$$

where $\hat{r}$ is a unit vector pointing away from the origin and α and β are positive constants.

In order to escape to infinity, the particle must have sufficient kinetic energy so that the work done by the radial force can never be as large as the initial kinetic energy carried by the particle at the origin. Recall that you learned about the connection between force, work, kinetic energy, and potential energy in your introductory physics courses.

Assume that the particle is non-relativistic and calculate the minimum speed the particle must have at the origin in order to escape to infinity.

Solution

$$\text{Work} = \Delta K = \int_0^\infty -\alpha m r e^{-\beta r^2}\mathrm{d}r = \frac{\alpha m}{2\beta}\int_0^\infty e^{-\beta r^2}\mathrm{d}(-\beta r^2)$$

$$= \frac{\alpha m}{2\beta}e^{-\beta r^2}\bigg|_0^\infty = -\frac{\alpha m}{2\beta} = -\frac{mv^2}{2}$$

$$v = \sqrt{\frac{\alpha}{\beta}}$$

Paris, France

Unit 5

Doppler Shifts, World Lines, Energy–Momentum Four-Vector

Introduction: The non-relativistic Doppler shift

The (non-relativistic) Doppler shift associated with sound can change the frequency of sound heard by a listener relative to the frequency emitted by a sound source. The effect arises from three things:

1. In still air of uniform pressure, temperature, and humidity, the speed of sound is constant, independent of any motion of the source. (It's about $331\,\mathrm{m/s}$ in dry air at $0°\mathrm{C}$.)

91

2. Motion of the listener might cause the listener to intercept more (or fewer) waves per second than would be the case for a stationary listener.
3. Motion of the source will cause the wave peaks to bunch up in some directions and be more widely spaced in other directions.

In the following diagram, wave fronts emitted by a sound source mounted on a subsonic train travel outward from their point of emission.

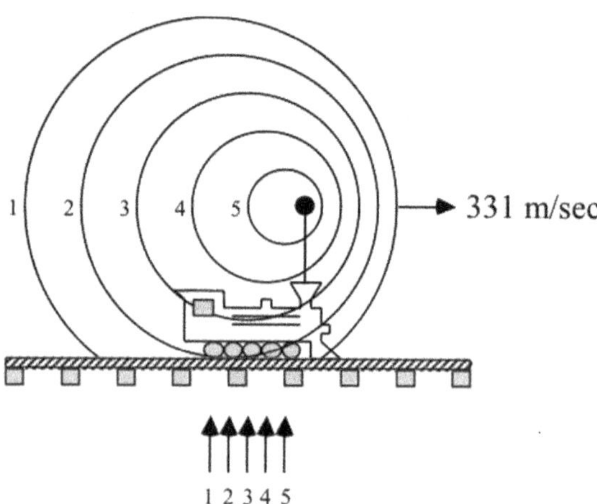

For a listener at rest, the distance from the emission point of successive wave fronts changes as the train moves. For example, if the listener is at an angle θ relative to the direction of travel of the train, we will have the situation shown in the following diagram.

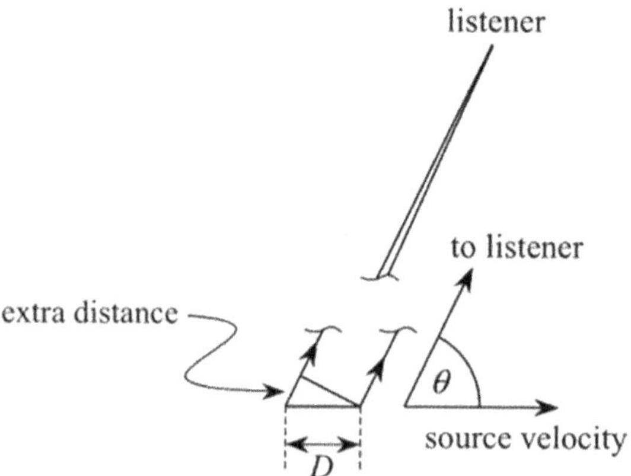

The distance D corresponds to the distance the train moves between the emission of successive wave fronts. The change in the distance to the stationary listener for successive wave fronts depends on the angle θ, of course.

If the period dialed into the train's sound generator is T (so that the frequency is $f = 1/T$), and the train is traveling with speed v, then $D = vT$.

Exercise 5.1: Doppler details

(a) Let's refer to the speed of sound in still air as c_{sound}. In terms of v, T, c_{sound}, and θ, what is T', the difference in arrival time of successive wave fronts at the listener?

Note that this will be the period heard by the listener, who will determine the frequency of the arriving sound to be $f' = 1/T'$.

Solution

The extra distance traveled by the previous wave front, relative to the most recent wave front (see the above diagram), is $D \cos \theta$. As a result, the most recent wave front takes $D \cos \theta / c_{\text{sound}}$ less time to arrive than it would have if the train had been stationary. Since $D = vT$, the time between arriving wavefronts is reduced from T to $T - vT \cos \theta / c_{\text{sound}}$ so that

$$T' = T - \frac{vT \cos \theta}{c_{\text{sound}}} = T \left(1 - \frac{v \cos \theta}{c_{\text{sound}}} \right)$$

(b) The ratio of apparent frequency and source frequency is f'/f. Find an expression for this ratio in terms of c_{sound}, v, and θ.

Solution

Since $f = 1/T$, we have

$$\frac{f'}{f} = \frac{T}{T'} = 1 \left/ \left(1 - \frac{v \cos \theta}{c_{\text{sound}}} \right) \right. = \frac{c_{\text{sound}}}{c_{\text{sound}} - v \cos \theta}$$

(c) What is f'/f when $\theta = 0°$? When $\theta = 180°$?

Solution

For $0°$, we get $1/(1 - v/c_{\text{sound}})$. For $180°$, $1/(1 + v/c_{\text{sound}})$.

Exercise 5.2: Relativistic Doppler shift

If we replace the train with a relativistic starship that is broadcasting an electromagnetic signal, we will still have the geometrical effects of the changing distance from the source to the observer. But an additional factor arises because the clocks on the starship tick slowly so that the apparent frequency of the broadcast signal is reduced by an additional factor of $1/\gamma$. (Of course, the speed of propagation of the signal that enters the Doppler shift formula is c instead of c_{sound}.)

Recall that

$$c^2 - v^2 = (c - v)(c + v)$$

so that

$$1/\gamma = \sqrt{c^2 - v^2}/c = \sqrt{(c - v)(c + v)}/c$$

Consider the cases $\theta = 0°$ and $\theta = 180°$. Making use of the algebraic convenience I just described, derive (and simplify as much as you can) expressions for the ratios $f_{\text{observed}}/f_{\text{source}}$ in these two cases. By f_{source} I mean the frequency that the crew of the starship believes their signal generator to be using.

Solution

On the starship, f_{source} is influenced, relative to that seen by external observers, by time dilation as well as Doppler shifts. But everything else will behave the same way.

Imagine that the ship emits a wave front every ship's time interval of T_{source}. The period in the observer frame will be longer so that a wavefront is seen to leave the ship every $T_{\text{source}}/\sqrt{1 - \beta^2}$ seconds, during which time the ship is seen to move by $\beta c T_{\text{source}}/\sqrt{1 - \beta^2}$. If the ship is traveling directly toward the observer, the second wavefront travels less distance than the first by the amount the ship moved, so it arrives at the observer later by $T_{\text{source}}/\sqrt{1 - \beta^2} - \frac{1}{c}\beta c T_{\text{source}}/\sqrt{1 - \beta^2}$ where the second term is the light travel time across the change in the ship's location. With a bit of algebra, we find that the delay between the observer's receipt of the first and second pulses is $T_{\text{source}}/\sqrt{1 - \beta^2} - \frac{1}{c}\beta c T_{\text{source}}/\sqrt{1 - \beta^2}$ so that, for $\theta = 0°$,

$$T_{\text{source}}/\sqrt{1 - \beta^2} - \frac{1}{c}\beta c T_{\text{source}}/\sqrt{1 - \beta^2}$$

$$= T_{\text{source}}/\sqrt{1 - \beta^2} - \beta T_{\text{source}}/\sqrt{1 - \beta^2}$$

$$= T_{\text{source}}\frac{1-\beta}{\sqrt{1-\beta^2}} = T_{\text{source}}\frac{1-\beta}{\sqrt{(1-\beta)(1+\beta)}}$$

$$= T_{\text{source}}\sqrt{\frac{1-\beta}{1+\beta}} = T_{\text{observer}}$$

As a result,

$$\frac{f_{\text{observer}}}{f_{\text{source}}} = \frac{T_{\text{source}}}{T_{\text{observer}}} = \sqrt{\frac{1+\beta}{1-\beta}}$$

For $\theta = 180°$, the pulse travels a greater distance and, after some algebra, can be seen to arrive

$$T_{\text{source}}\frac{1+\beta}{\sqrt{1-\beta^2}} \text{ later} = T_{\text{source}}\frac{1+\beta}{\sqrt{(1-\beta)(1+\beta)}}$$

$$= T_{\text{source}}\sqrt{\frac{1+\beta}{1-\beta}} = T_{\text{observer}}$$

so that

$$\frac{f_{\text{observer}}}{f_{\text{source}}} = \sqrt{\frac{1-\beta}{1+\beta}}$$

Exercise 5.3: Doppler shifts, red shifts, and the expanding universe

In his 1927 paper "Un Univers homogène de masse constante et de rayon croissant rendant compte de la vitesse radiale des nébuleuses extragalactiques,"[1] Belgian physicist Georges Lemaître proposed that the universe was expanding and that the speed with which distant objects were receding increased with distance. A few years later, Lemaître proposed that the expanding universe was born from a "Primeval Atom" which exploded outwards.[2] His initial work on the expansion of the universe preceded Edwin Hubble's by two years.

---

[1] See https://en.wikipedia.org/wiki/Georges_Lema%C3%AEtre. The title of his paper translates as "A homogeneous Universe of constant mass and growing radius accounting for the radial velocity of extragalactic nebulae."

[2] *Ibid.*

The typical speed v that an object at a distance D from Earth will be receding is described by Hubble's law, $v = H_0 D$. The Hubble constant H_0 is approximately 68 kilometers per second per megaparsec. (One parsec is 3.26 light years.) There are some subtleties to this: exactly what do we mean by distance when we are describing objects that are billions of light years away in a universe of finite volume? Let's be cavalier and ignore this.

You should have found in the previous exercise that

$$\frac{f_{\text{source}}}{f_{\text{observed}}} = \sqrt{\frac{c \pm v}{c \mp v}}$$

with the upper choice of signs corresponding to $\theta = 180°$ and the lower to $\theta = 0°$.

Astronomers define an object's **red** shift (indicated by the symbol z) this way:

$$1 + z = \frac{f_{\text{source}}}{f_{\text{observed}}}$$

It can be determined for a star (or a galaxy) by measuring the shift in spectral lines—both emission lines from the star and stellar atmospheric absorption lines—relative to their unshifted positions for a star that is not moving with respect to the Earth.

The coalescing black holes detected in recent years by the LIGO experiment are believed to be approximately 1.3 billion light years (399 megaparsecs) from Earth. Assuming the Hubble law is accurate, what is the red shift z for stars in the vicinity of these black holes?

Solution

We have

$$v = (399 \, \text{mpc}) \times (68 \, \text{km sec}^{-1} \, \text{mpc}^{-1}) = 2.71 \times 10^4 \, \text{km sec}^{-1}$$

As a result,

$$\frac{f_{\text{source}}}{f_{\text{observer}}} = \sqrt{\frac{3 \times 10^8 + 2.71 \times 10^7}{3 \times 10^8 - 2.71 \times 10^7}} = 1.0948$$

$$= 1 + z \text{ so that}$$

$$z = 0.0948$$

World lines

A world line is the trajectory of an object in space–time. We generally graph world lines using geometrical units: the vertical axis will be ct, while the horizontal axis (axes) will describe x (x and y). Here are a couple of examples.

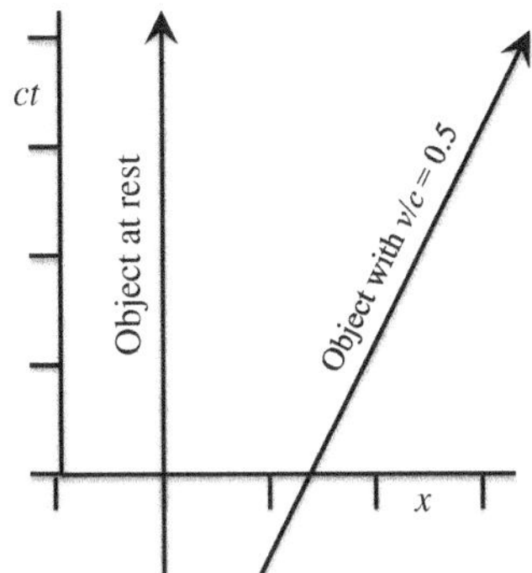

The Earth's world line as it orbits the Sun is a good approximation to a helix, with the Sun moving straight up the ct axis along the center of the helix. The Earth's orbital eccentricity is only 0.017, so it's not too far off the mark to call its world line a helix of radius 150 million kilometers.

The distance up the ct axis that the Earth travels in one year is a light year, of course, or 9.46 *trillion* kilometers. So, the pitch of the helix—the distance along ct to execute one full revolution in the x, y plane—is about 63,000 times larger than its radius. It doesn't corkscrew around very rapidly, does it?

Light always travels with speed c. Here is the world line for the laser beam in Mrs. Blepon's light clock (see the first unit's material), as measured in her rest frame.

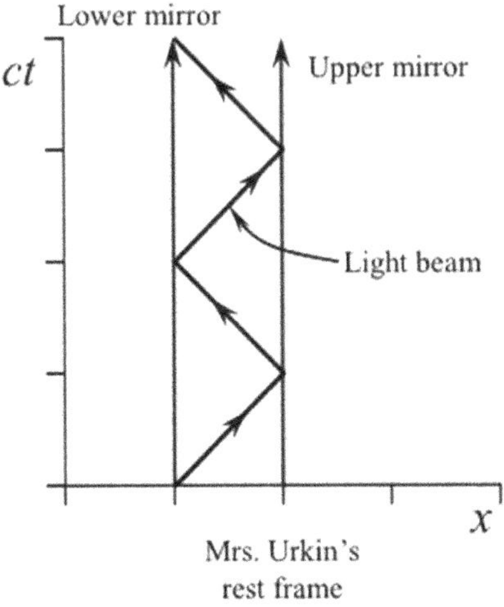

It's more complicated to draw this for a frame in which she is moving in the y-direction, while the light beam is bouncing in the x-direction. The following plot is for $v/c = 0.6$.

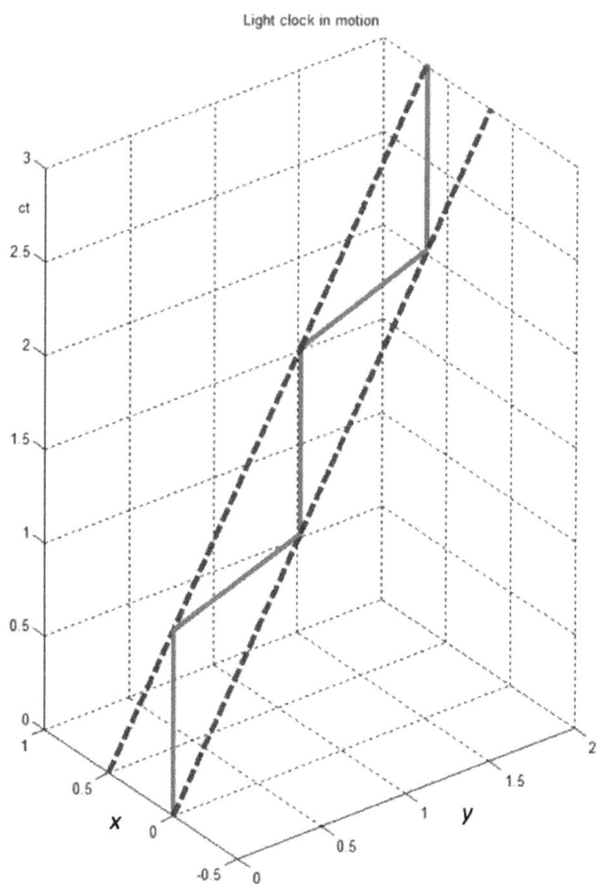

An object not subject to forces will, when viewed by observers in an inertial frame, always move so that its world line is straight.

Imagine that a clock travels from point A to point B, as shown in the following diagram. Note that I am specifying *both* the position and time of A and B.

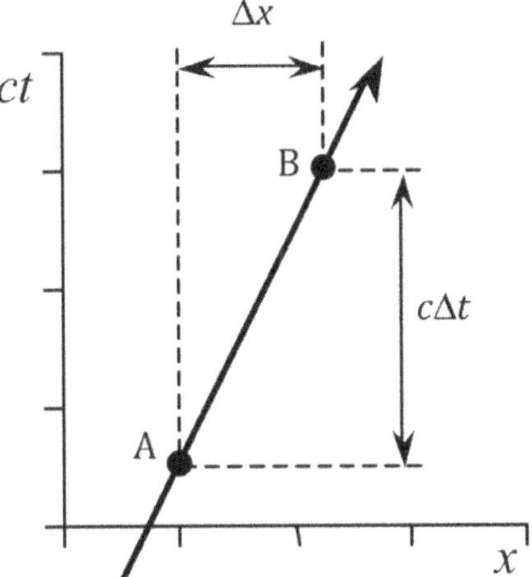

It is easy to figure out how much time passes on the clock: in its rest frame, it is just $\Delta\tau$. And since the proper time is a Lorentz invariant, we can calculate it this way:

$$\Delta T = \sqrt{\Delta t^2 - \Delta x^2/c^2}$$

If the object changed speed en route but still did it in a way to go from space–time point A to point B (by that I mean the departure and arrival times were unchanged from before), it might have a world line like this:

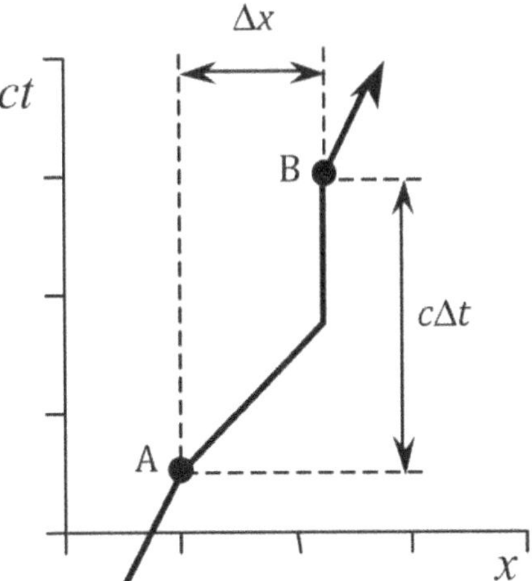

It's a simple matter to show that the elapsed time on the clock would be less than it is for the straight world line. Keep in mind that the time that elapses on the clock is an additive quantity: you can only do a simple calculation to derive proper time along a straight world line segment, but you can calculate separately the elapsed proper time along each segment, then add them together at the end. It's very much like measuring the lengths of a number of individual straight paths, then summing them to get the total path length for a complicated trajectory.

Perhaps the easiest comparison of this sort for the clock's elapsed time en route is for an object at rest in comparison to one traveling at nearly the speed of light:

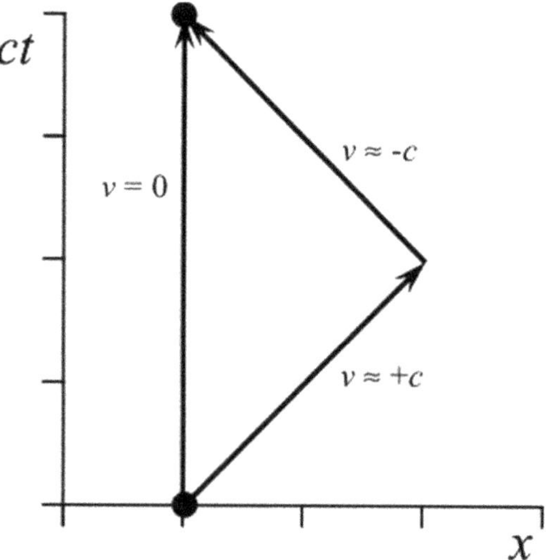

In this case, the clock traveling along the path with the kink has proper time that is almost zero.

Taylor and Wheeler refer to this as the "principle of maximal aging": an object that moves along a straight world line between two points in space–time will age more than an object whose world line follows any other path than this. Another way of saying it is that an object not subject to forces will move along a straight world line since this is the path of maximal aging.

In more sophisticated terms, objects not subject to forces move along world lines that are space–time geodesics.

Exercise 5.4: Twin paradox, via world lines

You surely have heard of this before. One twin stays at home on Earth while the other flies at high speed to a distant star and then turns around and returns home. How do their ages compare upon the return of the rocket twin? The apparent paradox arises from the fact that each twin sees the other as moving rapidly. Let's analyze this using world lines.

The rocket twin flies to Barnard's Star, a low-mass red dwarf 6 light years from Earth. The twin travels with speed $v = \frac{3}{4}c$, reverses course, and returns to Earth.

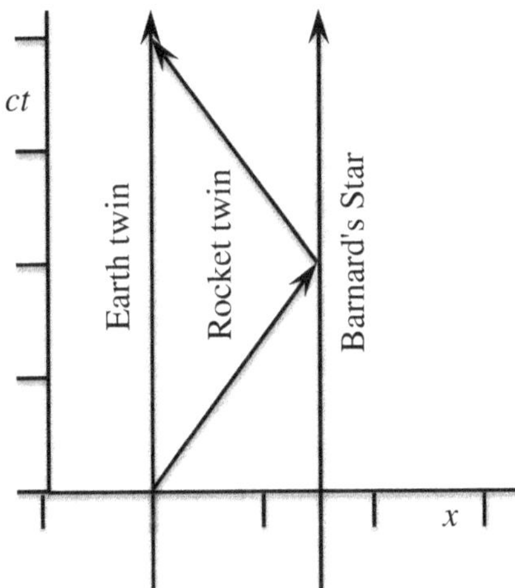

(a) What is the elapsed proper time experienced by the Earth twin during the rocket twin's voyage?

Solution

The rocket twin goes out 6 LY at $\frac{3}{4}$ LY/year, so the outbound trip takes 8 years, according to the Earth twin. The return trip also takes 8 years, so the Earth twin's proper time between the two events is 8 years.

(b) What is the proper time that elapses according to the rocket twin? Determine this using the information gleaned from the world line diagram.

Solution

By "proper time that elapses..." I mean the amount of time that passes on the rocket twin's clock. We can calculate this from the information measured by the Earth twin: the first half of the trip is seen to take 8 years; the end points of the half-trip are 6 LY apart. As a result,

$$\Delta\tau^2 = (\Delta t)^2 - (\Delta x/c)^2 = 8^2 - 6^2$$

$$= 64 - 36 \quad \text{so } \Delta\tau = \sqrt{28}$$

for the outbound trip and the same for the return trip. Doing the calculator work yields a round-trip time of about 10.58 years total.

(c) Why are the two numbers different?

Solution

The numbers are different because the rocket twin experiences the voyage in *two* different inertial frames, while the Earth twin only experiences it in one.

Energy–momentum four-vector

In the last unit, we discussed the space–time four-vector x^μ and the four-vector velocity u^μ. We constructed the four-velocity by dividing the space–time interval between two events by the Lorentz scalar proper time interval between the events. We found that $u^v = \gamma(c, \vec{v})$.

We can define a new four-vector by multiplying the four-velocity by the rest mass of the object that flew from the first event to the second. The rest mass of an object is, by definition, an invariant: regardless of frame, its value is the mass of the object *in the object's rest frame*. As a result,

$$p^\mu \equiv m u^\mu = \gamma m(c, \vec{v})$$

is automatically a four-vector. As it happens, the zeroth component is the object's total energy divided by c, while the space components are its linear momentum components.

Let's take a closer look.

Exercise 5.5: Binomial approximation to p^μ

Recall that the binomial expansion of $(1 + x)^n$ is $1 + nx + n(n - 1)x^2/2! + \cdots$. When x is very small, it is sometimes sufficiently accurate to keep only the first few terms.

(a) Write the first three terms in the binomial expansion of $\gamma = (1 - v^2/c^2)^{-1/2}$, then use this to calculate the small-velocity approximation to p^0, the time component of p^μ. The first term will be independent of velocity, while the second will be proportional to v^2/c.

Solution

$$\gamma = \left(1 - \frac{v^2}{c^2}\right)^{-1/2} \approx 1 - \frac{1}{2}\left(-\frac{v^2}{c^2}\right) + \left(-\frac{1}{2}\right)\left(-\frac{3}{2}\right)\frac{1}{2!}\left(-\frac{v^2}{c^2}\right)^2$$

$$= 1 + \frac{1}{2}\left(\frac{v^2}{c^2}\right) + \frac{3}{8}\left(\frac{v^4}{c^4}\right)$$

$$p^0 = \gamma mc \approx mc + \frac{1}{2}\left(\frac{mv^2}{c}\right) + \frac{3}{8}\left(\frac{mv^4}{c^3}\right) + \cdots$$

$$= \frac{1}{c}\left[mc^2 + \frac{mv^2}{2} + \frac{3}{8}mv^2\left(\frac{v^2}{c^2}\right) + \cdots\right]$$

(b) Multiply the second term by c to obtain something that should be familiar to you from introductory physics and high school physics. What is its usual name?

Solution

$$p^0 c = mc^2 + \frac{1}{2}mv^2 + \cdots$$

It's the usual (non-relativistic) kinetic energy!

Exercise 5.6: Relativistic vector momentum[3]

This is just a read-along section without any calculations for you to perform.

Imagine that an object of mass M, at rest at the origin, decays into two lighter fragments of mass m_0. Each fragment is equipped with an internal clock that will cause it to explode after a proper time $\Delta\tau$ elapses. Naturally, since momentum is conserved, the two fragments travel equal distances before exploding. The fragments are

---

[3]This analysis of the relativistic vector momentum is based on Dr. Paul Bamberg's 1973 material used in Unit 10 of Physics 1, Harvard University.

seen to travel nearly parallel to the x'-axis so that their displacements in the y'-direction at the time each explodes are small. Of course, the y' displacements are equal in magnitude: $\Delta y_1' = -\Delta y_2'$.

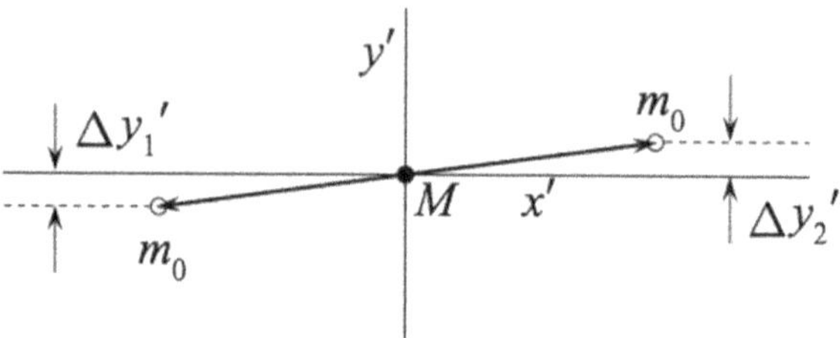

Now, shift to a frame of reference moving rapidly in the negative x'-direction so that the left fragment is moving directly along the $-y$-axis, as shown in the following. The angles between the trajectories and the x'-axis in the previous frame are so small that in the present frame the velocity with which the left mass moves is very very small compared to c.

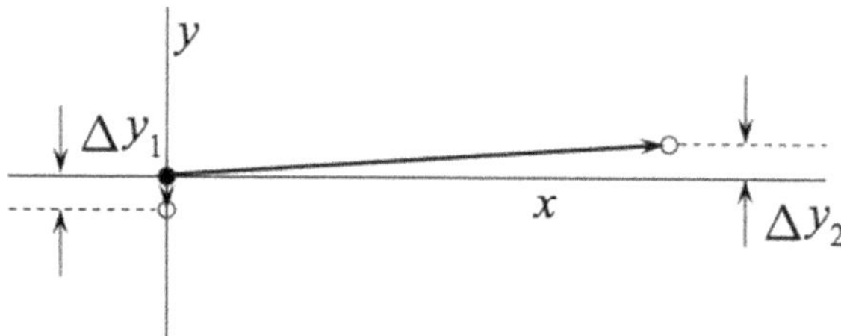

Since boosts along x do not change y distances, it is true that $\Delta y_1' = \Delta y_1$ and $\Delta y_2' = \Delta y_2$. As a result, $\Delta y_1 = \Delta y_2$. In this new frame, the explosions are no longer simultaneous—both still take place after a time $\Delta \tau$ passes on a mass's internal clock—since the right mass is relativistic and its clock ticks slowly. But when it does eventually explode, its displacement in the y-direction will be the same as that of the other mass, which has already blown up. And it will still be true that the two-mass system's y component of momentum will be zero. In this frame, the left mass blows up at time Δt_1 while the right mass blows up at time Δt_2.

The (non-relativistic) left mass has momentum $m_0 \Delta y_1 / \Delta t_1 = m_0 \Delta y_1 / \Delta \tau$ since it is moving so slowly that the time of its explosion in the unprimed frame is almost exactly the same as the reading on its slowly moving clock.

The right mass must have momentum $m_0 \Delta y_1 / \Delta t_1 = -m_0 \Delta y_2 / \Delta t_1$. Taking into account time dilation, we have $\Delta t_2 / \gamma = \Delta t_1$ since it takes longer for the right object's clock to bring about the explosion.

We have, as a result, $-m_0 \Delta y_2 / \Delta t_1 = -\gamma m_0 \Delta y_2 / \Delta t_2$. Consequently,

$$\left| m_0 \frac{\Delta y_1}{\Delta t_1} \right| = |m_0 v_{1y}| = \left| \gamma m_0 \frac{\Delta y_2}{\Delta t_2} \right| = |\gamma m_0 v_{2y}|$$

$$|m_0 v_{1y}| = |\gamma m_0 v_{2y}|.$$

In other words, if we define the y component of momentum of the relativistic object to be $\gamma m_0 v_y$, then it will be conserved in the breakup of the original mass M.

There is a similar argument that lets us conclude that the total energy of an object is what you saw earlier today: $E = \gamma m c^2$.

In conclusion,

$$p^\mu \equiv m u^\mu = \gamma m(c, \vec{v}) = \left(\frac{E}{c}, \vec{p} \right)$$

Exercise 5.7: Rest mass, of course

Since the four-momentum is a four-vector, it must be true that $\sum_\mu p^\mu p_\mu$ is a Lorentz invariant. Calculate what it is.

Solution

$$p^\mu \equiv m u^\mu = \gamma m(c, \vec{v})$$

$$p_\mu = m u_\mu = \gamma m(c, -\vec{v})$$

$$\sum_\mu (p^\mu p_\mu) = \gamma^2 m^2 \left(c^2 - v^2 \right) = \gamma^2 m^2 c^2 (1 - v^2/c^2)$$

$$= \frac{\gamma^2 m^2 c^2}{\gamma^2} = m^2 c^2$$

Chapter summary

In comparison to the non-relativistic case, Doppler shifts show the additional effects of time dilation: the (moving) signal source broadcasts with a down-shifted frequency since its internal time base ticks slowly, according to observers in a different frame of reference.

Sometimes it is illustrative to plot the space–time trajectory (the world line) of an object on axes that include time but expressed in "geometrical" units of ct. These world line diagrams can sometimes simplify analyses of complicated motions, such as the rocket twin in the twin paradox.

One of the most useful four-vectors is the four-momentum, whose components include an object's total energy and its vector momentum. We will make extensive use of the four-momentum when we work out the kinematics of relativistic particles next unit.

Problem 1: When the moon is full

A high-temperature gas cell in a physical chemistry laboratory contains a vapor of 1,2diphenylwolfsbane molecules. It is known that stationary molecules of this compound can absorb light of frequencies 7.000×10^{17} Hz and 7.300×10^{17} Hz. A laser beam with a frequency of 7.200×10^{17} Hz shines into the cell through a transparent window. The experimenters find that the gas is able to absorb some of the light, even though the laser is not tuned to one of the absorption frequencies. They correctly interpret their result to mean that some of the gas molecules are moving so rapidly that Doppler shifts play a significant role in the interaction of the gas molecules with the laser light. Calculate the ***minimum*** speed that will allow 1,2diphenyl-wolfsbane molecules to absorb light from the laser.

Solution

Absorption lines: we'll need to "blue shift" 7.2×10^{17} Hz up to 7.3×10^{17} Hz or "red shift" it down to 7.0×10^{17} Hz:

$$\sqrt{\frac{c+v}{c-v}} = \frac{7.2}{7.0} \quad \text{or} \quad \sqrt{\frac{c-v}{c+v}} = \frac{7.2}{7.3}.$$

Square both sides, plug in for c and do a little algebrato find $v = 0.0281c$ or $v = 0.01379c$. The second value is the smaller of the two.

Problem 2: Getting me about

Three clocks, each reading 0 initially, follow the world lines labeled A, B, and C in the following figure. (Units are light years.) What do each of the clocks read upon arrival at $ct = 8\,\text{LY}$?

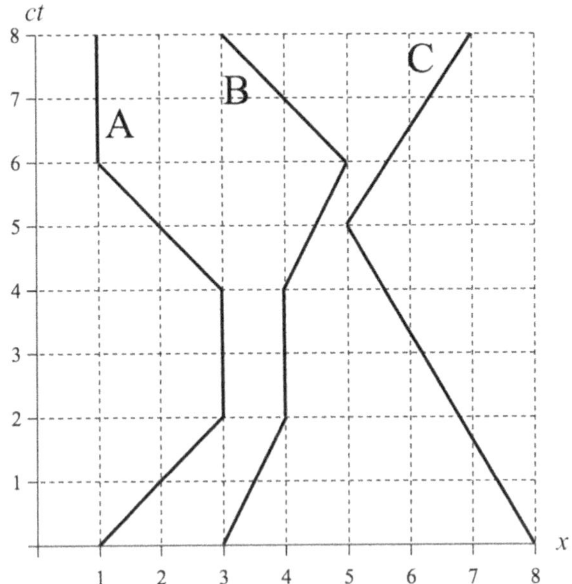

Solution

Add up the proper times along each segment:

$$A: \sqrt{2^2 - 2^2} + 2 + \sqrt{2^2 - 2^2} + 2 = 4\,\text{years}$$

$$B: \sqrt{2^2 - 1^2} + 2 + \sqrt{2^2 - 1^2} + \sqrt{2^2 - 2^2} = 2 + 2\sqrt{3} \approx 5.464$$

$$C: \sqrt{5^2 - 3^2} + \sqrt{3^2 - 2^2} = 4 + \sqrt{5} \approx 6.236$$

Problem 3: Mrs. Rumpdock's Silver Conveyance

Before its confiscation by the Department of Justice, Mrs. Rumpdock's hyperdrive-equipped sports car could attain relativistic speeds on straight portions of Interstate 94.

Forensic reconstruction of her collision with a stationary moose revealed that, at the time of the collision, the vector momentum of her Silver Conveyance (including Mrs. Rumpdock, contained therein) was $1{,}600c\,\text{kg} \cdot \text{m/sec}$. The total energy of the Jaguar and driver was $2{,}000c^2$ J.

What was the combined rest mass of Rumpdock and her car? Find a numerical answer with units of kg.

Solution

Recall that the four-momentum can be used to construct a Lorentz invariant quantity, namely $p^\mu \equiv mu^\mu = \gamma m(c, \vec{v}) = \left(\frac{E}{c}, \vec{p}\right)$:

$$\sum_\mu (p^\mu p_\mu) = \frac{E^2}{c^2} - \vec{p} \cdot \vec{p} = m^2 c^2$$

so that $E^2 - (\vec{p} \cdot \vec{p})c^2 = m^2 c^4$

Plug in:

$$(2000c^2)^2 - (1600c)^2 = m^2 c^4$$

$$(400c^2)^2[5^2 - 4^2] = m^2 c^4$$

$$1200c^2 = mc^2$$

$$1200\,\text{kg} = m$$

Problem 4: Discovering the pablon

Imagine that a Fermilab experiment discovers in its archives an overlooked signal that suggests the presence of a previously unknown neutral particle in the experiment's data. The collaboration proposes the name "pablon" for the new particle. The pablon is found to decay into pairs of charged pions.

Assume that the rest mass for a charged pion is $0.140\,\text{GeV}/c^2$.

In one event, a relativistic pablon decays in flight producing pions whose vector momenta in the lab frame are

$$\vec{p}_{\pi+} = (0, 0.3\,\text{GeV}/c, 7\,\text{GeV}/c)$$

$$\vec{p}_{\pi-} = (0, -0.3\,\text{GeV}/c, 12\,\text{GeV}/c)$$

(a) Calculate the total energy of π^+ in the lab frame. Also, calculate the total energy of the π^-.

Solution

$$E_+ = \sqrt{m^2 c^4 + p^2 c^2} = \sqrt{.140^2 + .09 + 49} = 7.008\,\text{GeV}$$

$$E_- = \sqrt{m^2 c^4 + p^2 c^2} = \sqrt{.140^2 + .09 + 144} = 12.005\,\text{GeV}$$

(b) Since each component of the system's four-momentum is conserved during the decay, you are able to use the information at hand to calculate the four-momentum of the pablon immediately before it decayed. Do so now.

Solution

Add the two four-momenta:

$$E = 7.008 + 12.005 = 19.013, \quad \vec{p} = (0, 0, 19\ GeV/c), \text{ so}$$

$$p^{\mu}_{\text{pablon}} = (19.013, 0, 0, 19)\ GeV/c.$$

(c) What is the rest mass of a pablon?

$$mc^2 = \sqrt{E^2 - p^2 c^2} = \sqrt{19.013^2 - 19^2}\, E \approx 0.686\ GeV$$

so that $m_{\text{pablon}} = 0.686\ GeV/c^2$.

Cassis, France

Unit 6

Conservation Laws, Relativistic Kinematics, and a Start on Dynamics

Λ baryon decay

A proton is made of two up quarks and one down quark: $p = uud$, while a neutron is made of one up quark and two down quarks: $n = udd$. A negative pi meson is made of a down quark and an anti-up quark: $\pi^- = \bar{u}d$.

111

The "lambda baryon" is a sort of first cousin of protons and neutrons. It is uncharged, and made from up, down, and strange quarks: $\Lambda = uds$. It is a little heavier than a proton and decays into a proton and a negative pion about 64% of the time. It can also decay into a neutron and a neutral pi meson, which it does about 36% of the time.

Let's analyze the decay $\Lambda \to p\pi^-$. The masses of the particles participating in this decay are as follows: $m_\Lambda = 1115.63\,\text{MeV}/c^2$, $m_p = 938.27\,\text{MeV}/c^2$, and $m_\pi = 139.57\,\text{MeV}/c^2$.

Kinematic tools at our disposal, in the Λ rest frame:

- Momentum conservation: $\gamma_p m_p |\vec{v}_p| = \gamma_\pi m_\pi |\vec{v}_\pi|$. Call this p.
- Energy conservation: $\gamma_p m_p c^2 + \gamma_\pi m_\pi c^2 = m_\Lambda c^2$.

We have two equations in the two unknowns $|\vec{v}_p|, |\vec{v}_\pi|$. In principle, we can solve this!

What we'll find is that the proton and $\bar{\pi}$ momenta are about $101\,\text{MeV}/c$ in the Λ rest frame. We can then boost back to the lab frame to describe what we see coming from our Λ beam as particles in it decay in flight. Since the speeds in the lab frame of the proton and pion aren't all that different, the Lorentz terms we use to transform to the lab frame are fairly similar for both particles.

Recall that $p \equiv |\vec{p}_\pi| = |\vec{p}_p|$ in the Λ rest frame. Solve for p:

$$E_p = \sqrt{p_\pi^2 c^2 + m_\pi^2 c^4} = \sqrt{p^2 c^2 + m_\pi^2 c^4}$$

$$E_p = \sqrt{p_p^2 c^2 + m_p^2 c^4} = \sqrt{p^2 c^2 + m_p^2 c^4}$$

$$E_\pi + E_p = m_\Lambda c^2$$

Square the last equation:

$$E_\pi^2 + E_p^2 + 2E_\pi E_p = m_\Lambda^2 c^4$$

or

$$p^2 c^2 + m_\pi^2 c^4 + p^2 c^2 + m_p^2 c^4 + 2\sqrt{(p^2 c^2 + m_\pi^2 c^4)(p^2 c^2 + m_p^2 c^4)} = m_\Lambda^2 c^4$$

Divide both sides by 2, rearrange terms, then square both sides:

$$(p^2 c^2 + m_\pi^2 c^4)(p^2 c^2 + m_p^2 c^4) = \left[\frac{(m_\Lambda^2 - m_\pi^2 - m_p^2)c^4}{2} - p^2 c^2\right]^2$$

or

$$p^4 c^4 + m_\pi^2 m_p^2 c^8 + (m_\pi^2 c^4 + m_p^2 c^4) p^2 c^2$$

$$= \left[\frac{(m_\Lambda^2 - m_\pi^2 - m_p^2) c^4}{2} \right]^2 + p^4 c^4 - 2 \left(\frac{m_\Lambda^2 - m_\pi^2 - m_p^2}{2} \right) c^4 p^2 c^2$$

$$m_\pi^2 m_p^2 c^8 + \left(m_\pi^2 c^4 + m_p^2 c^4 \right) p^2 c^2$$

$$= \left[\frac{(m_\Lambda^2 - m_\pi^2 - m_p^2) c^4}{2} \right]^2 + (-m_\Lambda^2 + m_\pi^2 + m_p^2) c^4 p^2 c^2$$

$$m_\pi^2 m_p^2 c^8 = \left[\frac{(m_\Lambda^2 - m_\pi^2 - m_p^2) c^4}{2} \right]^2 + (-m_\Lambda^2) c^4 p^2 c^2$$

so that

$$p^2 c^2 = \frac{\left(\frac{m_\Lambda^2 - m_\pi^2 - m_p^2}{2} c^4 \right)^2 - m_\pi^2 m_p^2 c^8}{m_\Lambda^2 c^4} = (100.5065 \,\text{MeV})^2$$

This is straightforward but slightly messy algebra.

Let's figure out what this decay looks like in the lab frame for a highly relativistic Λ. Say the Λ has 250 GeV energy: its Lorentz γ is

$$\gamma = \frac{E_\Lambda}{m_\Lambda c^2} = \frac{250}{1.11563} = 224.09$$

and its β is 0.99999004, which we can approximate as 1.

A quick comment about units

For a highly relativistic particle, $E \approx pc$ when $pc \gg mc^2$. It is convenient to express energy in GeV or MeV or eV or... and three-momentum in GeV/c or MeV/c or eV/c or.... That way, you don't divide by c when working with expressions involving a mix of p and E/c : the units absorb the c.

Exercise 6.1: Proton, pion four-momenta

The minimum π energy occurs for a decay in which the pion goes backwards in the Λ rest frame relative to the direction the Λ is traveling in the lab frame. What are the four-momenta for the proton and pion in the Λ ***rest frame*** in this case of minimum pion lab frame energy?

Solution

The magnitudes of the vector momenta in the Λ rest frame are both $100.5065\,\text{MeV}/c$. So, in the Λ rest frame,

$$p_\pi^\mu = (E_\pi/c, \vec{p}_\pi) = \left(\frac{\sqrt{p^2 c^2 + m^2 c^2}}{c}, -100.5065, 0, 0 \right) \text{MeV}/c$$

Do the calculator work to find $E_\pi = 171.99\,\text{MeV}$. What's left of the energy goes to the proton: $E_p = 1115.63 - 171.99\,\text{MeV} = 943.64\,\text{MeV}$. $p_p^\mu = (E_p/c, \vec{p}_p) = (943.64, +100.5065, 0, 0)\text{MeV}/c$.

Exercise 6.2: The necessary Lorentz transformation

Write down the Lorentz transformation matrix needed to transform to the lab frame. Keep in mind that β refers to the velocity of the new frame's origin from the perspective of the original frame. In order to enter a frame in which the Λ is moving rapidly in the positive x-direction, you'll need to boost from the Λ rest frame in the *negative* x-direction.

Solution

Easy: see a few units back. We have

$$\beta \approx -1, \quad \gamma = \frac{E_\Lambda}{m_\Lambda c^2} = 224.09 \quad \text{so that}$$

$$\begin{bmatrix} \gamma & -\gamma\beta & 0 & 0 \\ -\gamma\beta & \gamma & 0 & 0 \\ 0 & 0 & 1 & 0 \\ 0 & 0 & 0 & 1 \end{bmatrix} = \begin{bmatrix} 224.09 & +224.09 & 0 & 0 \\ +224.09 & 224.09 & 0 & 0 \\ 0 & 0 & 1 & 0 \\ 0 & 0 & 0 & 1 \end{bmatrix}$$

Exercise 6.3: Into the lab frame

Now, determine the minimum and maximum possible pion energies in the lab frame for decays of $250\,\text{GeV}\ \Lambda$ baryons. You should find that the proton always carries off most of the energy in the lab frame.

Solution

In the lab frame, the minimum pion energy will be this:

$$\begin{bmatrix} 224.09 & +224.09 & 0 & 0 \\ +224.09 & 224.09 & 0 & 0 \\ 0 & 0 & 1 & 0 \\ 0 & 0 & 0 & 1 \end{bmatrix} \begin{bmatrix} 171.99 \\ -100.5065 \\ 0 \\ 0 \end{bmatrix} = \begin{bmatrix} 16{,}018.7 \\ +16{,}018.7 \\ 0 \\ 0 \end{bmatrix}$$

so minimum $E_\pi = 16.02\,\text{GeV}$. For the maximum, flip the sign on the vector momentum before transforming to find $E_\pi = 61.06\,\text{GeV}$.

Exercise 6.4: When the final state particles have the same mass

The various formulas above also work fine for the two-body decays of other particles.

Imagine that you are working with a beam of neutral K mesons (made of down quarks and anti-strange quarks) and are observing their decays into pairs of charged pions: $K^0 \rightarrow \pi^+\pi^-$. The kaon mass is $497.67\,\text{MeV}/c^2$, while the positive and negative pion masses are identical.

(a) Determine pc for a pion in the kaon's rest frame.

Solution

$$p^2 c^2 = \frac{\left(\frac{m_K^2 - m_\pi^2 - m_\pi^2}{2} c^4 \right)^2 - m_\pi^2 m_\pi^2 c^8}{m_K^2 c^4} = \frac{\left[\frac{497.67^2}{2} - 139.57^2 \right]}{497.67^2} \quad \text{so}$$

$$pc = 206.007\,\text{MeV}$$

(b) If the energy of kaons in the beam corresponds to the same Lorentz γ as was the case for the Λ beam, what are the minimum and maximum pion energies in the lab for the decay products of beam kaons?

Solution

Start in the kaon rest frame, where each pion has half the kaon's rest energy. The two (pion) four-momenta of interest are $(248.835, \pm 206.007, 0, 0)$, so we have, in the lab frame,

$$E_\pi = 224.09(248.835, \pm 206.007) = 101.925 \quad \text{and} \quad 9.597.$$

Upgrading Newton's second law

We need a replacement for $\vec{F} = m\vec{a} = \frac{md^2\vec{x}}{dt^2}$.

An obvious problem with $\vec{F} = m\vec{a}$ is that $v < c$ for a massive object. Even though we might continue to apply a constant force to something, its rate of acceleration will decrease as $v \to c$.

How might we develop a proper connection between the applied force and an object's change in velocity? We need to preserve the connection between force, time, and impulse, namely that $\vec{F}dt = d\vec{p}$. And we'd also like to preserve the work-kinetic energy theorem: $\vec{F} \cdot d\vec{x} = dK$

So, let's forgo writing Newton's second law as $\vec{F} = m\vec{a} = md\vec{v}/dt$ for now but instead write it as $\vec{F} = d\vec{p}/dt$ where $\vec{p} = \gamma m\vec{v}$ is the kinematic quantity that is conserved in collisions and so forth. That automatically gives us $\vec{F}dt = d\vec{p}$: the induced change of momentum equals the product of the applied force and the duration of application.

Imagine we build an electromagnetic cannon that looks like this:

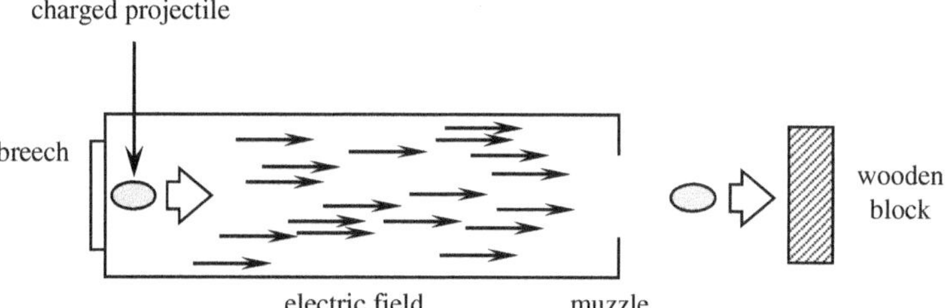

The projectile has mass m and carries charge Q. When we turn the cannon on, a uniform electric field is created everywhere inside the cannon which persists for a brief time Δt.

The projectile "feels" a force $Q\vec{E}$, receives a three-momentum impulse $\Delta\vec{p} = Q\vec{E}\Delta t$, and strikes a block of wood so that the wood and embedded projectile have total momentum $\vec{p} = Q\vec{E}\Delta t$ after the cannon is fired.

We expect the momentum of the final state to be independent of the mass of the projectile, even if the projectile is so light that it is relativistic.

Let's poke at $\vec{F} = d\vec{p}/dt$ where $\vec{p} = \gamma m \vec{v}$ for a bit, then see if we can use it to find a generalization of $\vec{F} = m\vec{a}$ that will work for relativistic systems. We'll need to come up with an equation that uses four-vectors and Lorentz scalars so that we'll know how it behaves when we change frames of reference.

We might find that $f^\mu \equiv dp^\mu/d\tau$ (where p^μ is a particle's four-momentum and τ is the proper time) will work! But f^μ defined this way has four components, not three, so let's postpone a discussion of the four-vector version of force for the time being.

How does $\vec{v}$ change when we apply a force in the relativistic case? Let's assume there's a force acting and that we measure the rate of change of three-momentum in response to this force. Take a derivative:

$$\frac{d\vec{p}}{dt} = \frac{d(\gamma m \vec{v})}{dt}$$

$$= m\frac{d}{dt}(\gamma \vec{v})$$

$$= m\vec{v}\left(\frac{d\gamma}{dt}\right) + m\gamma\frac{d\vec{v}}{dt}$$

Differentiate γ (which depends on speed, not velocity):

$$\frac{d\gamma}{dt} = \frac{d}{dt}\left(1 - |\vec{v}|^2/c^2\right)^{-\frac{1}{2}}$$

$$= -\frac{1}{2}\left(1 - |\vec{v}|^2/c^2\right)^{-\frac{3}{2}}\left(-\frac{2|\vec{v}|}{c^2}\right)\frac{d|\vec{v}|}{dt}$$

$$= \gamma^3\left(\frac{|\vec{v}|}{c}\right)\frac{d\left|\frac{\vec{v}}{c}\right|}{dt}$$

Now, re-express $\vec{v} = \vec{\beta}c$ to write this as

$$\frac{d\gamma}{dt} = \gamma^3|\vec{\beta}|\frac{d|\vec{\beta}|}{dt}$$

Let's define $\hat{\beta}_\perp$ and $\hat{\beta}_\parallel$ to be **unit vectors** perpendicular and parallel to $\vec{\beta}$ and let's say $\vec{\beta}$ changes to $\vec{\beta} + d\vec{\beta} = \vec{\beta} + (d\beta_\perp)\hat{\beta}_\perp + (d\beta_\parallel)\hat{\beta}_\parallel$ as shown in the following diagram.

The change in the length of $\vec{\beta}$ (call this $d|\vec{\beta}|$, though I suppose I could have said $d\beta$ instead) will be

$$d|\vec{\beta}| = \sqrt{(\beta + d\beta_{\parallel})^2 + (d\beta_{\perp})^2} - \sqrt{\beta^2}$$

$$= \sqrt{\beta^2 + 2\beta d\beta_{\parallel} + (d\beta_{\parallel})^2 + (d\beta_{\perp})^2} - \sqrt{\beta^2}$$

$$= \beta\sqrt{1 + \frac{2d\beta_{\parallel}}{\beta} + \left(\frac{d\beta_{\parallel}}{\beta}\right)^2 + \left(\frac{d\beta_{\perp}}{\beta}\right)^2} - \beta$$

$$\approx \beta + \frac{\beta}{2}\left[\frac{2d\beta_{\parallel}}{\beta} + \left(\frac{d\beta_{\parallel}}{\beta}\right)^2 + \left(\frac{d\beta_{\perp}}{\beta}\right)^2\right] - \beta$$

$$\approx d\beta_{\parallel} + \frac{\beta}{2}\left[\frac{2d\beta_{\parallel}}{\beta} + \left(\frac{d\beta_{\parallel}}{\beta}\right)^2 + \left(\frac{d\beta_{\perp}}{\beta}\right)^2\right]$$

$$\approx d\beta_{\parallel} + \frac{\beta}{2}\left[\left(\frac{d\beta_{\parallel}}{\beta}\right)^2 + \left(\frac{d\beta_{\perp}}{\beta}\right)^2\right] \approx d\beta_{\parallel}$$

as long as $d|\vec{\beta}| \ll |\vec{\beta}| = \beta$. (I used the binomial approximation $(1 + \varepsilon)^n \approx 1 + n\varepsilon$ to see this.) As a result,

$$\frac{d\gamma}{dt} = \gamma^3 \beta \frac{d|\vec{\beta}|}{dt} \approx \gamma^3 \beta \frac{d\beta_{\parallel}}{dt}$$

OK, that takes care of the first term in $d\vec{p}/dt = m\vec{v}d\gamma/dt + m\gamma d\vec{v}/dt$

The derivative in the second term on the right side (that we still need to figure out) is $d\vec{v}/dt$:

$$\frac{d\vec{v}}{dt} = c\frac{d\vec{\beta}}{dt} = c\left(\frac{d\beta_{\parallel}}{dt}\hat{\beta}_{\parallel} + \frac{d\beta_{\perp}}{dt}\hat{\beta}_{\perp}\right)$$

As a result, we can rewrite $d\vec{p}/dt = m\vec{v}\,d\gamma/dt + m\gamma\,d\vec{v}/dt$ as

$$\frac{d\vec{p}}{dt} = m\vec{v}\left(\frac{d\gamma}{dt}\right) + m\gamma\frac{d\vec{v}}{dt}$$

$$\approx m\vec{v}\gamma^3\beta\frac{d\beta_{\parallel}}{dt} + mc\gamma\left(\frac{d\beta_{\parallel}}{dt}\hat{\beta}_{\parallel} + \frac{d\beta_{\perp}}{dt}\hat{\beta}_{\perp}\right)$$

$$= m(\beta c\hat{\beta}_{\parallel})\gamma^3\beta\frac{d\beta_{\parallel}}{dt} + mc\gamma\left(\frac{d\beta_{\parallel}}{dt}\hat{\beta}_{\parallel} + \frac{d\beta_{\perp}}{dt}\hat{\beta}_{\perp}\right)$$

Grouping terms gives

$$\frac{d\vec{p}}{dt} = m(\beta c\widehat{\beta}_{\parallel})\gamma^3\beta\frac{d\beta_{\parallel}}{dt} + mc\gamma\left(\frac{d\beta_{\parallel}}{dt}\widehat{\beta}_{\parallel} + \frac{d\beta_{\perp}}{dt}\widehat{\beta}_{\perp}\right)$$

$$= \gamma mc\left[(\beta^2\gamma^2 + 1)\frac{d\beta_{\parallel}}{dt}\widehat{\beta}_{\parallel} + \frac{d\beta_{\perp}}{dt}\widehat{\beta}_{\perp}\right]$$

$$= \gamma mc\left[\gamma^2\frac{d\beta_{\parallel}}{dt}\widehat{\beta}_{\parallel} + \frac{d\beta_{\perp}}{dt}\widehat{\beta}_{\perp}\right]$$

since

$$\beta^2\gamma^2 + 1 = \frac{\beta^2}{1-\beta^2} + \frac{1-\beta^2}{1-\beta^2}$$

To recover some sort of comparison with the non-relativistic case, I can factor the leading c back into the derivatives to rewrite this:

$$\vec{F} = \frac{d\vec{p}}{dt} = \gamma m\left[\gamma^2\frac{dv_{\parallel}}{dt}\hat{\beta}_{\parallel} + \frac{dv_{\perp}}{dt}\hat{\beta}_{\perp}\right]$$

Here, I've used $\vec{v} = v_{\parallel}\hat{\beta}_{\parallel}$ (since the velocity points in the same direction as $\vec{\beta}$), with $d\vec{v} = (dv_{\parallel})\hat{\beta}_{\parallel} + (dv_{\perp})\hat{\beta}_{\perp}$. Note the extra factor of γ^2 in front of the parallel term.

The non-relativistic case has $\beta \to 0$ and $\gamma \to 1$ (but bear in mind that $\hat{\beta}_{\parallel}$ and $\hat{\beta}_{\perp}$ are unit vectors and do not change), so we recover, in the non-relativistic limit,

$$\frac{d\vec{p}}{dt} \xrightarrow[\lim \beta \to 0]{} m\left(\frac{dv_{\parallel}}{dt}\hat{\beta}_{\parallel} + \frac{dv_{\perp}}{dt}\hat{\beta}_{\perp}\right) = m\frac{d\vec{v}}{dt}$$

If we apply a force perpendicular to $\vec{v}$ so that $\vec{F} = F_{\perp}\hat{\beta}_{\perp}$, $\vec{p}$ will not change in magnitude, only direction. Since $\vec{p} = \gamma m\vec{v} = \gamma mv_{\parallel}\hat{\beta}_{\parallel}$, a force along $\hat{\beta}_\perp$ will only change $v_\perp$, not $v_\parallel$. We have

$$F_\perp = \gamma m \frac{dv_\perp}{dt}$$

If we apply a force along $\vec{v}$ so that $\vec{F} = F_\parallel \hat{\beta}_\parallel$, $v_\perp$ won't change, but $v_\parallel$ will change so that

$$F_\parallel = \gamma^3 m \frac{dv_\parallel}{dt}$$

How does the velocity change with time if we apply a constant force along the direction of motion to an object initially at rest?

Since we are defining force so that it is the rate of change of momentum, we can use $p(t) = Ft$ and solve

$$p = \gamma m v = Ft$$

so

$$\frac{mv}{\sqrt{1 - v^2/c^2}} = Ft$$

$$m^2 v^2 = F^2 t^2 (1 - v^2/c^2)$$

$$m^2 v^2 + F^2 t^2 v^2/c^2 = F^2 t^2$$

$$(mv)^2 \left(1 + \frac{F^2 t^2}{m^2 c^2}\right) = F^2 t^2$$

$$v = \frac{Ft/m}{\sqrt{1 + \frac{F^2 t^2}{m^2 c^2}}}$$

We can also show that, in general, the work-energy theorem still works. Start by using $E^2 = p^2 c^2 + m^2 c^4$ and differentiating both sides, assuming that $\vec{F} \parallel \vec{v}$:

$$2E \frac{dE}{dt} = 2pc^2 \frac{dp}{dt}$$

Since $E = \gamma m c^2$ and $p = \gamma m v$, dividing both sides by $2E$ gives

$$\frac{dE}{dt} = \frac{pc^2}{E} \frac{dp}{dt}$$

$$\frac{dE}{dt} = \frac{\gamma m v c^2}{\gamma m c^2}\frac{dp}{dt}$$

$$= v\frac{dp}{dt}$$

so that

$$dE = vdp$$

and therefore

$$\frac{dE}{dx} = v\frac{dp}{dx}$$

$$= \frac{dx}{dt}\frac{dp}{dx}$$

$$= \frac{dp}{dt}$$

As a result, since $F = dp/dt$,

$$F = \frac{dp}{dt} = \frac{dE}{dx}$$

so

$$\Delta E = F\Delta x$$

for small enough Δx that we can treat the force as constant during the interval Δx. This is the work-energy theorem. It still works!

Also (assuming straight line motion), we can integrate over a larger distance:

$$\Delta E = \int_{\text{start}}^{\text{finish}} dE = \int_{\text{start}}^{\text{finish}} F dx$$

If the object is initially at rest, its change in energy will be $\Delta E = \gamma m c^2 - m c^2 = (\gamma - 1)m c^2$.

Exercise 6.5: Free fall

The despicable Mrs. Rumpdock falls towards a black hole of mass $M = 2 \times 10^{30}$ kg. Her initial distance from the black hole is so great that it can be taken to be infinite; her initial velocity is almost zero but pointed toward the center of the black hole. Naturally, as she descends into oblivion, her speed increases.

Ignoring complications introduced by general relativity, we can take the force acting on her to be the classical Newtonian gravitational force of GmM/r^2. The value of G is $6.674 \times 10^{-11}\,\mathrm{N \cdot m^2 \cdot kg^{-2}}$.

How fast is she moving as she falls through a point that is $3\,\mathrm{km}$ from the center of the black hole? (Your answer won't agree with a fully general relativistic calculation, but we won't worry about that here.)

Solution

Let's use the work-energy theorem. The change in energy is $(\gamma - 1)mc^2$ so that

$$\int_{\infty}^{3\,\mathrm{km}} \frac{GMm}{r^2}\,\mathrm{d}r = (\gamma - 1)mc^2$$

$$\left. \frac{GM}{r} \right|_{\infty}^{3000} = (\gamma - 1)c^2$$

$$\frac{6.674 \times 10^{-11}}{3 \times 10^3 (3 \times 10^8)^2} = (\gamma - 1)/M$$

$$\gamma = 1.49437 = 1/\sqrt{1 - v^2/c^2} \text{ so } v = 0.74c$$

Problem 1: The Greisen-Zatsepin-Kuzmin (GZK) upper limit on cosmic ray energies

The contemporary universe is filled with low-energy photons left over from the time when the early universe cooled sufficiently for the omnipresent plasma of electrons and protons to form neutral hydrogen. These photons, born as visible- and ultraviolet-wavelength electromagnetic radiation, have lost energy as the universe has expanded. The current "cosmic microwave background" (CMB) spectrum is well described by a black body spectrum with temperature $2.725\,\mathrm{K}$. This corresponds to a peak wavelength of 1.9 millimeters or a peak energy per photon of about $6.5 \times 10^{-4}\,\mathrm{eV}$.

In 1966, Kenneth Greisen, Vadim Kuzmin, and Georgiy Zatsepin realized that cosmic ray protons of sufficiently high energy could, in collisions with these photons, have their internal quark structure scrambled to produce heavier, unstable particles such as the Δ^+

through the process $\gamma p \to \Delta^+ \to p\pi^0$ and $\gamma p \to \Delta^+ \to n\pi^+$. After the Δ^+ decays, the proton (or neutron) would have significantly less energy in the "lab" frame than it had before the collision. (By lab frame, I mean a frame in which the CMB spectrum is not Doppler shifted away from its peak wavelength of $1.9\,\mathrm{mm}$.)

The rest masses of the proton and Δ^+ are $0.93827\,\mathrm{GeV}/c^2$ and $1.232\,\mathrm{GeV}/c^2$, respectively.

At what proton energy should the GZK effect begin to make itself felt? You can assume that all CMB photons have energy $6.5 \times 10^{-4}\,\mathrm{eV}$. Feel free to use a calculator.

Solution

The protons are so energetic that we can approximate β as 1. But note that we can't use this when calculating γ, which becomes infinite at $\beta = 1$! In a head-on collision, we have (in the "lab" frame, with the proton moving in the $+x$-direction, the photon in the $-x$-direction)

$$p^\mu_{\text{proton}} = \left(\frac{E_{\text{proton}}}{c}, p_{\text{proton}}, 0, 0 \right)$$

$$p^\mu_{\text{photon}} = \frac{1}{c}(E_{\text{photon}}, -E_{\text{photon}}, 0, 0)$$

$$p^\mu_{\text{system}} = \frac{1}{c}(E_{\text{proton}} + E_{\text{photon}}, E_{\text{proton}} - E_{\text{photon}}, 0, 0)$$

$$(m_{\text{system}}c^2)^2 = (E_{\text{proton}} + E_{\text{photon}})^2 - (E_{\text{proton}} - E_{\text{photon}})^2$$

$$= 4E_{\text{proton}}E_{\text{photon}} = (m_\Delta c^2)^2$$

$$4E_{\text{proton}} \times 6.5 \times 10^{-4}\,\mathrm{eV} = \left(1.232 \times 10^9 \mathrm{eV}\right)^2$$

$$E_{\text{proton}} = 5.838 \times 10^{20}\,\mathrm{eV} = 5.838 \times 10^{11}\,\mathrm{GeV}$$

Problem 2: This is why we build them as colliders now

Some years ago, Fermilab used to extract its high-energy proton beam for use by "fixed target" experiments situated at the ends of external beamlines a mile north of the Tevatron ring. The energy available for the production of (unstable) heavy particles in collisions between a high-energy proton and a stationary proton is very different from that available when two beams with the full Tevatron energy collide head-on.

Calculate the most massive single particle that can be produced in a collision of a proton with total energy $1000\,\mathrm{GeV}$ with a stationary proton with total energy $mc^2 = 0.93827\,\mathrm{GeV}$. Also, calculate the most massive particle that can be produced in a head-on collision of two $1000\,\mathrm{GeV}$ beams. (It's fine to use the approximation $\beta \approx 1$ when appropriate.)

Solution

Let's do the "fixed target" version first: one proton is at rest:

$$p^\mu_{\text{system}} = \frac{1}{c}\left((1000 + .938), 1000, 0, 0\right)$$

$$mc^2 = \sqrt{(1000 + .938)^2 - 1000^2}\,\mathrm{GeV} = 43.32\,\mathrm{GeV}$$

Now, for the colliding beams version:

$$p^\mu_{\text{system}} = \frac{1}{c}\left((1000 + 1000), 1000 - 1000, 0, 0\right)$$

$$mc^2 = \sqrt{(2000)^2}\,\mathrm{GeV} = 2000\,\mathrm{GeV}$$

Problem 3: Mu2e at Fermilab

We know that the "Standard Model," which successfully describes nearly all phenomena involving strong, electromagnetic, and weak interactions, is incomplete. It is curious that a theory so strong in predictive power, and so accurate in nearly everything it covers, is entirely unable to provide insight into the nature of dark matter and dark energy. How can it provide accurate predictions when it is blind to 95.6% of everything that there is in the universe?

One experiment that might provide insight into some of the Standard Model's missing pieces is called *Mu2e*: the collaboration will search for neutrinoless transitions into electrons of muons that have been captured by aluminum nuclei. The reaction they are seeking to observe is $\mu^-\mathrm{Al} \to \mathrm{e}^-\mathrm{Al}$, where the initial state can be thought of as a stationary muon bound to a stationary aluminum nucleus.

The masses of the muon and aluminum nucleus are approximately $105.66\,\mathrm{MeV}/c^2$ and $25.13\,\mathrm{GeV}/c^2$, respectively; the electron mass is $0.511\,\mathrm{MeV}/c^2$.

If they are lucky enough to observe a few $\mu^-\text{Al} \to e^-\text{Al}$ transitions, they'll be measuring the energy of the outgoing electron in their apparatus. They will *not* be able to observe the aluminum nucleus as it recoils in the opposite direction from the muon. Since energy and momentum must be conserved, they will identify a monochromatic electron energy signal as coming from these neutrinoless transitions.

Calculate the energy of the electron produced in the process $\mu^-\text{Al} \to e^-\text{Al}$. Also, calculate the *kinetic* energy of the aluminum after the electron is produced, using the non-relativistic approximation $K = p^2/(2m)$. (Keep in mind that the aluminum nucleus carries off momentum and a small amount of kinetic energy in order to balance the momentum imparted to the electron, and neglect any effects associated with the Coulomb binding energy of the muon/electron to the nucleus.)

Solution

Use conservation of four-momentum. The initial energy is $25.13 + 0.10566\,\text{GeV}$, while the initial vector momentum is zero. The final state four-momenta are

$$p^u_{Al} = \gamma_{Al}(m_{Al}c, m_{Al}\vec{v}, 0, 0) = (\gamma_{Al}m_{Al}c, \vec{p}, 0, 0)$$

$$p^{\mu}_e = (\gamma_e m_e c, -\vec{p}, 0, 0)$$

We can use the algebra developed earlier to describe Λ baryon decay into a proton and a pion, but replacing the initial state (stationary) Λ with an initial bound state of a muon and an aluminum nucleus, treated as a single particle. Plugging into the expression for p^2c^2 and doing the calculator work eventually yields $p^2c^2 = 0.011117$ so that $pc = 0.105438\,\text{GeV}$. So, the electron is still relativistic! Calculate the energies now, using the non-relativistic approximation for the Al nucleus kinetic energy:

$$E_{\text{electron}} = \sqrt{p^2c^2 + m^2c^4} = 0.105438\,\text{GeV}$$

$$KE_{Al} = p^2/2M_{Al} = \frac{0.011117}{2 \times 25.13} = 2.21 \times 10^{-4}\,\text{GeV}.$$

The nucleus carries off very little energy!

Problem 4: Proton and electron linacs

Superconducting accelerating structures in modern linear accelerators are able to support electric fields in excess of 5×10^7 volts per meter.

The magnitude of the electron and proton charges is 1.6×10^{-19} C. In S.I. units, the electron and proton masses are 9.11×10^{-31} kg and 1.67×10^{-27} kg.

How long is a linac (assuming an accelerating field of 50 megavolts per meter) that's intended to produce an electron beam traveling at $0.9c$? How long would a proton linac need to be to accelerate its beam to $0.9c$?

Solution

The work-energy theorem still holds. An electron going $0.9c$ has $\gamma = 1/\sqrt{1 - .9^2} \approx 2.294157$ with an energy change of $(\gamma - 1)mc^2 = 0.66134$ MeV.

An electron gains 5×10^7 eV per meter in a 50 megavolt per meter field, so we only need an accelerating structure that is 0.66134 MeV/50 MV/m $= 0.013$ m.

The proton's energy change will be much larger: $\Delta K = (\gamma - 1)mc^2 = 1.213919$ GeV so we need 1.213919 GeV/.05 GeV per meter $= 24.28$ m.

Note that it is much easier to do this using eV and the work-energy theorem than working, instead, with forces.

Cassis, France

Unit 7

Massless Particles, Relativistic Dynamics, the Electromagnetic Field

Zero-mass particles

There are such things: photons. We used to think that neutrinos were massless, but now we know that's not actually the case!

Massless particles carry *both* energy and momentum. The equation $E^2 = p^2c^2 + m^2c^4$ is still valid, reducing to $E^2 = p^2c^2$ or $E = pc$.

127

From introductory physics, recall that

$$p = \frac{h}{\lambda}, \quad E = \frac{hc}{\lambda}$$

for a photon.

How fast do massless particles travel?

Let's say a photon is traveling with speed v and total energy E (with momentum $p = E/c$) in the $+x$-direction. Its four momentum is

$$p = (E/c, 0, 0, E/c)$$

If the photon were to have $v < c$, we could boost to a frame in which the photon was stationary:

$$p'^{\mu} = \sum_{v} \Lambda^{\mu}_{v} p^{v}$$

or

$$p'^{v} = \left(\gamma(1 - \beta)\frac{E}{c}, 0, 0, \gamma(1 - \beta)\frac{E}{c} \right)$$

where the photon's vector momentum must be zero in this frame. That means

$$\gamma(1 - \beta)E/c = 0$$

The only way for this to work is for β to be unity! (Even though γ diverges, $\gamma(1 - \beta)$ goes to zero.)

Exercise 7.1: Proof

Recall that $a^2 - b^2 = (a - b)(a + b)$. This might help you do the following: prove that $\lim_{\beta \to 1} \gamma(1 - \beta) = 0$.

Solution

$$\gamma(1 - \beta) = \frac{(1 - \beta)}{\sqrt{(1 - \beta^2)}} = \frac{1 - \beta}{\sqrt{(1 + \beta)(1 - \beta)}} = \sqrt{\frac{1 - \beta}{1 + \beta}}$$

$$\lim_{\beta \to 1} \gamma(1 - \beta) = \lim_{\beta \to 1} \sqrt{\frac{1 - \beta}{1 + \beta}} = \lim_{\beta \to 1} \sqrt{\frac{1 - \beta}{2}} = 0$$

We can't ever get to the same velocity as the photon by doing a boost—the photon must be traveling with $v = c$. Massless things necessarily move at the speed of light. Calculations are particularly simple for massless objects since it's always true that $E = pc$.

An example: a π^0 decays (with an average lifetime of 8.4×10^{-17} seconds) into a pair of photons 98.8% of the time. What are the energies carried off by the photons?

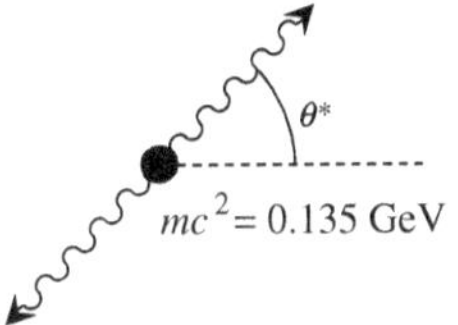

Look at the four-momenta in the pion's rest frame: before decaying, the π^0 has $p^\mu = (m_{\pi^0}c, 0, 0, 0)$.

Let's say the photons make angles θ^* and $\theta^* + 180°$ with respect to the x-axis and travel in the $x - y$ plane. The right-side photon has energy $m_{\pi^0}c^2/2$ and three-momentum of magnitude $m_{\pi^0}c/2$. The same is true for the other photon. The photons' four-momenta are

$$p_{right} = \frac{m_{\pi^0}c}{2}(1, \cos \theta^*, \sin \theta^*, 0)$$

$$p_{left} = \frac{m_{\pi^0}c}{2}(1, -\cos \theta^*, -\sin \theta^*, 0)$$

Exercise 7.2: In the lab frame

Lorentz transformed the photons' four momenta in the above decay into a frame in which the π^0 is moving along $+\hat{x}$ with velocity $v = \beta c$ at the time it decays. By doing this, calculate the angles between the two photons' velocities and the x-axis in this new frame.

Solution

Keep in mind that the frame you're going into is moving *to the left* (in this frame, the pion is seen to be moving in the positive direction), so its origin's velocity is negative:

$$\begin{bmatrix} E/c \\ p_x \\ p_y \\ p_z \end{bmatrix} = \begin{bmatrix} \gamma & -\gamma\beta & 0 & 0 \\ -\gamma\beta & \gamma & 0 & 0 \\ 0 & 0 & 1 & 0 \\ 0 & 0 & 0 & 1 \end{bmatrix} \frac{m_\pi c}{2} \begin{bmatrix} 1 \\ \pm\cos\theta \\ \pm\sin\theta \\ 0 \end{bmatrix}$$

so

$$p_x = \frac{m_\pi c}{2}(-\gamma\beta \pm \gamma\cos\theta) = \frac{m_\pi c\gamma}{2}(-\beta \pm \cos\theta)$$

$$p_y = \pm\sin\theta\frac{m_\pi c}{2}$$

$$\text{angle} = \tan^{-1}(p_y/p_x) = \tan^{-1}\left(\frac{1}{\gamma}\frac{\sin\theta}{\cos\theta \pm |\beta|}\right)$$

Relativistic dynamics

We want equations that are candidates for consideration as "laws of nature" to have the same form in all inertial frames. A statement such as $\vec{E} = 0$ (true outside an electrically neutral current-carrying wire at rest, in the absence of externally supplied electric fields) won't be true in a frame in which the wire is seen to be moving. So, that's not going to cut it as a law of nature.

As long as we build our equations from objects that behave well under Lorentz transformations—four-vectors, Lorentz scalars, and Lorentz tensors—we'll do fine. An equation of this sort that is true in one frame of reference will *necessarily* be true in any other frame of reference, without changing form.

We saw last unit that Newton's second law, written as $\vec{F} = \frac{d\vec{p}}{dt}$ (with $\vec{p} \equiv \gamma m\vec{v}$), is true even for a fast-moving object. But we can't really call this a "law of nature" in Einstein's sense since the time interval in the "denominator" of the derivative mixes with an associated space interval dx upon change of reference frame.

Some general principles concerning Lorentz scalars, four-vectors, and tensors

1. Lorentz scalars

- The sum, difference, product, or quotient of two Lorentz scalars is a Lorentz scalar.

- The scalar product of any two four-vectors is a Lorentz scalar:

$$s = \sum_\mu A^\mu B_\mu = \sum_\mu A^\mu \sum_\rho B^\rho g_{\mu\rho} = \sum_{\mu,\rho} A^\mu B^\rho g_{\mu\rho}$$

- The "contraction" of both indices of any two tensors is a Lorentz scalar:

$$b = \sum_{\mu,v} S^{\mu v} T_{\mu v} = \sum_{\mu,v} \left[S^{\mu v} \sum_{\rho,\sigma} g_{\mu\rho} g_{v\sigma} T^{\rho\sigma} \right] = \sum_{\mu,v,\rho,\sigma} S^{\mu v} g_{\mu\rho} g_{v\sigma} T^{\rho\sigma}$$

2. Four-vectors

- The product of a Lorentz scalar and a four-vector is a four-vector.
- The sum or difference of two four-vectors is a four-vector.
- The "contraction" of a four-vector and one index of a tensor is a four-vector:

$$K^v = \sum_\mu S^{\mu v} A_\mu = \sum_\mu \left[S^{\mu v} \sum_\rho g_{\mu\rho} A^\rho \right] = \sum_\mu \sum_\rho S^{\mu v} g_{\mu\rho} A^\rho$$

3. Tensors

- The product of a Lorentz scalar and a tensor is a tensor.
- The sum or difference of two tensors is a tensor.
- The outer product of two four-vectors is a tensor: $T^{\mu v} = A^\mu B^v$.
- A product of two tensors done with the contraction of one index is a tensor: $A^{\mu v} = \sum_\rho [B^\mu_\rho C^{\rho v}] = \sum_\rho \{[\sum_\sigma g_{\rho\sigma} B^{\mu\sigma}] C^{\rho v}\}$.

4. General covariance of "good" equations

- An equation written as a relationship among Lorentz scalars, four-vectors, and Lorentz tensors that is known to be true in one inertial frame will be true in all inertial frames.

5. Behavior under Lorentz transformations

- Scalars: $s' = s$.
- Four vectors: $A'^\mu = \sum_v \Lambda^\mu_v A^v$.
- Tensors: $T'^{\mu v} = \sum_{\rho,\sigma} \Lambda^\mu_\rho \Lambda^v_\sigma T^{\rho\sigma}$.

Where's the physics?

There's actually not very much physics in this: the speed of light is constant, so space–time intervals transform from frame to frame according to the Lorentz transformations. The rest is just slick, powerful machinery that enables rapid calculations, as well as easy recognition of a system's essential features. Try to keep that in mind: there's much less "there" there than meets the eye. It's all about how space and time transform when shifting frames of reference and is all contained in the Lorentz transformations.

Definition of four-force

Let's turn $\vec{F} = m\vec{a}$ into an equation that behaves sensibly under Lorentz transformations. We'd like to recast it as an equation involving four-vectors and Lorentz scalars so that it keeps its form under Lorentz transformations.

An obvious thing to try is this: cook up some kind of four-force f^μ that will satisfy

$$f^\mu \Delta\tau = \Delta p^\mu$$

where $\Delta\tau = \left[\sqrt{c^2\Delta t^2 - \Delta x^2 - \Delta y^2 - \Delta z^2}\right]/c$ and $\Delta p^\mu = \Delta[\gamma m\,(c, \vec{v})]$ are the changes in proper time and four-momentum. (We want to try using this equation since our non-four-vector equation that works so well for non-relativistic systems is $\vec{F}\Delta t = \Delta\vec{p}$.)

Can we actually cook up a true equation of this form? If so, what should f^μ look like? And how do we deal with Δt in $\vec{F}\Delta t = \Delta\vec{p}$ compared to $\Delta\tau$ in the $f^\mu \Delta\tau = \Delta p^\mu$ equation?

Let's be very explicit here. Say, we apply a (three-vector) force $\vec{F}$ to an object between times t and $t+\Delta t$ as measured by clocks in our frame. We have two events:

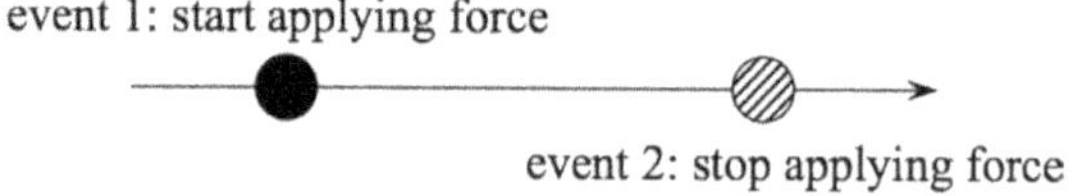

Let's restrict ourselves to the case $\vec{F}\|\vec{v}$ for the time being.

We apply our force $\vec{F}$, measure the three-vector velocity $\vec{v}$ at t (event 1) then at $t+\Delta t$ (event 2). We then calculate $\vec{p} = \gamma m\vec{v}$ at the two times t and $t+\Delta t$.

Due to our definitions of force, momentum, etc., we find $\vec{F} = \Delta\vec{p}/\Delta t$. That's not surprising.

Let's say the interval Δt is small enough so that the change in velocity is small compared to $|\vec{v}| : |\Delta\vec{v}|/|\vec{v}| << 1$.

In this case, γ is nearly constant between events 1 and 2 so that $\Delta\tau = \Delta t/\gamma$.

$\Delta\tau$ would be the amount of time that would pass on a clock moving from event 1 to event 2, traveling with speed $|\vec{v}|$. As a result, we can rewrite $\vec{F} = \Delta\vec{p}/\Delta t$ as $\gamma\vec{F} = \Delta\vec{p}/\Delta\tau$.

If the object were initially at rest at time t, it would be moving so slowly at $t + \Delta t$ that we'd have $\gamma \approx 1$ and $t \approx \Delta\tau$, as one would expect.

Recall that $\Delta\vec{p} \equiv \vec{p}_2 - \vec{p}_1$ populates the last three components (the "space components") of the four-vector $\Delta p^\mu \equiv p_2^\mu - p_1^\mu$. As a result, the right side of $\gamma\vec{F} = \Delta\vec{p}/\Delta\tau$ looks just like the last three components of a four-vector. If we take $\gamma\vec{F}$ to be the last three components of a four-force f^μ, then we'll have three of the four components sorted out for a definition that works in $f^\mu\Delta\tau = \Delta p^\mu$.

How about the first component? Let's investigate $f^0 = \Delta p^0/\Delta\tau$ now. We know that $\Delta p^0 = \Delta E/c$. Also, since the work–energy theorem still holds, we have $\Delta E = \vec{F} \cdot \Delta\vec{x} = F\Delta x$ since $\vec{F}\|\Delta\vec{x}$ and both the force and the displacement are in the $\hat{x}$-direction. As a result,

$$\frac{\Delta p^0}{\Delta\tau} = \gamma\frac{\Delta p^0}{\Delta t} = \frac{\gamma}{\Delta t}\frac{\Delta E}{c}$$

$$= \frac{\gamma}{\Delta t}\frac{F\Delta x}{c}$$

$$= \gamma F\frac{\Delta x}{c\Delta t}$$

$$= \gamma F\beta$$

Combine what we have found so far: $\Delta\vec{p}/\Delta\tau = \gamma\vec{F}$ and $\Delta p^0/\Delta\tau = \beta\gamma F$. We have

$$\frac{\Delta p^\mu}{\Delta\tau} = \left(\frac{\Delta p^0}{\Delta\tau}, \frac{\Delta\vec{p}}{\Delta\tau}\right) = (\gamma F\beta, \gamma\vec{F}) = (\gamma F\beta, \gamma F, 0, 0)$$

Since we want $f^\mu \Delta\tau = \Delta p^\mu$, or $f^\mu = \frac{\Delta p^\mu}{\Delta\tau}$, it looks like a good definition of the four-force is

$$f^\mu = \gamma(\beta F, F, 0, 0)$$

What's f'^μ in a frame in which the particle is at rest at time t? We should find that the four-force corresponds to the entirely non-relativistic form of Newton's law. Boost to the particle's rest frame to write

$$f'^\mu = \begin{bmatrix} \gamma & -\beta\gamma & 0 & 0 \\ -\beta\gamma & 1 & 0 & 0 \\ 0 & 0 & 1 & 0 \\ 0 & 0 & 0 & 1 \end{bmatrix} \begin{bmatrix} \gamma\beta F \\ \gamma F \\ 0 \\ 0 \end{bmatrix} = \begin{bmatrix} 0 \\ \gamma^2(1-\beta^2)F \\ 0 \\ 0 \end{bmatrix}$$

Since $\gamma^2(1-\beta^2) = 1$, we find $f'^\mu = (0, F, 0, 0) = (0, \vec{F})$, as we'd expect.

In a coordinate-independent fashion, in a frame in which a particle is at rest, we have

$$f'^\mu = (0, \vec{F})$$

In a different frame, in which the particle is seen to be moving with velocity $\vec{v}$, we'll have

$$f^\mu = \gamma(\vec{\beta} \cdot \vec{F}, \vec{F})$$

We'll also have, in these two frames,

$$f^\mu = \frac{dp^\mu}{d\tau} \quad \text{and} \quad f'^\mu = \frac{dp'^\mu}{d\tau}$$

Note that the equation has the same form in the two frames. This is what we want to be the case for a candidate "law of physics."

What does it mean to have the first component of the four-force be zero in the particle's rest frame? Investigate the equation

$$f^0 = \frac{dp^0}{d\tau} = \frac{1}{c}\frac{dE}{dt}$$

in the particle's rest frame, where $dt = d\tau$.

Before we shift to the rest frame, we have

$$p^0 = \frac{E}{c}$$

$$= \frac{1}{c}\sqrt{|\vec{p}|^2 c^2 + m^2 c^4}$$

$$p^0 \approx \frac{1}{c}\left(mc^2 + \frac{1}{2}mv^2\right)$$

$$\approx \frac{1}{c}\left(mc^2 + \frac{|\vec{p}|^2}{2m}\right)$$

In the particle's rest frame, this reduces to $p^0 = mc$.

After a time interval Δt, our particle will no longer be at rest, instead having $v = (F/m)\Delta t$ so that $\vec{p} = \vec{F}\Delta t$ and $\frac{\Delta E}{c} \approx \frac{|\vec{p}|^2}{2mc} = \frac{F^2}{2mc}(\Delta t)^2$.

As $\Delta t \to 0$, $\Delta E/c$ goes to zero quadratically in Δt, much more rapidly than does Δp.

Equivalently, for our slow-moving particle, we can write

$$\Delta p^\mu \approx m((\Delta|\vec{v}|)^2/c, \Delta\vec{v})$$

$$= m(\Delta\vec{\beta}\cdot\Delta\vec{v}, \Delta\vec{v})$$

Since $|\Delta\vec{\beta}| <<< 1$ (the particle is nonrelativistic), the change in energy is negligible compared to the change in momentum in the frame in which the particle is initially at rest. And that's why the first component of the four-force is zero when the particle is, for an instant, at rest as it undergoes (possibly uniform) acceleration.

A summary so far

1. The Newtonian expressions for changes in energy and vector momentum still hold for a particle undergoing relativistic motion:

$$\vec{F}\Delta t = \Delta\vec{p}$$

$$\vec{F}\cdot\Delta\vec{x} = \Delta E$$

    but we must redefine energy and vector momentum as $E = \gamma mc^2$ and $\vec{p} = \gamma m\vec{v}$.

2. The four-velocity and four-momentum are defined this way:

$$u^\mu = \gamma(c, \vec{v})$$

$$p^\mu = mu^\mu = \gamma m(c, \vec{v}) = \left(\frac{E}{c}, \vec{p}\right)$$

The four velocity and four momentum transform in the expected way under Lorentz transformations:

$$u'^{\mu} = \sum_{v} \Lambda^{\mu}_{v} u^{v}, \quad p'^{\mu} = \sum_{v} \Lambda^{\mu}_{v} p^{v}$$

3. The connection between the four-force and the four-momentum for a particle in motion is this:

$$f^{\mu}\Delta\tau = \Delta p^{\mu}$$

while the connection between the (Newtonian) force and four-force for a particle moving with velocity βc is this:

$$f^{\mu} = \gamma(\vec{F}\cdot\vec{\beta}, \vec{F})$$

An example

It's time for an example. Usually, you'll want to make use of

$$\vec{F} = \frac{d\vec{p}}{dt}$$

with $\vec{p} = \gamma m\vec{v}$.

Exercise 7.3: Tevatron magnets

Let's figure out the strength of the bending magnets in Fermilab's Tevatron, in which relativistic protons and antiprotons circulated.

The force exerted by electric and magnetic fields on a charge Q is

$$\vec{F} = Q(\vec{E} + \vec{v}\times\vec{B})$$

Say the protons had total energy 1000 GeV (it was slightly less than this) and (vector) momentum magnitude $p \approx 1000$ GeV/c. The accelerator was a circular ring 1000 m in radius.

(a) Calculate the beam's orbital angular frequency ω as it traveled around the ring.

Solution

Since $\omega = v/r$, and since they're going so close to the speed of light, we have $\omega \approx c/r = 3\times 10^8/10^3 = 3\times 10^5$ radians per second.

(b) For an object traveling in a circle with constant speed, the rate of change of its (vector) momentum is

$$\frac{d\vec{p}}{dt} = \vec{\omega} \times \vec{p}$$

The magnet force on a proton in the beam is perpendicular to the proton's velocity and of magnitude $Q|\vec{v}|B$, where the proton's charge is $Q = 1.6 \times 10^{-19}$ C. From this, derive an expression for the average magnetic field in terms of the relevant kinematic quantities and solve for its value in Tesla.

Solution

We need to use S.I. values to get an answer in Tesla:

$$|\vec{F}| = \left|\frac{d\vec{p}}{dt}\right| = |\vec{\omega} \times \vec{p}\,| = \omega p = QvB$$

$$\omega p = \omega \gamma m v \approx \omega \gamma m c$$

$$\gamma = E_{\text{proton}}/m_{\text{proton}}, \quad c^2 = 1000\,\text{GeV}/0.938\,\text{GeV} = 1066.1$$

$$\left|\frac{d\vec{p}}{dt}\right| = \omega \gamma m c = (3 \times 10^5)(1066.1)(1.67 \times 10^{-27}\,\text{kg})(3 \times 10^8)$$

$$= 1.602 \times 10^{-10}$$

$$= QvB = (1.6 \times 10^{-19})(3 \times 10^8)B$$

so

$$B = \frac{1.602 \times 10^{-10}}{(1.6 \times 10^{-19})(3 \times 10^8)} = 10/3\,\text{T}$$

The actual magnets were about twice as strong as this since not all of the accelerator's beam pipe was surrounded by dipole magnets. There were also focusing (quadrupole) magnets, vacuum pump inlets, and so forth.

A curious fact: it's not true in general that $\vec{F}$ and $\vec{a}$ are parallel. This is easy to see since something going close to the speed of light can't speed up appreciably. The acceleration induced by a force parallel to the velocity will be minimal. However, forces perpendicular to the velocity will succeed in changing the velocity's direction, causing a non-zero acceleration.

Since $\vec{F}\Delta t = \Delta\vec{p}$ (the time interval Δt is very short), a force *perpendicular* to the velocity (which doesn't change the magnitude of the velocity) will cause a change in velocity given by

$$\vec{F}\Delta t = \gamma m \Delta\vec{v}$$

or an acceleration $\vec{a} = \frac{\Delta\vec{v}}{\Delta t} = \frac{\vec{F}}{\gamma m}$.

A force *parallel* to $\vec{v}$ will change γ as well as $\vec{v}$; from an equation some pages back, we have, in this case,

$$\vec{a} = \frac{\vec{F}}{\gamma^3 m}$$

If we define $\vec{F} \equiv \begin{bmatrix} F_\| \\ \vec{F}_\perp \end{bmatrix}$ and $\vec{a} \equiv \begin{bmatrix} a_\| \\ \vec{a}_\perp \end{bmatrix}$, then we have $\vec{a} = \begin{bmatrix} a_\| \\ \vec{a}_\perp \end{bmatrix} = \frac{1}{\gamma m}\begin{bmatrix} 1/\gamma^2 & 0 \\ 0 & 1 \end{bmatrix}\begin{bmatrix} F_\| \\ \vec{F}_\perp \end{bmatrix}$.

The electric and magnetic fields

In elementary physics courses, we tend to treat the electric and magnetic fields as distinct things, which transform as vectors under rotations. In the context of special relativity, it might be tempting to think of them as the last three components of four vectors, for example:

$$(E_0, E_x, E_y, E_z) \quad \text{and} \quad (B_0, B_x, B_y, B_z)$$

If that were the case, then we would transform them under boosts using a single Lorentz transformation matrix, just as we do for four velocities and four momenta.

But that would be completely wrong! As you know from an earlier unit, changes of frame will mix together the components of the electric and magnetic fields as observed in one frame when we shift to another frame.

The proper description of the fields is as the components of a rank two Lorentz tensor. The electromagnetic field strength tensor is this:

$$F^{\mu\nu} \equiv \begin{bmatrix} 0 & -E_x/c & -E_y/c & -E_z/c \\ E_x/c & 0 & -B_z & B_y \\ E_y/c & B_z & 0 & -B_x \\ E_z/c & -B_y & B_x & 0 \end{bmatrix}$$

It behaves this way under Lorentz transformation:

$$\underset{\approx}{F'} = \underset{\approx}{\Lambda}\,\underset{\approx}{F}\,\underset{\approx}{\Lambda}^{T} = \underset{\approx}{\Lambda}\,\underset{\approx}{F}\,\underset{\approx}{\Lambda}$$

Writing it out explicitly for boosts in the x-direction,

$$F'^{\mu\nu} \equiv \begin{bmatrix} \gamma & -\gamma\beta & 0 & 0 \\ -\gamma\beta & \gamma & 0 & 0 \\ 0 & 0 & 1 & 0 \\ 0 & 0 & 0 & 1 \end{bmatrix} \begin{bmatrix} 0 & -E_x/c & -E_y/c & -E_z/c \\ E_x/c & 0 & -B_z & B_y \\ E_y/c & B_z & 0 & -B_x \\ E_z/c & -B_y & B_x & 0 \end{bmatrix}$$

$$\times \begin{bmatrix} \gamma & -\gamma\beta & 0 & 0 \\ -\gamma\beta & \gamma & 0 & 0 \\ 0 & 0 & 1 & 0 \\ 0 & 0 & 0 & 1 \end{bmatrix}$$

How might we construct a four-force associated with the electromagnetic field? We would want to build it out of quantities with good Lorentz properties and end up with something that satisfies what we already know, namely that

$$\vec{F} = Q(\vec{E} + \vec{v} \times \vec{B})$$

The most obvious candidates to use are the field strength tensor and the four-velocity. Perhaps this will work? Can we replace the "$\sim$" with an equal sign in the following expression?

$$f^{\mu} \sim Q \sum_{\nu} F^{\mu\nu} u_{\nu} = Q \sum_{\nu} \left[F^{\mu\nu} \sum_{\rho} g_{\nu\rho} u^{\rho} \right]$$

Note that this is just the matrix multiplication of F and the covariant version of u:

$$Q \sum_{\nu} F^{\mu\nu} u_{\nu} = Q \begin{bmatrix} 0 & -E_x/c & -E_y/c & -E_z/c \\ E_x/c & 0 & -B_z & B_y \\ E_y/c & B_z & 0 & -B_x \\ E_z/c & -B_y & B_x & 0 \end{bmatrix} \begin{bmatrix} \gamma c \\ -\gamma v_x \\ -\gamma v_y \\ -\gamma v_z \end{bmatrix}$$

Exercise 7.4: Electromagnetic force

Perform the above matrix multiplication. How does the product compare to what you would expect, if this really is the right expression for the four-force? In other words, is it true that

$$
f^\mu \equiv Q
\begin{bmatrix}
0 & -E_x/c & -E_y/c & -E_z/c \\
E_x/c & 0 & -B_z & B_y \\
E_y/c & B_z & 0 & -B_x \\
E_z/c & -B_y & B_x & 0
\end{bmatrix}
\begin{bmatrix}
\gamma c \\
-\gamma v_x \\
-\gamma v_y \\
-\gamma v_z
\end{bmatrix}
$$

$$
=
\begin{bmatrix}
\gamma \vec{\beta} \cdot Q(\vec{E} + \vec{v} \times \vec{B}) \\
\gamma Q(\vec{E} + \vec{v} \times \vec{B})_x \\
\gamma Q(\vec{E} + \vec{v} \times \vec{B})_y \\
\gamma Q(\vec{E} + \vec{v} \times \vec{B})_z
\end{bmatrix} \, ?
$$

Solution

Just do the multiplications:

$$
f^\mu = Q
\begin{bmatrix}
0 & -E_x/c & -E_y/c & -E_z/c \\
E_x/c & 0 & -B_z & B_y \\
E_y/c & B_z & 0 & -B_x \\
E_z/c & -B_y & B_x & 0
\end{bmatrix}
\begin{bmatrix}
\gamma c \\
-\gamma v_x \\
-\gamma v_y \\
-\gamma v_z
\end{bmatrix}
$$

$$
= Q
\begin{bmatrix}
\gamma v_x E_x/c + \gamma v_y E_y/c + \gamma v_z E_z/c \\
\gamma E_x + \gamma v_y B_z - \gamma v_z B_y \\
\gamma E_y - \gamma v_x B_z + \gamma v_z B_x \\
\gamma E_z + \gamma v_x B_y - \gamma v_y B_x
\end{bmatrix}
= Q
\begin{bmatrix}
\gamma \vec{v} \cdot \vec{E}/c \\
\gamma(E_x + (\vec{v} \times \vec{B})_x) \\
\gamma(E_y + (\vec{v} \times \vec{B})_y) \\
\gamma(E_z + (\vec{v} \times \vec{B})_z)
\end{bmatrix}
$$

$$
= Q
\begin{bmatrix}
\gamma \vec{\beta} \cdot \vec{E} \\
\gamma(E_x + (\vec{v} \times \vec{B})_x) \\
\gamma(E_y + (\vec{v} \times \vec{B})_y) \\
\gamma(E_z + (\vec{v} \times \vec{B})_z)
\end{bmatrix}
$$

Since the cross product is perpendicular to the velocity, we can add a "propitious zero" to the uppermost term to rewrite what we have as

$$f^\mu = \gamma Q \begin{bmatrix} \vec{\beta} \cdot (\vec{E} + \vec{v} \times \vec{B}) \\ (E_x + (\vec{v} \times \vec{B})_x) \\ (E_y + (\vec{v} \times \vec{B})_y) \\ (E_z + (\vec{v} \times \vec{B})_z) \end{bmatrix}$$

Since $\vec{F} = (\vec{E} + \vec{v} \times \vec{B})$ we have

$$f^\mu = \gamma Q \begin{bmatrix} \vec{\beta} \cdot \vec{F} \\ F_x \\ F_y \\ F_z \end{bmatrix}$$

This is what we were hoping to find! (From before, we had $\vec{F} = Q(\vec{E} + \vec{v} \times \vec{B})$ and $f^\mu = \gamma(\vec{\beta} \cdot \vec{F}, \vec{F})$.)

Quick summary

We discussed the basic facts about transformations of scalars, four-vectors, and tensors. These are important and very useful in all sorts of calculations.

The motions of massless particles are particularly simple to analyze, as long as you don't try to use expressions that diverge as v approaches c.

We discussed (and contrasted) forces and four-forces and how to use them. We also began to work with the electromagnetic field strength tensor.

Problem 1: The remembrance of things not completed

You might not have made it all the way through the very last exercise in this unit. So, try it again, and see the solution to that last exercise.

Problem 2: NASA propaganda and other things

(a) A few years ago, NASA released a long movie about a hypothetical interstellar mission to an imaginary star six light years from Earth, about which an earthlike planet had been found in the star's

"Goldilocks" zone. The film described the fully automated probe as being the size of an attack submarine and reaching a maximum speed of 0.2c.

The mass of a Los Angeles-class attack submarine is about 6×10^6 kg.[1] The worldwide consumption of energy a dozen years ago was estimated[2] to have been approximately 6×10^{20} joules. Neglecting ohmic losses in the electrical grid, the annual worldwide production of energy is the same as the annual consumption of energy.

If *all* of the world's energy production were diverted into bringing the NASA probe up to speed, how long would it take?

Solution

We need to calculate the starship's kinetic energy:

$$\mathrm{KE} = (\gamma - 1)mc^2 == (\gamma - 1)6 \times 10^6 \times 9 \times 10^{16}$$
$$= (\gamma - 1) \times 5.4 \times 10^{23} \text{ Joules}$$

$(\gamma - 1) = 0.020621$ so the starship's KE is 1.1135×10^{22} Joules

The exact length of a year is 3.15566×10^7 seconds. (A useful approximation that's easy to remember: a year is slightly more than $\pi \times 10^7$ seconds.) As a result, it'll take *all* of the world's energy production for about 18.56 years to accelerate the starship to 0.2c. The NASA propaganda is nonsense!

(b) The record for the most intense laser beam some years ago was held by a research group in Osaka, Japan.[3] Their device fired a 2 petawatt (2×10^{15} watts) pulse of duration 10^{-12} seconds. Imagine that the beam is directed at a perfectly reflecting object initially at rest. How large a force does the object experience while being illuminated by the laser?

Solution

Recall that force is the time derivative of (vector) momentum and that the beam carries momentum E/c. Neglecting the change in the speed of the target as it's struck by the laser, the reflection will

---

[1]https://en.wikipedia.org/wiki/Los_Angeles-class_submarine.
[2]https://en.wikipedia.org/wiki/World_energy_consumption.
[3]http://www.dailymail.co.uk/sciencetech/article-3179045/The-Death-Star-weapon-Japan-just-fired-world-s-powerful-laser.html.

transfer twice as much momentum to the target as was originally carried by the beam. It takes place over 10^{-12} seconds, so we have

$$F = \frac{\Delta p}{\Delta t} = \frac{2(E_{beam}/c)\Delta t}{\Delta t} = \frac{2(2 \times 10^{15}/c)10^{-12}}{10^{-12}}$$

$$= 1.33 \times 10^7 \, \text{N}$$

That's a large force! Since a one-pound object weighs about 4.46 N, the laser exerts 3.5 million lbs. of thrust on its target. Recall that the Saturn V first stage, used to launch moon-bound Apollo astronauts, was rated for about 7.5 million lbs. of thrust.

Problem 3: Putting Col. Stapp to shame

On the run after her conviction for mail fraud, Mrs. Rumpdock drives her Silver Conveyance at relativistic speed v eastward on Interstate 90. As she crosses into Montana, she sees a moose blocking the road and slams on her brakes. Her brakes are able to apply a (Newtonian, not four) force F to the car as it hurtles toward the unlucky animal.

Assuming the combined mass of Rumpdock and her car is m, derive an expression for her initial deceleration.

Solution

In a non-relativistic problem, we'd just say that $a = F/m$ and be done with it. But in the relativistic case, a force parallel to $\vec{v}$ will change γ as well as $\vec{v}$; from an equation some pages back, we have, in this case,

$$\vec{a} = \frac{\vec{F}}{\gamma^3 m}$$

If you'd like to derive this yourself, here is my version. There's a lot of algebra:

$$f^\mu = \gamma(\vec{\beta} \cdot \vec{F}, \vec{F}) = \frac{dp^\mu}{d\tau}$$

In the highway rest frame, $dt = \gamma d\tau$.

Looking at the space components of the first equation, we can write $\gamma \vec{F} = \frac{d\gamma m\vec{v}}{d\tau}$ and, for the force parallel to the velocity,

$F = \frac{dmv}{d\tau} = \frac{d\gamma mv}{dt}$. Do the differentiation:

$$\frac{d\gamma mv}{dt} = m\left[\frac{d\gamma}{dt}v + \gamma\frac{dv}{dt}\right] = m\frac{dv}{dt}\left[\gamma + \gamma\left(\frac{v\gamma}{c}\right)^2\right] \quad \text{after some algebra}$$

$$= \gamma m\frac{dv}{dt}\left[1 + \left(\frac{v\gamma}{c}\right)^2\right] = \gamma m\frac{dv}{dt}\left[1 + \frac{v^2/c^2}{1 - v^2/c^2}\right]$$

$$= \gamma m\frac{dv}{dt}\left[\frac{1}{1 - v^2/c^2}\right]$$

$$= \gamma^3 m\frac{dv}{dt} = F \quad \text{or} \quad a = \frac{F}{\gamma^3 m}$$

Problem 4: Some geometry

A small area patch on the surface of a sphere of radius r is shown in the first figure, in the following. (The sphere is not drawn.) The patch extends from θ to $\theta + d\theta$ and ϕ to $\phi + d\phi$. A patch on a second sphere, of slightly larger radius $r + dr$, but subtending the same angular ranges as the first patch, is shown in the second figure.

The patches can be thought of as defining two sides of a volume whose six faces are nearly rectangular. I have marked two vertices of the six-sided figure with black dots.

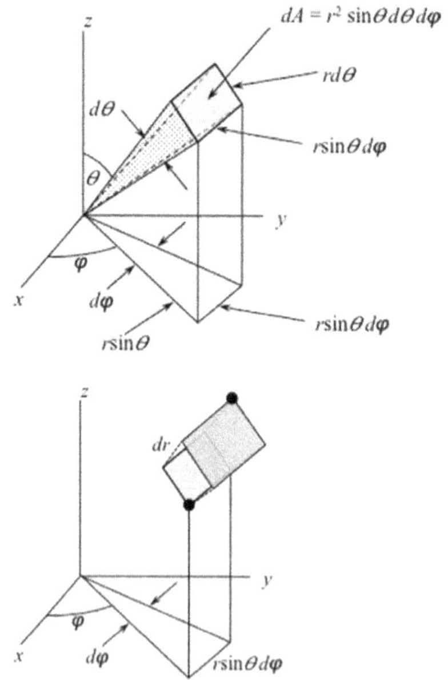

The square of the distance between the dots can be written in the form

$$D^2 = \begin{bmatrix} dr & d\theta & d\varphi \end{bmatrix} \begin{bmatrix} \eta_{11} & \eta_{12} & \eta_{13} \\ \eta_{21} & \eta_{22} & \eta_{23} \\ \eta_{31} & \eta_{32} & \eta_{33} \end{bmatrix} \begin{bmatrix} dr \\ d\theta \\ d\varphi \end{bmatrix}$$

where η_{ij} are the components of a matrix that takes care of the fact that we are using spherical coordinates. (In Cartesian coordinates, the analog of η would be the identity matrix.) Recall that the length of a vector does not change when we switch from Cartesian coordinates to another kind of coordinate system or rotate our coordinate axes. (Lengths are invariant under coordinate transformations.) The matrix η helps us form the quantity that is invariant under coordinate transformations.

What are the nine elements η_{ij}?

An aside: in special relativity, we know that the contraction of any two four-vectors is a Lorentz scalar and therefore invariant under Lorentz transformations. The metric tensor $g_{\mu\nu}$ is the relativistic analog of η_{ij}.

Solution

For a rectangular solid with side lengths a, b, c, the length of the body diagonal is $\sqrt{a^2 + b^2 + c^2}$. The side lengths of our little box are $dr, r\sin\theta d\phi$, and $rd\theta$. So, the distance between the marked points is

$$D = \sqrt{(dr)^2 + (r\sin\theta d\varphi)^2 + (rd\theta)^2}$$

To write D^2 as

$$D^2 = \begin{bmatrix} dr & d\theta & d\varphi \end{bmatrix} [\eta] \begin{bmatrix} dr \\ d\theta \\ d\varphi \end{bmatrix}$$

requires η to be diagonal since there are no terms like $d\theta d\phi$:

$$D^2 = \begin{bmatrix} dr & d\theta & d\varphi \end{bmatrix} \begin{bmatrix} \eta_{11} & 0 & 0 \\ 0 & \eta_{22} & 0 \\ 0 & 0 & \eta_{33} \end{bmatrix} \begin{bmatrix} dr \\ d\theta \\ d\varphi \end{bmatrix}$$

$$= \eta_{11}(dr)^2 + \eta_{22}(d\theta)^2 + \eta_{33}(d\varphi)^2$$

so

$$\eta_{11} = 1, \quad \eta_{22} = r^2, \quad \eta_{33} = (r\sin\theta)^2 \text{ so that}$$

$$\eta = \begin{bmatrix} 1 & 0 & 0 \\ 0 & r^2 & 0 \\ 0 & 0 & r^2\sin^2\theta \end{bmatrix}$$

Les Îles d'Hyères, Porquerolles, France

Unit 8

An Introduction to General Relativity: Non-Euclidean Geometry, the Metric Tensor, Space-time Curvature

A different way of thinking about gravity

General Relativity is naturally viewed as a geometrical theory: the presence of matter and energy in the universe alters the geometry of space–time so that it is non-Euclidean.

147

Rather than describing gravity as an action-at-a-distance force that pulls on massive objects (which makes it something very much like electromagnetism and other—often quantum mechanical—forces), we remove it entirely from the roster of possible forces influencing the observable universe. Instead, we say that an object in free fall experiences no force at all, moving through space–time along a world line that maximizes the amount of time passing on a clock attached to the object as it travels from space–time point A to space–time point B.

The presence of matter and energy in the universe changes how the space and time intervals associated with a clock's voyage from A to B combine to yield the proper time that passes on the clock.

This might actually sound a little familiar if you've learned about the "principle of least action" in classical mechanics. Imagine that an object is subject only to forces that allow the definition of a position-dependent potential energy. If the object moves from point A to point B, beginning its trip at time t_A and ending at time t_B, the classical action is defined as

$$S = \int_{t_A}^{t_B} L\,dt \quad \text{for } L \equiv T - V$$

with T and V the object's kinetic and potential energy and L its Lagrangian. The object will always move in a way that minimizes S as long as all trial paths are constrained to begin at t_A and end at t_B. (For the purists, the exact statement is that S will be an extremum, not necessarily a minimum.)

Back to relativity. Even without geometry-changing matter, an object's correct world line from A to B will always yield a larger proper time than an unphysical world line running from A to B.

Here's an example of what I mean.

Exercise 8.1: Maximizing the proper time

Consider the following two world lines, proposed for an object that is not experiencing any external forces. It should be obvious that the path on the right is impossible: it violates Newton's second law. (How do we know this is the case?) But let's pretend for the moment that we don't know about Newton's laws.

At $ct = 1$ light year, the object is at $x = 2$ light years; this is space–time point A. Five years later, according to observers at rest in this inertial frame, the object passes through the space–time point $ct = 6$ light years, $x = 5$ light years. Calculate the time that passes on a clock that follows each of these world lines. (For the path on the right, you'll need to calculate the time that passes during each of the straight segments, then add them together.)

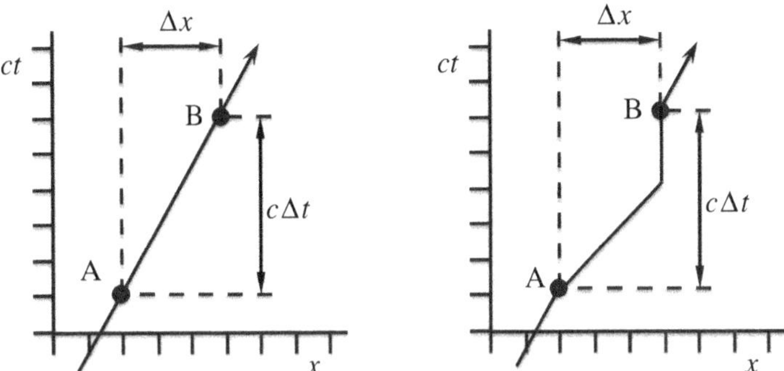

You should find that less time passes on the clock following the kinked world line.

Solution

On the left: $\Delta x = 3\,\mathrm{LY}$, $c\Delta t = 5\,\mathrm{LY}$, so $(c\Delta t)^2 - (\Delta x)^2 = 25 - 9 = 16$. As a result, the proper time is $\Delta\tau = 4$ years.

On the right: calculate the proper time separately for the two parts of the trajectory, on either side of the kink. We get

$$\Delta\tau = \sqrt{3^2 - 3^2} + \sqrt{2^2 - 0} = 2$$

The metric tensor $g_{\mu\nu}$

For small intervals dx, dy, dz, dt, we calculate the square of the (Lorentz invariant) proper time in the usual way:

$$(d\tau)^2 = (dt)^2 - \left(\frac{dx}{c}\right)^2 - \left(\frac{dy}{c}\right)^2 - \left(\frac{dz}{c}\right)^2 = (dt)^2 - \left(\frac{dL}{c}\right)^2$$

where dL is the length of the vector with components (dx, dy, dz).

A more compact way to write this is to take advantage of the covariant notation we've been developing. The contravariant vector dx^μ is

$$dx^\mu = (dx^0, dx^1, dx^2, dx^3)$$

with $dx^0 = cdt$, $dx^1 = dx$, $dx^2 = dy$, and $dx^3 = dz$. We define the covariant version of this as follows:

$$dx_\mu = (cdt, -dx, -dy, -dz)$$

$$= (dx_0, dx_1, dx_2, dx_3) = (dx^0, -dx^1, -dx^2, -dx^3)$$

$$= \sum_\nu g_{\mu\nu} dx^\nu$$

We can write

$$(d\tau)^2 = (dt)^2 - \left(\frac{dx}{c}\right)^2 - \left(\frac{dy}{c}\right)^2 - \left(\frac{dz}{c}\right)^2$$

$$= \frac{1}{c^2} \sum_\mu dx^\mu dx_\mu$$

$$= \frac{1}{c^2} \sum_\mu dx^\mu \sum_v g_{\mu\nu} dx^\nu$$

$$= \frac{1}{c^2} \sum_{\mu,\nu} dx^\mu g_{\mu\nu} dx^\nu = \frac{1}{c^2} \sum_{\mu,\nu} g_{\mu\nu} dx^\mu dx^\nu$$

The metric tensor $g_{\mu\nu}$ is the mathematical tool that knows how we combine pairs of components of four-vectors to form a Lorentz invariant. In Cartesian coordinates, all it does is to flip the signs on the space components of a contravariant four-vector to produce a covariant four-vector. As a result, $g_{\mu\nu}$ is

$$g_{\mu\nu} = \begin{bmatrix} 1 & 0 & 0 & 0 \\ 0 & -1 & 0 & 0 \\ 0 & 0 & -1 & 0 \\ 0 & 0 & 0 & -1 \end{bmatrix}$$

It is very much akin to the matrix I called η in the last unit. In that case, we were describing the length of a short vector, either in

Cartesian or spherical coordinates. In the Cartesian case, we formed (the square of) the length this way:

$$(dL)^2 = (dx_1)^2 + (dx_2)^2 + (dx_3)^2$$

$$= \sum_{i,j} \eta_{i,j} dx_i dx_j$$

with

$$\eta = \begin{bmatrix} 1 & 0 & 0 \\ 0 & 1 & 0 \\ 0 & 0 & 1 \end{bmatrix}$$

If we built a long straight line out of short segments, we could either add up all the lengths of the short segments or just calculate the length of the line directly, as long as we are using Cartesian coordinates:

$$L = \sqrt{(x_{\text{final}} - x_{\text{initial}})^2 + (y_{\text{final}} - y_{\text{initial}})^2 + (z_{\text{final}} - z_{\text{initial}})^2}$$

But (even in Cartesian coordinates) if we want to calculate the length of a curve, we'll need to break the curve up into tiny segments and add the lengths of the individual segments.

In spherical coordinates, for the moment defining q_1 as r, q_2 as θ, and q_3 as ϕ, you should have found in a recent problem that

$$(dL)^2 = \sum_{i,j} \eta_{i,j} dq_i dq_j$$

with

$$\eta = \begin{bmatrix} 1 & 0 & 0 \\ 0 & r^2 & 0 \\ 0 & 0 & r^2 \sin^2 \theta \end{bmatrix}$$

Note that $dr, d\theta$, and $d\phi$ need to be tiny for this to be accurate. And even if you are moving along a straight line, you'll need to add up the lengths of lots of little segments unless θ and ϕ are constant so that $d\theta$ and $d\phi$ are zero.

Exercise 8.2: Distance traveled

A robotic explorer on the surface of Mars sets out from a point at (Martian) latitude 60° North (this corresponds to the spherical coordinate $\theta = 30°$) and longitude 45° East ($\phi = 45°$). It drives along the shortest possible route (which is nearly a straight line) to a point with latitude 61° North and longitude 46° East. How long was the drive? (The mean radius of Mars is 3,389 km.)

Solution

We'll need to convert all the angles to radians, of course. That's not a big deal. Here we go:

$$\{dL\}^2 = \sum_{i,j} \eta_{ij} dq_i dq_j = \eta_{22} d\theta^2 + \eta_{33} d\phi^2$$

$$= r^2 d\theta^2 + r^2 \sin^2 \theta d\phi^2$$

$$dL = 3{,}389 \, \text{km} \times \sqrt{\left(\frac{\pi}{180}\right)^2 + \sin^2 30° \left(\frac{\pi}{180}\right)^2}$$

$$dL = 66.13 \, \text{km}$$

Special relativity and $g_{\mu\nu}$

The three space–space components of the metric tensor (g_{11}, g_{22}, g_{33}) tell us how to combine dx, dy, and dz (or $r, \theta, \phi, dr, d\theta$, and $d\phi$) to calculate a length, just as they do in Euclidean geometry.

We use all four of the diagonal components—including the time–time component (g_{00})—to calculate proper time intervals and other Lorentz invariants.

An object moving with constant velocity will follow a straight world line. If we are working with Cartesian coordinates, we can calculate both the distance it has traveled and the amount of time that has passed on its clock by looking at the difference in its space–time coordinates at the end points of its trajectory, as you did in the first exercise for the left-hand world line.

If the object moves along a curved world line (perhaps it is equipped with a rocket engine), you will need to break the world line up into short segments.

In spherical coordinates, the metric tensor $g_{\mu\nu}$ of special relativity is

$$g_{\mu\nu} = \begin{bmatrix} 1 & 0 & 0 & 0 \\ 0 & -1 & 0 & 0 \\ 0 & 0 & -r^2 & 0 \\ 0 & 0 & 0 & -r^2\sin^2\theta \end{bmatrix}$$

That's not at all surprising: the last three diagonal elements tell us how to calculate the ordinary measure-it-with-a-tape-measure distance and (except for the sign flip) must be the same as the diagonals of η.

So, $g_{\mu\nu}$ is really telling us about two things: how to calculate the length of a short line segment in a particular frame of reference (that's just the Pythagorean theorem of Euclidean geometry) and how to calculate the Lorentz invariant quantity associated with a space–time interval.

We know that both space and time intervals change when boosting to a different frame, but we still have Euclidean geometry and the Pythagorean theorem at our disposal for calculating lengths in the new frame.

Under General Relativity, the presence of matter and energy changes the geometry of space–time so that measure-it-with-a-ruler distances are no longer reliably calculated using the Pythagorean theorem. GR also changes the rate of passage of time.

We capture these effects by solving the Einstein Equations to determine the components of $g_{\mu\nu}$ in our no-longer-empty universe. That's not an easy task, and we won't attempt it ourselves! We'll just use the results found by 20th-century physicists, such as Einstein and Karl Schwarzschild.

Euclidean geometry

The geometry of our universe in the vicinity of the Earth is, to an excellent approximation, Euclidean. From growing up in a quasi Euclidean world, we have come to expect the following:

- The separation between two straight lines which are initially parallel remains constant.
- We can "enlarge" geometrical shapes without changing their basic properties (e.g. the sum of an equilateral triangle's interior angles

will remain $180°$; a sphere's ratio of volume to surface area will always be $r/3$).

- The circumference of a circle is $2\pi r$, no matter how small or large the radius r.
- The surface area and volume of a sphere are always $4\pi r^2$ and $4\pi r^3/3$, respectively.

It is not so easy for us to visualize a 3-dimensional non-Euclidean world. That doesn't mean we can't do the math, though. Just as we're comfortable with the idea that the distance between two points in a five-dimensional Euclidean space is

$$L = \sqrt{(\Delta x_1)^2 + (\Delta x_2)^2 + (\Delta x_3)^2 + (\Delta x_4)^2 + (\Delta x_5)^2},$$

we can work with geometries in which our intuitions about circumferences, areas, volumes, and parallel lines are no longer true. We'll use mathematics to avoid being misled by faulty intuition.

It isn't hard to visualize a non-Euclidean geometry in 2-space: we can embed it in a Euclidean 3-space and draw pictures for ourselves.

But we don't need to embed a (possibly non-Euclidean) space in a higher-dimensional space to learn something about its geometry.

Determining whether or not our geometry is Euclidean

Some *gedanken* (thought) experiments for 3-space:

1. Fill all of space with polystyrene foam, take a rope of length s, and anchor one end to a point in the plastic foam. Stretch the rope to its full length and remove all the plastic that comes into contact with the rope as you move the rope through the full range of (spherical coordinate) angles θ and φ.

   Obtain a very large number of small cubes of known volume. (Alphabet blocks?) See how many can fit inside the hollowed-out space in the polystyrene, determining the volume of the void you've created by counting how many alphabet blocks were required. (You can carve up the alphabet blocks as necessary to fill the void all the way out to the polystyrene-void boundary.)

   Repeat the process for different values of rope length s.

   If the volume deviates from $4\pi s^3/3$ for large s, you're living in a non-Euclidean space. Note that the deviation could go in either direction: the volume might grow more rapidly than expected, or less rapidly.

2. Remove all the alphabet blocks from the holes you carved in the plastic foam. Make a VERY large number of small, identical paper squares. Use the squares to cover the (interior) surface of the hollowed-out regions. Identify the surface area of each sphere as the total area of paper squares used as wallpaper.

   If the area deviates from $4\pi s^2$ for large s, you're living in a non-Euclidean space. Again, the sense of the deviation could go in either direction.

3. Measure the ratio of the circumference and radius of circles of different sizes. It is always 2π in a Euclidean space, of course. There's a small complication: we'll need to make sure that the circular path we measure actually lies in a plane.

   Here's one way to do this: go into one of the hollowed-out voids, station ourselves at the very center of the void, put a mass on one end of the rope we used to carve out this sphere, anchor the other at the center of the void, and start the mass moving perpendicular to the line made by the taut string. The rope will sweep out a planar disc, and the mass on the end of the rope will move in a circle, as long as there are no external forces acting on the mass.

   If we find the circumference of the circle to be $2\pi r$, independent of radius, our world's geometry is Euclidean.

   If we are testing the properties of a 2-space, our volume and area thought experiments drop down by one dimension to determinations of the area and circumference of a circle as a function of radius.

If we are residents of a non-Euclidean universe, there's generally going to be some length scale that characterizes when the geometry begins to differ noticeably from Euclidean geometry. At distances that are small compared to this scale, the geometry will be approximately Euclidean.

As I said earlier, General Relativity is naturally viewed as a geometrical theory: the presence of matter and energy in the universe alters the geometry of space–time so that it is non-Euclidean.

Geodesics in Euclidean geometry

We know exactly what we mean by "straight line" in a Euclidean universe, but we'll need to generalize it so that we can discuss force-free motion in a universe with gravitationally induced space–time curvature.

The world line of an object not experiencing external forces follows a *geodesic*. Note that gravity is no longer considered a force: it is just a consequence of the linkage between matter, energy, and alterations to the geometry of space–time.

A geodesic is a curve that is an extremum (shortest or longest) path connecting two points. For this to mean anything, we'll need to have a sensible definition of distance so that "shortest" and "longest" are meaningful concepts.

In 3-space, the definition is just what you'd think: run a string from point A to point B and pull it taut. The string lies on a geodesic, where we are using the physical length of the string as the measure of path length: the geodesic is the path of minimum length. In equations, we write

$$dL^2 = dx_1^2 + dx_2^2 + dx_3^2$$

$$= \sum_{i,j} \eta_{i,j} dx_i dx_j$$

I am being a little cavalier here, leaving out the parentheses. When I write dL^2, I really mean $(dL)^2$. For convenience, I'll keep doing that from this time forward.

When we are discussing space–time world lines, however, the geodesic linking points A and B is the path which maximizes the (proper) time that passes on a clock which travels from A to B.

Back to 3-space. When using spherical coordinates, we use a different η and write the square of the separation between points at (r, θ, φ) and $(r + dr, \theta + d\theta, \varphi + d\varphi)$ in about the same way:

$$dL^2 = \eta_{11} dr^2 + \eta_{22} d\theta^2 + \eta_{33} d\varphi^2$$

This particular η for a Euclidean 3-space is

$$\eta_{ij} = \begin{bmatrix} 1 & 0 & 0 \\ 0 & r^2 & 0 \\ 0 & 0 & r^2 \sin^2 \theta \end{bmatrix}$$

Not surprisingly, the length of a macroscopic curve is

$$L = \int_{\text{beginning}}^{\text{end}} dL$$

In Euclidean geometry, geodesics are straight lines.

Elliptic geometry

For the moment, let's continue to consider spatial geometries, not space–time geometries.

An elliptic geometry is one in which lines that are initially parallel—the paths followed by a pair of perfectly aligned laser beams, for example—will decrease in separation as they propagate. An example of a 2-space with elliptic geometry is something that, when viewed from higher dimensions, looks like the surface of a sphere. Perhaps we have embedded our space in a Euclidean 3-space, but that is not necessary.

The geodesics on the sphere are sections of great circles. (A great circle on the surface of a sphere is a circle whose center coincides with the sphere's center.) Imagine stretching a string between two points on the circle. When the string is pulled taut, it follows the geodesic path from one point to the other.

Note that inhabitants of the 2-space can only be aware of directions that are parallel to the sphere's surface. They do not see a horizon: the curve of their "planet" (as we see it from our 3-space vantage point) does not block their view of places far away. For them (and for "light" propagating in their 2-space), there is no up or down, only directions along the surface. With a telescope, an inhabitant looking forward can see the back of their head.

Curvature in an elliptic space

We can perform our *gedanken* experiments to determine whether or not our geometry is Euclidean. Use a string of length s to draw a circle, as shown in the following figure. There's a bit of algebra ahead; consider following along by rewriting the equations on a piece of scratch paper as you read through the next few pages.

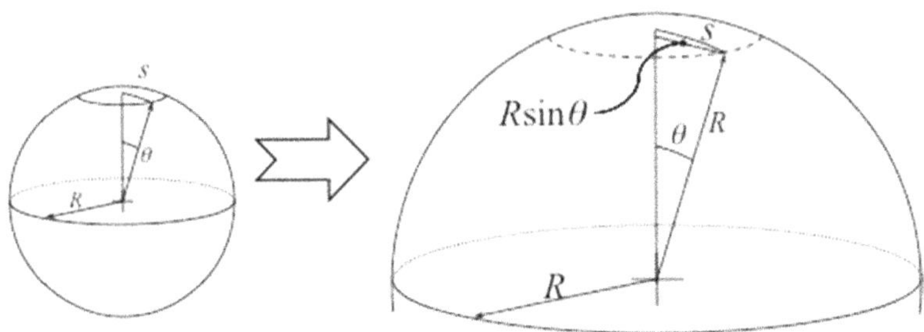

The geometry is simple: the length of the arc that runs from the "north pole" to the circle we will draw is $s = R\theta$ (because that's how angles in radians, arc lengths, and radii work); the length of an arc depends on the radius and subtended angle. The radius of the circle we drew on the surface of our sphere is $r = R\sin\theta$, which is smaller than s. (The center of the circle is inside the sphere, not at the top of the sphere.)

The inhabitants of the 2-space might naively expect their circle to have circumference $2\pi s$ but will measure it to be $2\pi r$, which is smaller, since $2\pi r = 2\pi R\sin(\theta) < 2\pi s = 2\pi R\theta$. And by "measure," we mean the usual, familiar thing: grab a tape measure from our toolbox and run it around the perimeter of our circle.

Using the approximation $\sin\theta \approx \theta - \theta^3/3!$, the fractional disagreement between the inhabitants' expected and measured values can be written as

$$\frac{\Delta\text{circumference}}{\text{expected circumference}} = \frac{2\pi s - 2\pi r}{2\pi s} = \frac{R(\theta - \sin\theta)}{R\theta}$$

$$\approx \frac{\theta^2}{3!} = \frac{1}{6}\left(\frac{s}{R}\right)^2$$

They'll also overestimate the surface area of the circle, expecting it to be πs^2. That's not correct: consider a band on the surface of the sphere that extends from θ' to $\theta' + d\theta'$ and goes all the way around.

Its width is $Rd\theta'$ and its circumference is $2\pi R\sin(\theta')$, so it will have area

$$dA = (Rd\theta') \times 2\pi R\sin\theta'$$

$$= 2\pi R^2 \sin\theta' d\theta'$$

The area of the patch extending from the "north pole" to the polar angle θ is

$$A = \int_{\theta'=0}^{\theta'=\theta} dA = \int_{\theta'=0}^{\theta'=\theta} 2\pi R^2 \sin\theta' d\theta'$$

$$= 2\pi R^2(1 - \cos\theta) \approx 2\pi R^2 \left(\frac{\theta^2}{2!} - \frac{\theta^4}{4!}\right)$$

(I've just used the series expansion for the cosine.) Since $s = R\theta$, the true area they measure (by tiling their circle with pieces of paper, for example) can be reexpressed as

$$A \approx 2\pi R^2 \left(\frac{\theta^2}{2!} - \frac{\theta^4}{4!} \right)$$

$$= \pi s^2 - \pi s^2 \frac{\theta^2}{12}$$

$$= \pi s^2 \left(1 - \frac{1}{12} \left(\frac{s}{R} \right)^2 \right)$$

The fractional disagreement between their expectations and their actual measurement is

$$\frac{\Delta A}{A_{\text{expected}}} = \frac{\pi s^2 - \pi s^2 \left(1 - \frac{1}{12} \left(\frac{s}{R} \right)^2 \right)}{\pi s^2} = \frac{\pi s^2 \frac{1}{12} \left(\frac{s}{R} \right)^2}{\pi s^2} = \frac{1}{12} \left(\frac{s}{R} \right)^2$$

When s/R is not miniscule, we'll see disagreement between the measured area and the naïve, Euclidean expectation.

The *scalar curvature* S for a 2-space with the elliptical geometry of a sphere of radius R is defined to be

$$S \equiv \frac{2}{R^2}$$

This gives us

$$\frac{\Delta A}{A_{\text{expected}}} = \frac{s^2 S}{24}$$

For some reason, geometers sometimes use the "Gaussian curvature" K instead of the scalar curvature; in an elliptical geometry 2-space

$$K = \frac{1}{R^2} = \frac{S}{2}$$

Keep in mind: large R means *small* curvature. And in this particular example, the curvature is constant everywhere. Note that I'm using upper-case letters as the symbols for curvature.

There's a lot of good information in texts on differential geometry (as well as Wikipedia!) about curvature and non-Euclidean geometries.

If we choose to test whether or not a 3 -space is Euclidean, we might want to compare the volume inside a sphere of radius s with our naïve Euclidean expectation of $4\pi s^3/3$. If our geometry is elliptic, we'll find that the true volume of the sphere is always less than the Euclidean volume.

To sum up, briefly: in an elliptic geometry, there is less empty space than we might expect inside the usual figures familiar to us from Euclidean geometry. Spheres don't hold as much, spheres and circles have less surface area than we'd expect, and both spheres and circles have smaller circumferences than we would have predicted based on intuition that is informed by life in a quasi-Euclidean environment.

Hyperbolic geometry

In a hyperbolic geometry 2-space, a pair of laser beams that are initially parallel (in the high school geometry sense) will increase in separation as they propagate. The curvature of a hyperbolic geometry is negative; in a hyperbolic geometry 3-space,[1] the surface area and volume of a sphere of radius r are larger than we would expect to find in a Euclidean space. It's the opposite of what we find in an elliptic geometry space.

Elliptic geometry, curvature, and the metric

The presence of matter and energy in the universe changes space from something well described by Euclidean geometry to something characterized by a non-Euclidean geometry that bears a small resemblance to an elliptic geometry whose curvature varies with position.

The closer you get to a concentration of mass, the more noticeable the departure from Euclidean geometry.

The Einstein equations relate the density of mass and energy to the induced space–time curvature. We'll talk about those later. First, let's do some more work with a 2-space described by an elliptic geometry.

Here's that diagram from before, in which we use a string of length s to draw a circle on the ground of what we discover, to our surprise, is a non-Euclidean world. Keep in mind that, living in the 2-space

---

[1]http://en.wikipedia.org/wiki/Hyperbolic_geometry.

which has the same geometry as the surface of the sphere, *we have no idea that there are spies in a 3-space watching us* or that it is even possible to find a vantage point in a space of more than two dimensions.

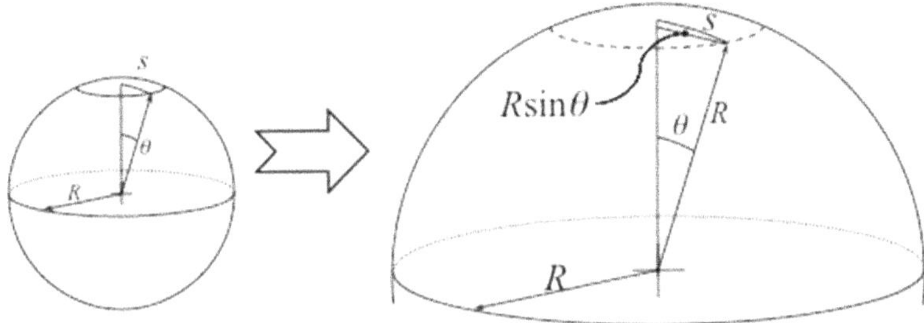

Let's describe the circle we've drawn using polar coordinates.

If our 2-space were Euclidean (it's not), it'd all be easy. (In that case, the radius of the circle we'd draw would be s rather than $R\sin\theta$.)

The circumference of the circle can always be calculated by following its perimeter and adding up the lengths of all the (small) arcs from which the circle is constructed, each of which subtends the (small) angle $d\varphi$.

If the space were Euclidean, we'd find, of course, that the circumference satisfied $C = 2\pi s$.

We'd also see (in a Euclidean space) that the square of the distance between the two points in the following figure is

$$(dL)^2 = (ds)^2 + (sd\varphi)^2$$

which I'll shorten to $dL^2 = ds^2 + s^2 d\varphi^2$ in a minor abuse of the mathematical notation.

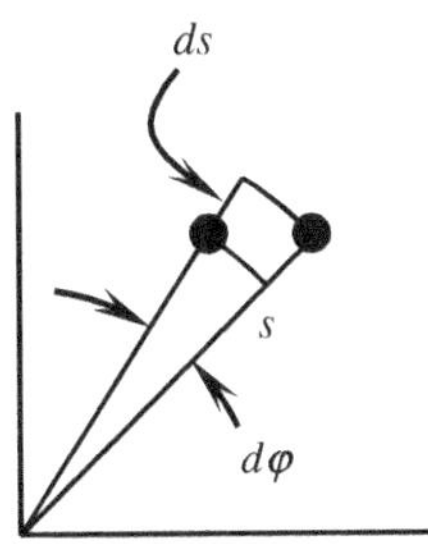

But if we are actually living in an elliptic geometry 2-space, we'll get a smaller value for the circumference C than we might have expected from multiplying the length of our string by 2π.

Since s is the length of our string (and $s = R\theta$ for the polar angle θ), in our elliptic geometry 2-space, we'll have

$$C = 2\pi R \sin\theta \approx 2\pi R \left(\theta - \frac{\theta^3}{3!}\right)$$

$$= 2\pi R\theta \left(1 - \frac{1}{6}\theta^2\right) = 2\pi s \left(1 - \frac{1}{6}\frac{s^2}{R^2}\right)$$

$$= 2\pi s \left(1 - \frac{K}{6}s^2\right)$$

with K the Gaussian curvature of our geometry. Note that I'm assuming that the radius of curvature R is much bigger than the distance $s : Ks^2 \ll 1$. That way, I can truncate the series representation of the sine after the second term.

Imagine we wanted to define a new coordinate in which we measure the circumference C of a circle (perhaps by using a tape measure), then divide this by 2π. It would be natural to name this new coordinate "r" where $r \equiv C/2\pi = s(1 - \frac{K}{6}s^2)$.

This variable r is sometimes called the "coordinate radius" or the "reduced circumference." But it's extracted from a measurement of the circumference, which has a funny relationship to the length of the string we used to draw our circle. In a non-Euclidean geometry, s and r do not have to be equal.

Distances

Let's say we want to calculate the distance between the two closely spaced points in the following figure, in our elliptic geometry 2-space.

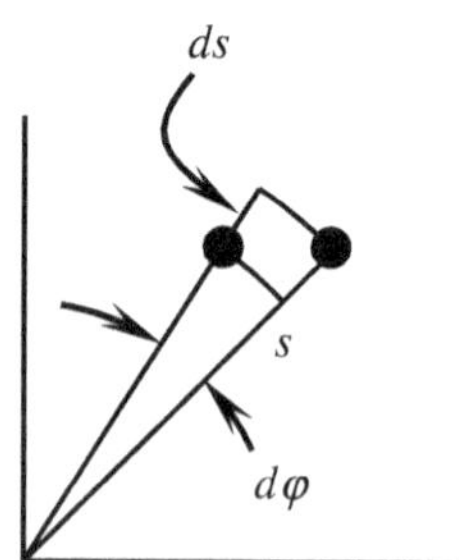

The radial component of their separation is just ds. That is unsurprising: it's the small difference in the lengths of our strings used to draw circular arcs through those two points.

The length of the orthogonal component, in the φ-direction, is NOT $sd\varphi$ since the circumference is *smaller* than $2\pi s$. What we really need to do is pay attention to what fraction of the entire circle the angle $d\varphi$ subtends, and multiply that fraction by the circumference of the circle. We've invented the coordinate radius r so that it satisfies $C = 2\pi r$, so the length of our short segment will be $rd\varphi$ (and not $sd\varphi$).

This way, an arc drawn with constant s (and therefore constant r) which subtends an angle φ will always have an arc length that is just $r\varphi$. And if we change the length of our string in a way that induces a change in coordinate radius to $r + dr$, the change in the circumference of our circle will be $dC = 2\pi dr$.

In the extreme case when our string runs all the way from the north pole to the equator, changing the length of the string a little bit doesn't change the circumference of the circle to first order in ds. Changing the value of s will not change the value of r.

In the figure above, instead of using $sd\varphi$ as the component of distance along the φ-direction (because it's not!), we'll need to use

$$rd\varphi = s\left(1 - \frac{K}{6}s^2\right)d\varphi$$

As a result, the square of the distance between the two points will satisfy

$$dL^2 = ds^2 + r^2d\varphi^2$$

$$= ds^2 + s^2\left(1 - \frac{K}{6}s^2\right)^2 d\varphi^2 \quad \text{(OK)}$$

instead of

$$dL^2 = ds^2 + s^2d\varphi^2 \quad \text{(NOT TRUE)}$$

If we'd prefer to express everything in terms of r and φ instead of s and φ, we'd use

$$r = s\left(1 - \frac{K}{6}s^2\right)$$

or, since $Ks^2 \ll 1$,

$$s = \frac{r}{(1 - \frac{K}{6}s^2)} \approx \frac{r}{(1 - \frac{K}{6}r^2)}$$

where I am only keeping terms to first order in Kr^2 and Ks^2. (Note that I replaced s^2 in the denominator in the rightmost term with r^2.)

Not surprisingly, for positive curvature, s is (a little bit) bigger than r. If we change s by the small amount ds, the associated change in r (call it dr) will be a little bit smaller than ds:

$$dr \approx ds \left(1 - \frac{K}{6}r^2\right) \quad \text{or} \quad \frac{dr}{(1 - \frac{K}{6}r^2)} \approx ds$$

Thanks to this, we can rewrite our equation from before

$$dL^2 = ds^2 + s^2 \left(1 - \frac{K}{6}s^2\right)^2 d\varphi^2$$

in terms of r, not s, as

$$dL^2 = \frac{dr^2}{(1 - \frac{K}{6}r^2)^2} + s^2 \left(1 - \frac{K}{6}r^2\right)^2 d\varphi^2 = \frac{d\varphi^2}{(1 - \frac{K}{6}r^2)^2} + r^2 d\varphi^2$$

$$\approx \frac{dr^2}{(1 - \frac{K}{3}r^2)} + r^2 d\varphi^2 \quad \text{since } s\left(1 - \frac{K}{6}r^2\right) \approx r$$

where I have used a binomial approximation on the denominator in the first term.

Either coordinate system is fine to use: s and ϕ are just as good as r and ϕ.

Keep in mind that dL is what we would measure with a tape measure as the distance between the two points, while s is what we would measure with a tape measure as the distance from the "north pole," which we think is the center of our circle, to the arc of the circle.

Also, bear in mind that r is a derived quantity, calculated by dividing the measured circumference by 2π.

As I did before, I can define the metric η so that we can express the square of the separation between points (in our 2-space elliptic geometry) this way:

$$dL^2 = \eta_{11}dr^2 + \eta_{22}d\varphi^2$$

We have

$$\eta_{ij} = \begin{bmatrix} \dfrac{1}{1 - \frac{K}{3}r^2} & 0 \\ 0 & r^2 \end{bmatrix}$$

What we're doing is jamming into the metric all the information about how to include the effects of curvature (and the use of a non-Cartesian coordinate system) when we calculate a length.

What we mean by "length" is just what you'd expect: if we walked over to those two points and connected them with a tape measure stretched taut, we'd find the same value for dL as we calculated above.

If we'd like to measure the separation between two events that are *not* close together, we'd find a geodesic that connected them (stretch a string between the two points to determine the geodesic: it's a great circle if we are in a 2-space with constant positive curvature) and then calculate

$$L = \oint_{\text{beginning}}^{\text{end}} dL$$

along this path.

This technique will also work fine in a 2-space where the curvature changes from place to place, as long as we are able to determine a geodesic by running a taut string from one point to the other. And the procedure generalizes to a 3-space elliptical geometry, but we won't discuss that here.

But what about space–time?

I haven't said anything concrete about what might cause the geometry of space to deviate from the Euclidean. And I haven't said anything yet about the behavior of time intervals, about the possibility that whatever influences the geometry also influences the rate of passage of time.

Even if the geometry gets weird, we expect that an observer in an inertial frame will see that a light beam zipping past his/her position always travels at speed c as it flashes past. As long as the curvature isn't so extreme that we are unable to find a small neighborhood around our position that seems Euclidean, this will be the case.

Imagine that we have a large number of observers at rest in an inertial frame in an elliptical geometry 2-space and that the clocks held by those observers are both synchronized *and* ticking at the same rate. A laser beam travels past point 1, then passes point 2 at a later time. The geodesic path followed by the laser beam is shown in the following figure. Observers at points 1 and 2, and many points in between, record the times when the light beam passes their positions.

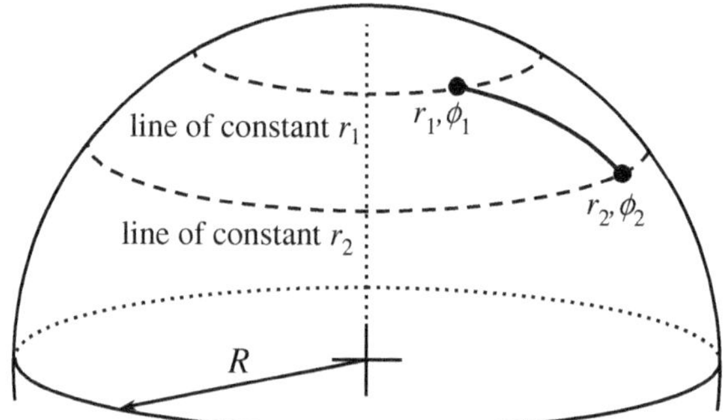

Two observers along the path of the light beam, separated by a small distance dL, will measure the times of passage of the beam as differing by dt, where $dt = dL/c$. The square of the proper time between the two events—the first observer's, then the second observer's observation of the beam—is zero, of course: we're watching the behavior of a light beam. We can calculate the (square of the) proper time as

$$d\tau^2 = g_{00}(dt)^2 + g_{11}\left(\frac{dr}{c}\right)^2 + g_{22}\left(\frac{d\varphi}{c}\right)^2$$

as we did earlier. And since the 11 and 22 elements of $g_{\mu\nu}$ are just the sign-flipped corresponding elements of η, we have (in our 2-space's elliptic geometry)

$$g_{\mu\nu} = \begin{bmatrix} 1 & 0 & 0 \\ 0 & \dfrac{-1}{1 - \frac{K}{3}r^2} & 0 \\ 0 & 0 & -r^2 \end{bmatrix}$$

It's not a big deal, just some algebra.

It is only slightly more complicated if the effect that pushes the geometry from Euclidean to elliptical also changes the rate of passage of time from place to place. We still expect that an observer—unaware that their apparently Euclidean neighborhood turns into something weird at larger distances—will see the laser beam traveling past their position at speed c. This is *always* true. And even in a region of strong gravitational fields, observers can cast themselves into (locally) inertial frames by allowing themselves to fall freely.[2]

We'll need to think carefully about what we mean by dt. It is this: a very distant observer, free from the influence of whatever is changing the geometry and the behavior of time intervals, provides us with the dt value by receiving signals from the observers separated by dL, then correcting from any difference in signal propagation time from the observers' positions.

Perhaps it would be more clear if I wrote

$$d\tau^2 = g_{00}(dt_{\text{distant observer}})^2 + g_{11}\left(\frac{dr}{c}\right)^2 + g_{22}\left(\frac{d\varphi}{c}\right)^2$$

If the modification to the flow of time is a function of position (perhaps the local observers' clocks tick slowly when they are close to the origin) so that

$$dt_{\text{local}} = f(r, \varphi)dt_{\text{distant observer}}$$

our requirement that the local observer sees the light beam traveling at c means

$$\frac{dL}{dt_{\text{local}}} = c$$

so that

$$\frac{dL}{dt_{\text{distant observer}}} = \frac{dL}{dt_{\text{local}}}f(r, \varphi)$$
$$= cf(r, \varphi)$$

---

[2] *Cf.* Tom Petty, "Free Fallin.," *Full Moon Fever*, track 1 (1989).

Now, the metric for our elliptical 2-space becomes

$$g_{\mu\nu} = \begin{bmatrix} f^2(r,\varphi) & 0 & 0 \\ 0 & \dfrac{-1}{1 - \frac{K}{3}r^2} & 0 \\ 0 & 0 & -r^2 \end{bmatrix}$$

There's nothing surprising about the extension of this to three spatial dimensions.

The Einstein equations

The master equations for general relativity are called the "Einstein equations." They establish the connection between the distribution of matter and energy in the universe and the deviation from Euclidean geometry and a universal rate of passage of time for all observers at rest in an inertial frame.

We will be working with the simplest of non-trivial cases: the behavior of space–time in the presence of a compact object of mass M at the origin.

As always, the proper time interval between two events will be the amount of time that passes on a clock that travels from the first event to the second. (Keep in mind that an event is [happens at] a single point in space–time.)

To Einstein's great surprise, it was only a few months after his principal paper on general relativity that Karl Schwarzschild solved the equations to produce the applicable metric.

Consider two points in 3-space, in which we use spherical coordinates to describe positions.

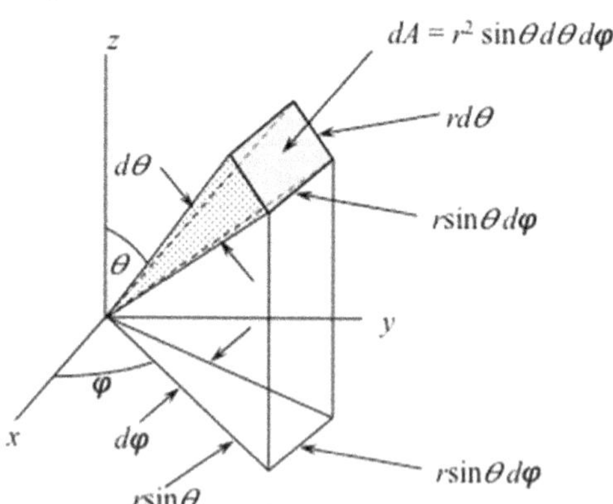

Schwarzschild found the (square of the) measure-it-with-a-ruler separation between points (when a mass M is at the origin) to be

$$dL^2 = \frac{dr^2}{1 - \frac{2GM}{rc^2}} + r^2 d\theta^2 + r^2 \sin^2 \theta d\varphi^2$$

where G is Newton's constant, $6.674 \times 10^{-11}\,\mathrm{N \cdot m^2/kg^2}$.

To compare this with our elliptical 2-space metric, consider what we have if we restrict ourselves to motion in the x, y plane so that $\theta = 90°$ and $d\theta = 0$:

$$dL^2 \to \frac{dr^2}{1 - \frac{2GM}{rc^2}} + r^2 d\varphi^2 \quad \text{(Schwarzschild, } \theta = 0)$$

instead of

$$dL^2 = \frac{dr^2}{1 - \frac{K}{3}r^2} + r^2 d\varphi^2 \quad \text{(elliptical 2-space)}$$

Including the effects of curvature on the flow of time, the fullblown Schwarzschild metric becomes

$$g_{\mu\nu} = \begin{bmatrix} \left(1 - \frac{2GM}{rc^2}\right) & 0 & 0 & 0 \\ 0 & -\left(1 - \frac{2GM}{rc^2}\right)^{-1} & 0 & 0 \\ 0 & 0 & -r^2 & 0 \\ 0 & 0 & 0 & -r^2 \sin^2 \theta \end{bmatrix}$$

Note the bad behavior as $r \to 2GM/c^2$!

The Schwarzschild radius

When $r = 2GM/c^2$, everything goes crazy. That's called the Schwarzschild radius for obvious reasons. Something with the sun's mass, but squished into an extraordinarily dense lump, will have a Schwarzschild radius of about $3\,\mathrm{km}$. The Earth's is a little less than $9\,\mathrm{mm}$.

But here's a point of confusion: what exactly do we mean by t, r, θ, and ϕ?

What the coordinates actually mean

Here's an engineering approach to establishing a coordinate system around a mass M at the origin: let's build a large number of gigantic hoops to create a coordinate grid. We proceed as follows. (See the following figure.)

Creating the spatial coordinate grid

1. Construct a large number of rigid circular hoops in a region of the universe that is so far from massive objects that its geometry is Euclidean. Build them so that, while under construction, they are parallel to the x, y plane and centered on the z-axis. Fabricate the jth hoop so that its circumference is $C_j = 2\pi j$.

   Recall that I defined the *coordinate radius* in terms of the circumference: $r = C/2\pi$. This way, the jth hoop has coordinate radius $r_j = C_j/2\pi = j$ meters. In this part of the universe, where the curvature is negligible, the measure-it-with-a-tape-measure radius is also j meters.

2. Paint fiducial (measurement) markings on each hoop at $0°$, $1°, \ldots, 359°$.

3. Install a clock near each of the fiducial marks on each hoop. Synchronize all of the clocks.

4. Transport the hoops along the z-axis as necessary so that they eventually lie in the x, y plane, centered on the origin.

   These hoops define the r, ϕ coordinate grid at $\theta = 90°$ and are to be used by all observers at rest with respect to the origin. If a meteor wipes out a clock on the $\phi = 15°$ mark on a ring with coordinate radius $12\,\mathrm{m}$, we'll say that the clock-destruction event happened at $r = 12, \theta = 90°$, and $\phi = 15°$.

5. Build another set of circular hoops by repeating steps 1 through 4, but this time construct them parallel to the x, z plane and centered on the y-axis, at a great distance from the origin.

   Once they are moved into place, they define an r, θ coordinate grid at $\phi = 0°$ and are to be used by all observers at rest with respect to the origin.

6. Repeat Step 5, but install the hoops at $\phi = 1°$. Keep installing sets of hoops at one degree ϕ increments up to $\phi = 179°$.

   These hoops define the r, θ, ϕ coordinate grid and are to be used by all observers at rest with respect to the origin. We can interpolate between grid points—we have a fiducial mark every meter in coordinate radius and every degree in both θ and ϕ.

The presence (or absence) of a mass at the center of our collection of concentric rings **does not** change any of the rings' circumferences and, consequently, does not alter any of their coordinate radii.

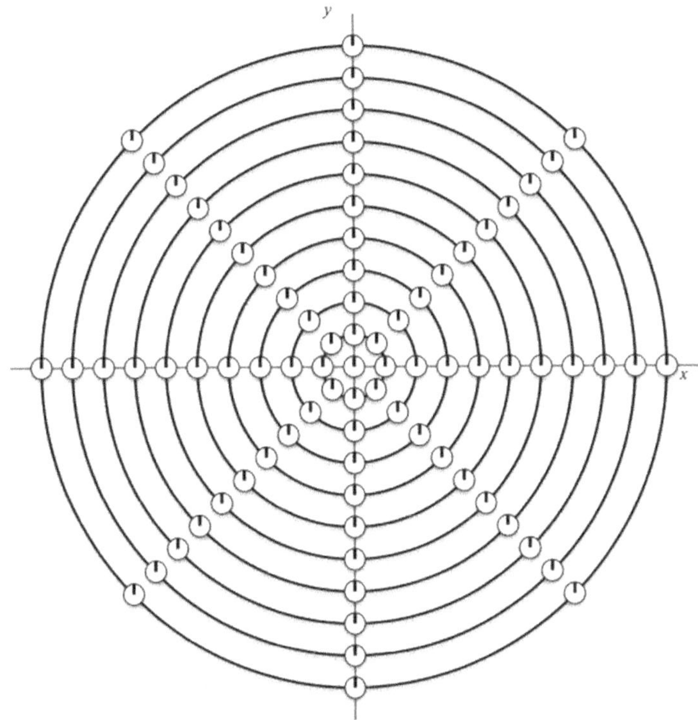

Defining the time interval dt

Imagine that the clocks we installed on our rings were manufactured to tick once per second. An observer at rest next to one of the clocks will see that it continues to tick once per second. This will be the case even if the clock and observer are close to a very massive object.

Here's a way to think about this: if the observer and clock are suddenly detached from the ring, they will begin to free fall toward the mass. For a very short time after entering free fall, they will be moving at negligible speed and will not feel the effects of any (non-gravitational) net force. It will seem to the observer that they live in a force-free inertial frame and that the clock behaves as designed by the manufacturer. The observer will conclude that the (proper) time interval between ticks is $\Delta\tau = 1$ second.

Each time it ticks, the clock sends a laser pulse toward the distant engineer who supervised the construction of all those rings. The transit time from the laser to the engineer might be substantial, but each pulse will take the same amount of time to make the trip.

The time interval between arriving pulses measured by the distant engineer is dt. This is the time interval that we use in a calculation of the proper time between events.

What we mean by a measurement of the space–time interval between two events

Our distant engineer maps out the transit time of a light beam from each clock to his/her position by sending a laser pulse toward each clock and measuring how long it takes for the reflection off the clock to return to his/her position. The one-way transit time is half this, of course.

The distant engineer observes that a clock at r, θ, ϕ is zapped by a meteor. Later, the engineer sees the explosion when the clock at $r + dr, \theta + d\theta, \phi + d\phi$ is struck by a second meteor. After correcting for the different transit times of light from the two explosions, the engineer concludes that the time interval between the explosions is dt, though an observer next to the clocks would measure the time interval to be $\Delta\tau$, which could be different from dt.

The distant engineer is unable to perform a direct measurement of the distance between the two clocks, but they see which hoops had supported the now-destroyed time pieces and can use the $dr, d\theta, d\phi$ information gleaned in this fashion to calculate dL using the Schwarzschild metric.

The speed of light

The observers **next to the now-destroyed clocks** will always, always, always measure the speed of a light beam that passes their positions as 299,792,458 meters per second.

Some quick examples

We place a neutron star whose Schwarzschild radius is 2 km at the origin. A clock at coordinate radius $r = 4$ km flashes a lamp once per second, according to its own sense of time (so $\Delta\tau = 1$). What is the interval dt between flashes received by the distant observer?

Here's the answer: since the flashes come from the same (stationary) clock, $dr, d\theta$, and $d\phi$ are all zero. As a result, we have

$$\Delta\tau^2 = 1 = g_{00}(dt)^2 = \left(1 - \frac{2GM}{rc^2}\right)(dt)^2$$

$$= \left(1 - \frac{2\text{ km}}{4\text{ km}}\right)(dt)^2. \text{ Since } \Delta\tau = 1 \text{ we have}$$

$$dt = \frac{1}{\sqrt{1/2}} = 1.414 \text{ seconds.}$$

Though it is stationary, the clock that is deep in the gravitational field is seen to tick slowly!

Here's another, with the same neutron star.

We run a tape measure from a ring with **coordinate radius** $r = 4\,\text{km}$ radially inward to a ring whose coordinate radius is $1\,\text{m}$ less. (This ring's circumference is 2π meters less, of course.) What is the distance ds determined by the tape measure?

The observations of the tape measure's ends are made at the same time so that $dt = 0$. The tape measure is oriented radially so that $d\theta$ and $d\phi$ are zero. We have

$$ds = \sqrt{-g_{11}(dr)^2} = \frac{dr}{\sqrt{1 - \frac{2GM}{rc^2}}} = \sqrt{2}$$

There is more space between the concentric rings than there would have been in the absence of the neutron star!

Summary

After an initial discussion of proper times and distances along the surface of a sphere, we launched into a more traditional approach to general relativity pedagogy, in which we put down our pens, laptops, pieces of chalk, smartphones, etc. and talked about the underlying concepts. These included

- curvature and non-Euclidean geometry;
- use of a metric;
- calculating finite intervals by integrating infinitesimal intervals;
- the Schwarzschild metric.

When using the Schwarzschild metric, the time coordinate t (used, for example, when we describe the time difference between two events) is the time as determined by a distant observer in a curvature-free region of space. The *coordinate radius* r is a derived quantity: we construct a circle centered on the origin that goes through the event under consideration, measure the circumference C of this circle, and then calculate $r = C/2\pi$. It might well be the case that r is smaller than the length s of the string that we pegged to the origin, then used the other end to draw the circle.

Problem 1: Strings

These are easy and are intended to give you a chance to do some simple calculations in a curved geometry. Assume that the radius of the Earth is 4,000 miles.

(a) You peg one end of a long string to the north pole and find that the string, when pulled taught, reaches to the equator. Ignoring problems with trees, mountains, and bodies of water, you walk due east, keeping the string taught, to trace a circle on the surface of the Earth. Answer the following:

1. How long is the string?
2. What is the circumference of the circle you've just traced?
3. If you hadn't known that the Earth was a sphere, you would have naïvely expected the circle's circumference to be 2π times the length of your string. What is the difference between the naïve circumference and the actual circumference of your circle?

Solution

1. The string reaches to the equator, so it is $\frac{1}{4}$ as long as the circumference of the Earth, namely, $(2\pi)(4000)/4 = 2000\pi$ miles.
2. It's just $(2\pi)(R) = (2\pi)(4000) = 8000\pi$ miles.
3. $C_{\text{naive}} - C_{\text{actual}} = (2\pi)(2000\pi) - 8000\pi = 14{,}346$ miles.

(b) Make a reasonably accurate graph of the quantity

$$\frac{\text{naïve circumference - true circumference}}{\text{naïve circumference}}$$

as a function of string length, with the string length ranging from zero to 4,000 miles. Also, make an accurate graph of the true circumference as a function of string length.

Solution

Here's a Python program to make the plots.

```python
"""
This file is unit_08.py
make a plot for the World Scientific text, unit 8, problem 1b
February 10, 2025
@author: George Gollin, University of Illinois
"""
# import libraries.
import numpy as np
import matplotlib
import matplotlib.pyplot as plt

# earth radius, in miles
R_earth = 4000
# string length
string_length_min = 1
string_length_max = R_earth

# arrays holding the ratio info
nbins = string_length_max - string_length_min + 1
s_array = np.array([0.0] * nbins)
ratio_array = np.array([0.0] * nbins)
true_circumference_array = np.array([0.0] * nbins)


ibin = 0
for s in range(string_length_min, string_length_max):
    # naive circle circumference
    circumference_naive = 2. * np.pi * s
    # polar angle made by the end of the string, in radians
    theta = s / R_earth
    # circumference of the actual circle made by sweeping the end of the
    # string through 2pi
    radius_true = R_earth * np.sin(theta)
    circumference_true = 2. * np.pi * radius_true
    s_array[ibin] = s
    ratio_array[ibin] = (circumference_naive - circumference_true) \
        / circumference_naive
    true_circumference_array[ibin] = circumference_true
    ibin = ibin + 1
# close any already-open graphics windows
matplotlib.pyplot.close("all")
first_bin = 0
last_bin = nbins - 1
# now plot that ratio
fig = plt.figure()
ax = fig.gca()
ax.set_xlabel("string length (miles)")
ax.set_ylabel("(naive - true) / naive ratio")
ax.set_title("(naive circumference - true circumference) / naive circumference")
ax.plot(s_array[first_bin:last_bin], ratio_array[first_bin:last_bin], '-', c = 'blue')
# also plot the true circumference
fig = plt.figure()
ax = fig.gca()
```

```
ax.set_xlabel("string length (miles)")
ax.set_ylabel("true circumference")
ax.set_title("true circumference vs string length")
ax.plot(s_array[first_bin:last_bin], true_circumference_array[first_bin:last_bin],
    '-', c = 'red')
```

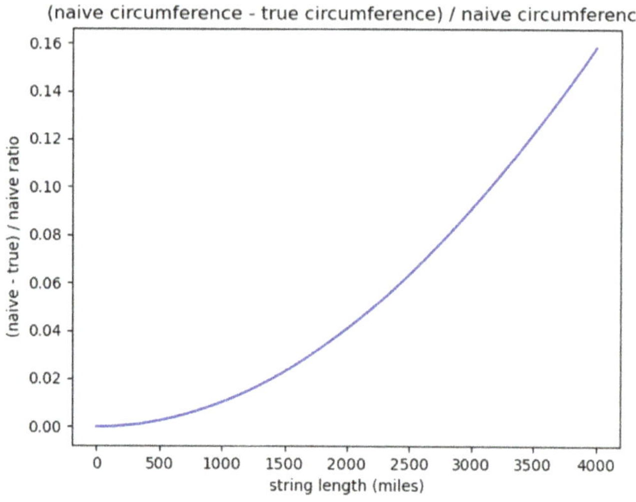

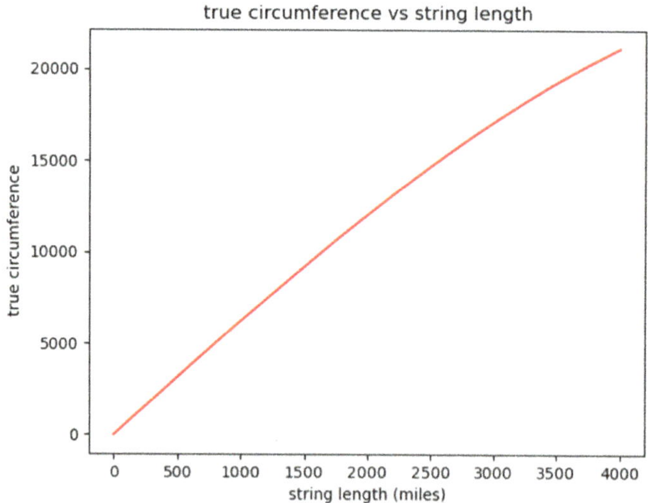

Problem 2: Looks familiar

In the following diagram, the angle θ refers to the latitude above the equator. As before, the radius of the Earth is to be taken as 4,000 miles. Note that with this definition of θ,

$$r = R\cos\theta, \quad s = \left(\frac{\pi}{2} - \theta\right) R$$

The quantity s is the length of a string fixed to the north pole that is used to draw a circle along a path of constant latitude, as shown in the figure. From a measurement of the circumference $C = 2\pi R\cos\theta$, one can define the coordinate radius r by dividing the circumference by 2π to obtain the same quantity r as indicated in the figure below, and in the first equation in this problem, above.

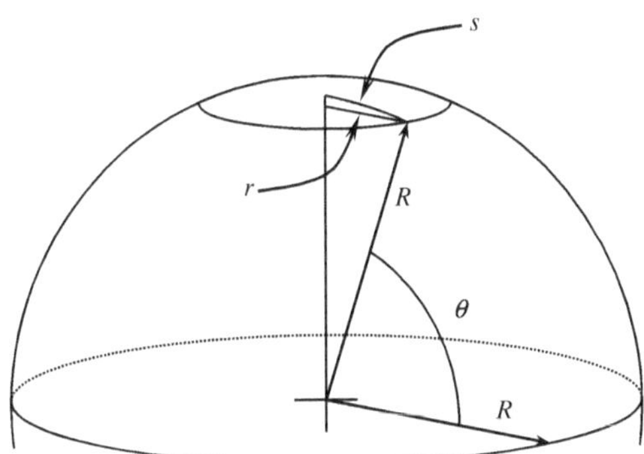

(a) Determine an algebraic expression for ds/dr, expressing your answer in terms of r and R *without* making explicit reference in your formula to the angle θ. Naturally, your expression will be singular at $\theta = 0$ since r does not change with infinitesimal changes in the length s of the string.

Solution

Use the chain rule plus a well-known trig identity:

$$\frac{ds}{dr} = \frac{ds}{d\theta}\frac{d\theta}{dr} = \frac{\frac{ds}{d\theta}}{\frac{dr}{d\theta}}$$

$$= \frac{\frac{d(\frac{\pi}{2}-\theta)R}{d\theta}}{\frac{dR\cos\theta}{d\theta}} = \frac{-R}{-R\sin\theta} = \frac{1}{\sin\theta}$$

Use

$$\sin\theta = \sqrt{1 - \cos^2\theta}$$

$$\text{so that } \frac{ds}{dr} = \frac{1}{\sin\theta} = \frac{1}{\sqrt{1 - \cos^2\theta}} = \frac{1}{\sqrt{1 - \left(\frac{r}{R}\right)^2}} \quad \text{since } \frac{r}{R} = \cos\theta$$

(b) Make a graph of ds/dr as a function of r for $0 < r < 3{,}500$ miles. Use Matlab, Excel, Python, or any other method you prefer, including by hand on graph paper.

Solution

More Python!

```
"""
This file is unit_08b.py
make a plot for the World Scientific text, unit 8, problem 2b
February 10, 2025
@author: George Gollin, University of Illinois
"""


# import libraries.
import numpy as np
import matplotlib
import matplotlib.pyplot as plt

# earth radius, in miles
R_earth = 4000
# string length
r_length_min = 1
r_length_max = 3500

# arrays holding the ratio info
nbins = r_length_max - r_length_min + 1
r_array = np.array([0.0] * nbins)
ratio_array = np.array([0.0] * nbins)

ibin = 0
for r in range(r_length_min, r_length_max):

    r_array[ibin] = r
    ratio_array[ibin] = 1. / np.sqrt(1 - (r * r) / (R_earth * R_earth))
    ibin = ibin + 1

# close any already-open graphics windows
matplotlib.pyplot.close("all")

first_bin = 0
last_bin = nbins - 1
```

```python
# now plot that ratio
fig = plt.figure()
ax = fig.gca()
ax.set_xlabel("r (miles)")
ax.set_ylabel("ds/dr")
ax.set_title("(ds/dr vs. r")
ax.plot(r_array[first_bin:last_bin], ratio_array[first_bin:last_bin], \
        '-', c = 'blue')
```

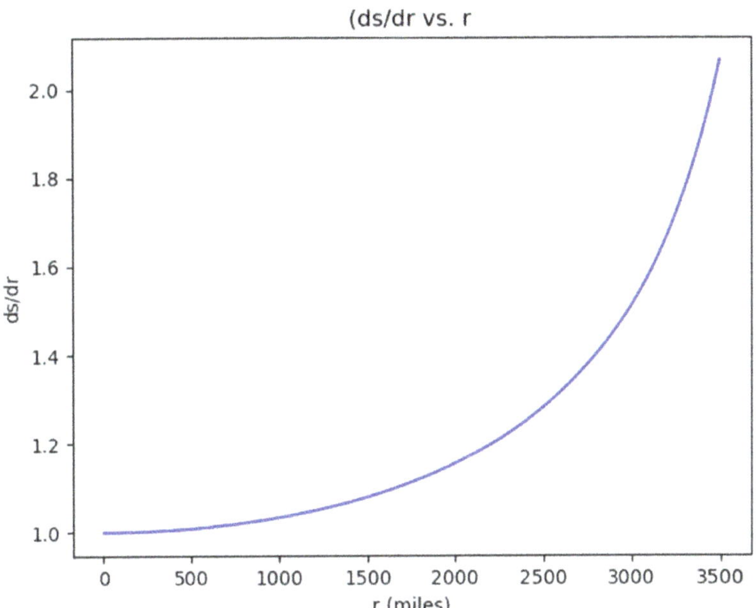

Problem 3: Metric stuff

Consider an observer at the origin in an elliptical geometry with constant Gaussian curvature $K = 10^{-6} \, \mathrm{m}^{-2}$. The observer constructs a circle, centered at the origin, with circumference (as measured by a tape measure) of 200π meters and marks a point on the circle at the angle $\varphi = 0$. The observer then constructs a larger circle (also centered on the origin) with a circumference of 202π meters and marks points at the angles $\varphi = 0$ and $\varphi = 0.01$ radians.

(a) The observer uses a tape measure to determine the distance between the two points at the angle $\varphi = 0$ on the pair of circles. What is the distance so measured?

Solution

We have

$$d\varphi = 0, \quad dr = \frac{202\pi}{2\pi} - \frac{200\pi}{2\pi} = 1\,\mathrm{m}$$

and, using the metrics (and a binomial approximation),

$$dL^2 = \eta_{11}dr^2 + \eta_{22}d\varphi^2$$

$$= \frac{dr^2}{1 - \frac{Kr^2}{3}}$$

$$\text{so that } dL = \frac{dr}{\sqrt{1 - \frac{Kr^2}{3}}} \approx dr\left(1 + \frac{Kr^2}{6}\right)$$

$$\text{since } K = 10^{-6}, \ dr = 1, \quad \text{and} \quad r = 100 \text{ so} \dots$$

$$dL \approx 1.000017$$

Note that this is slightly larger than $1\,\mathrm{m}$!

(b) The observer uses a tape measure to determine the distance along the arc of the circle between the points at $\varphi = 0$ and $\varphi = 0.01$ radians on the larger of the two circles. What is the distance so measured?

Solution

Now,

$$dL^2 = \eta_{11}dr^2 + \eta_{22}d\varphi^2$$

$$= r^2 d\varphi^2$$

$$\text{so that } dL = rd\varphi = 101 \times 0.01 = 1.01\,\mathrm{m}$$

(c) The observer uses a tape measure to determine the distance between the point at $\varphi = 0$ on the smaller circle and the point at $\varphi = 0.01$ radians on the larger circle. What is the distance so measured?

Solution

Now, we have

$$dL^2 = \eta_{11}dr^2 + \eta_{22}d\varphi^2 = (1.000017)^2 + (1.01)^2$$

$$dL = 1.42131\,\text{m}$$

Problem 4: Integrating it

Once again, consider an observer at the origin in an elliptical geometry 2-space with constant Gaussian curvature K. The metric in this space is

$$\eta_{ij} = \begin{bmatrix} \dfrac{1}{1 - \frac{K}{3}r^2} & 0 \\ 0 & r^2 \end{bmatrix}$$

so that the square of the separation between points is

$$dL^2 = \eta_{11}dr^2 + \eta_{22}d\varphi^2$$

As long as we restrict ourselves to regions where $r^2 \ll 1/K$, we can use a binomial approximation to rewrite the metric:

$$\eta_{ij} \approx \begin{bmatrix} 1 + \dfrac{K}{3}r^2 & 0 \\ 0 & r^2 \end{bmatrix}$$

The observer stretches a string from the point $r = 0$, $\phi = 0$ to the point $r = R, \phi = 0$. Assuming that $R^2 \ll 1/K$, how long is the string? Express your answer in terms of K and R.

Solution

We have

$$dL = \frac{dr}{\sqrt{1 - \frac{Kr^2}{3}}} \approx dr\left(1 + \frac{Kr^2}{6}\right)$$

$$\text{so } L = \int_0^R \left(1 + \frac{Kr^2}{6}\right) \mathrm{d}r = \left(r + \frac{K}{18}r^3\right)\Big|_0^R$$

$$= R\left(1 + \frac{K}{18}R^2\right)$$

So, that's the length!

La Calanque d'En Vau, Cassis, France

Unit 9

The Riemann Curvature Tensor, the Einstein Field Equations, the Schwarzschild and Kerr Metrics, and "Frame Dragging." And Just for Fun, LIGO

About Einstein's approach

This unit will mostly be a lecture-style presentation of the topics at hand.

Einstein's approach to gravity is conceptually different from Newton's, though the predictions of general relativity are nearly identical to those of Newtonian gravitation in the limit of weak fields.

Einstein identified the space–time curvature associated with gravitational effects as arising from the presence of matter and energy in space. The Einstein Field Equations define this relationship between curvature and mass/energy.

Here's a comparison of Newton's and Einstein's very different approaches.

Newton's view

1. An object of mass m under the influence of the gravitational field $\vec{G}$ generated by another mass will experience a force $m\vec{G}$. If no other forces act on the mass, it will undergo an acceleration $\vec{a} = m\vec{G}/m = \vec{G}$.
2. The gravitational field generated by a point mass falls inversely with the square of the distance from the mass.

Einstein's view

1. The world line of an object that is not subject to external (non-gravitational) forces is a geodesic: a straight line when viewed from an inertial frame in an empty universe, and (more generally) the space–time trajectory that maximizes the amount of proper time that passes on a clock carried by the particle between when it passes through point A and arrives at point B.
2. In the absence of space–time curvature, it is possible to find an inertial frame that is infinite in extent in which *all* objects will move along straight world lines.
3. Physics in an inertial frame in the absence of space–time curvature is *indistinguishable* from physics in a free-falling frame in a gravitational field, as long as measurements are confined to a sufficiently small region that curvature-induced (tidal) effects are negligible. An observer in free fall near a (possibly moving) test particle will see it moving in a straight line with constant velocity.

4. Concentrations of mass and/or energy influence the geometry of space–time, inducing curvature so that it is no longer possible to find a single inertial frame in which the world lines of all particles are straight lines.

An observer who is not subject to non-gravitational forces never experiences any force at all as they coast through (curved) space–time. The curvature business lets us stitch together the various local inertial frames used by observers at different positions.

The field equations are compactly stated this way:

$$G^{\mu v} = \frac{8\pi G}{c^4} T^{\mu v}$$

G is Newton's gravitational constant, $6.674 \times 10^{-11}\,\mathrm{m}^3/\mathrm{kg}/\mathrm{s}^2$.

$T^{\mu v}$ is the stress–energy tensor, a four-by-four rank-two tensor whose elements depend on the density of energy and momentum at points in space–time. We'll discuss it in more detail shortly.

$G^{\mu v}$ is the Einstein tensor, which gives a *partial* description of the curvature of space–time. (It is sometimes called the "Einstein curvature tensor.") More on this later.

The field equations are *local* equations: $T^{\mu v}$ and $G^{\mu v}$ are defined as functions of position and time, and the equations will hold for all points in space at all times.

The stress–energy tensor

Imagine that space is filled with a (possibly relativistic) fluid with position-dependent pressure P and "proper density" ρ (the density measured by an observer at rest with respect to the surrounding fluid). Perhaps the fluid is a diffuse gas, perhaps it is the dense plasma at the center of a star, or perhaps it is empty space, whose density and pressure are zero.

$T^{\mu v}$ is a function of position and time, of course (as are the pressure and density), so I suppose I could have written it as $T^{\mu v}(x^\sigma)$.

Recall from an earlier unit the definition of the four-velocity: $u^v = \gamma(c, \vec{v})$. We can assign a four-velocity to each point in space–time by determining the velocity of the fluid at that point. In the definition of $T^{\mu v}$, the fluid's four-velocity will always be multiplied by pressure or density, so we won't need to worry about how to assign four-velocities to locations in empty space.

The stress–energy tensor is

$$T^{\mu\nu} = Pg^{\mu\nu} + \left(\frac{P}{c^2} + \rho\right) u^\mu u^\nu$$

The first term includes the (contravariant) metric tensor $g^{\mu\nu}$, which carries information about space–time curvature.

Note the presence of both the pressure and the mass density in the second term. This has interesting consequences for the formation of a black hole when a star collapses! At a certain point in the collapse of a sufficiently massive star, the increasing pressure acts like an increase in the mass that is the source of the gravitational forces driving the collapse. The faster the pressure goes up, the more rapidly the star collapses!

The stress–energy tensor has units of energy per unit volume. This is easy to see: I've defined each of the terms, so the units are consistent throughout; the "time–time component" T^{00} contains $\rho u^0 u^0$, which is the mass per unit volume multiplied by γc^2. Including the other terms, we see that T^{00} is the energy density, including the effects of pressure.

The component T^{0i} (for $i = 1, 2, 3$) is the density of vector momentum multiplied by the speed of light. Since the metric tensor $g^{\mu\nu}$ is diagonal (its off-diagonal elements are zero), we have

$$T^{0i} = \left(\frac{P}{c^2} + \rho\right) \gamma c u^i \quad (i = 1, 2, 3)$$

$$= c \left(\frac{P}{c^2} + \rho\right) \gamma v^i$$

Recall that vector momentum is $\vec{p} = \gamma m \vec{v}$, so $\gamma \rho \vec{v}$ is the momentum per unit volume in special relativity. The general relativistic expression includes the pressure as well as the mass density.

The "space–space" elements T^{ij} ($i, j = 1, 2, 3$) tell us about the flow of momentum in the i direction that is transported by the fluid through a plane that is perpendicular to the j axis.

$T^{\mu\nu}$ is symmetric: it should be obvious to you from its definition that $T^{\mu\nu} = T^{\nu\mu}$.

Parallel transport and the Riemann curvature tensor

We need to develop a way to describe space–time curvature that lets us deal with the most general case—curvature that varies from point to point in space–time. That sounds like a daunting task, but there's an "experimental" technique we could use to map out the curvature everywhere. (Well, maybe not everywhere: that'd take an infinite amount of time.) The method is called "parallel transport."

For the moment, let's discuss parallel transport in an elliptic 2-space rather than an Einsteinian space–time.

Imagine we take the vectors $\vec{A}$, $\vec{B}$, and $\vec{C}$, each of which is short compared to the distance over which the effects of curvature become appreciable. Use $\vec{A}$ and $\vec{B}$ to create a parallelogram like this:

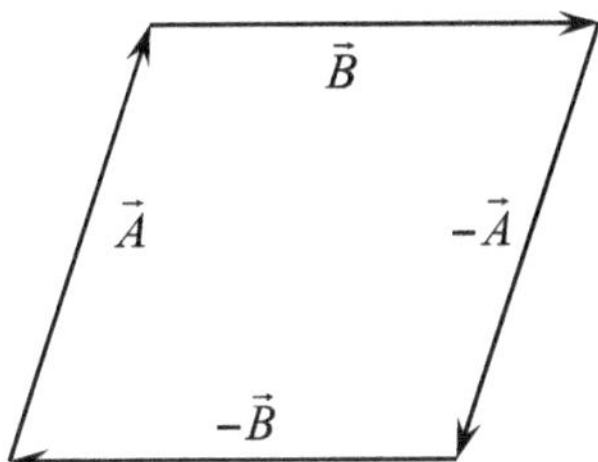

If our geometry isn't perfectly flat, it'll be necessary to fudge a little to make the figure close. That's OK: do this by adjusting the lengths of the sides labeled $-\vec{A}$ and $-\vec{B}$ slightly as needed.

Take the third vector $\vec{C}$ and parallel transport it around the parallelogram. By this, I mean start at the lower left corner and carry the vector along the four sides of the parallelogram, keeping the angle between the vector and the side of the parallelogram constant while traveling along that side. Let's call the vector after we bring it back to its starting point $\vec{C}'$. Here's a diagram:

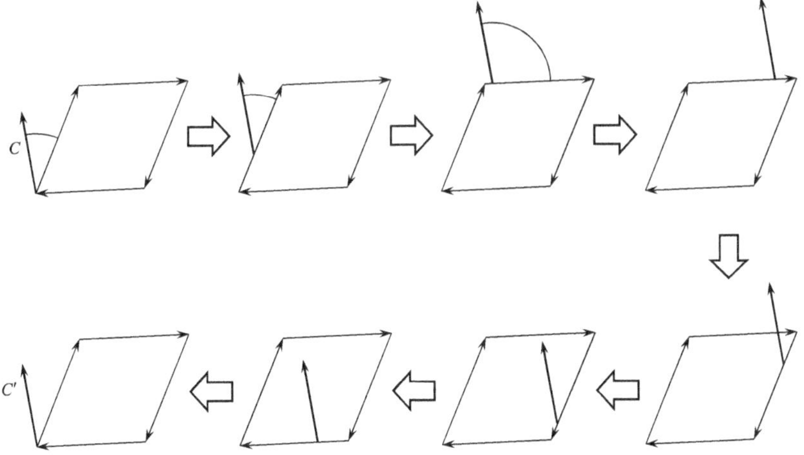

If our geometry is Euclidean, the orientation of the traveling vector will not change, so $\vec{C}' = \vec{C}$ and $\vec{\delta} \equiv \vec{C}' - \vec{C} = 0$.

In an elliptic geometry, this won't be the case, as shown in the following figure. (The size of the parallelogram is exaggerated!) The sides of the parallelogram are the elliptical geometry analogues of "straight lines," namely great circles.

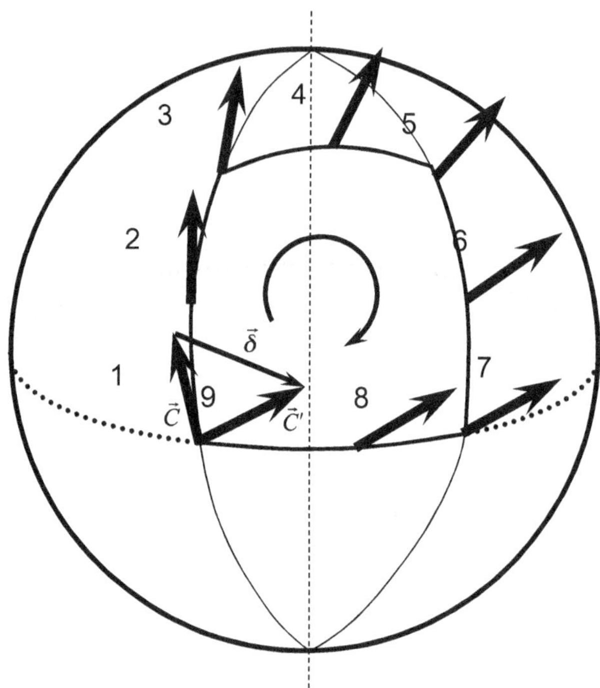

We'd expect that increasing the lengths of the sides of the parallelogram should increase the amount of rotation of the vector being transported. This is the case: the angle through which the transported vector rotates is proportional to $|\vec{A}||\vec{B}|$, and therefore the area of the parallelogram, as long as the vectors are very small compared to the curvature scale of the space. In addition, the longer the $\vec{C}$, the longer the $\vec{\delta}$ will be, though the angle of rotation of $\vec{C}$ is unaffected.

If we wanted to derive how $\vec{C}$ changes as we carry it around the not-quite parallelogram, we'd need to work out a procedure that takes the three vectors $\vec{A}$, $\vec{B}$, and $\vec{C}$ as inputs and produces the single vector $\vec{\delta}$ as its output. The "Riemann curvature tensor" is the key to this. In the vocabulary of some of my mathematics colleagues, it is a $(1,3)$-tensor since it produces one output vector after being fed three input vectors.

But it's just a piece of mathematical machinery: you put some stuff in and something useful comes out. As long as you've been provided with all of its components at each point in space (or space–time), you can treat the Riemann tensor like the bread machine that I used to use to make pizza dough: put five ingredients into it, push the start button, and (a couple of hours later) retrieve a ball of dough ready to form into a crust. Unless I need to open the machine to repair it, I don't have to think about what's inside. In the present case, the Riemann tensor only needs three ingredients and doesn't plug into an electrical outlet.

In a 2-space, we use it this way:

$$\delta_i = \sum_{j=1}^{j=2}\sum_{k=1}^{k=2}\sum_{l=1}^{l=2} R_{ijkl} A_j B_k C_l.$$

To parallel transport a four-vector in an Einsteinian universe with $3+1$ (space + time) dimensions, we'd use it this way:

$$\delta^\mu = \sum_{v=0}^{v=3}\sum_{\rho=0}^{\rho=3}\sum_{\sigma=0}^{\sigma=3} R^\mu_{v\rho\sigma} A^v B^\rho C^\sigma = \sum_{v,\rho,\sigma} R^\mu_{v\rho\sigma} A^v B^\rho C^\sigma$$

The notion of a curvature tensor may sound daunting, but the ideas aren't, at heart, so difficult: we're all capable of grinding out the calculation of $\vec{\delta}$ as a function of $\vec{A}, \vec{B}$, and $\vec{C}$ for an elliptical 2-space, in the spirit of what I did to calculate the radius of a circle drawn on the sphere last unit. It might take us all afternoon, but we could figure it out if it were absolutely necessary. We would just need to know enough about the geometry of our space to say how the vector $\vec{C}$ would change as we moved around the loop. The vectors of interest are small enough so that there'd be all sorts of small-angle approximations we could employ.

A very useful fact from differential geometry: if we are able to know the outcome at all points in space (or space–time) of carrying any (infinitesimal) vector around the quasi-parallelogram defined by any pair of (infinitesimal) vectors, we'll have enough information to know *everything* about the curvature of our geometry.

Why do we need something as complicated as a tensor to describe curvature? Lots of reasons, and it's not hard to see that there's more than just a scalar quantity kicking around. Think back to our discussion of elliptic geometry: we worked with a special case where the curvature resembled that of a 2-sphere. It didn't matter in which direction we shot off a couple of parallel lines; their rate of convergence would be the same. But I could have made the curvature vary with direction; this would amount to drawing an ellipsoid rather than a sphere to represent the geometry, and working up somewhat more complicated analyses of the circumference and area of circles (ellipses, really), we might draw on the surface of the ellipsoid that represented our 2-space.

In three-dimensions, there are even more handles needed to characterize what is going on.

The Einstein tensor

$G^{\mu\nu}$ is a contravariant rank-2 tensor with 16 elements, while the Riemann tensor is a mixed (one contravariant, three covariant indices) rank-4 object with 256 elements. So, Einstein couldn't just set $G^{\mu\nu}$ to be the same as $R^{\mu}_{\nu\rho\sigma}$. There's too much information in the Riemann tensor for it to be equal to the stress–energy tensor, the right side of the Einstein Equations.

In a world with $(3+1)$-dimensions, there are *two* distinct rank-two curvature tensors that can be built from the Riemann tensor. They're called the Ricci curvature tensor and the Weyl curvature tensor. At least one of them must be non-zero for gravitational effects to be possible.

The Ricci tensor "represents the amount by which the volume element of a geodesic ball in a curved Riemannian manifold deviates from that of the standard ball in Euclidean space."[1] That's actually not a complicated idea: think back to our discussion of elliptic geometry. The Ricci tensor lets us figure out how much the volume of a sphere of radius r differs from the Euclidean volume of $4\pi r^3/3$. But it doesn't tell us about tidal effects associated with spatial variations in the strength and direction of the gravitational force.

There's a good, concise description of the Weyl tensor on Wikipedia. "In differential geometry, the Weyl curvature tensor,

---

[1]See a text on differential geometry or else the Wikipedia article http://en.wikipedia.org/wiki/Ricci_curvature for a brief description of this.

named after Hermann Weyl, is a measure of the curvature of space–time or, more generally, a pseudo-Riemannian manifold. Like the Riemann curvature tensor, the Weyl tensor expresses the tidal force that a body feels when moving along a geodesic. The Weyl tensor differs from the Riemann curvature tensor in that it does not convey information on how the volume of the body changes, but rather only how the shape of the body is distorted by the tidal force."[2]

The Einstein tensor $G^{\mu\nu}$ is built from the Ricci tensor $R^{\mu\nu}$, the (contravariant) metric tensor $g^{\mu\nu}$, and the scalar curvature R. There's nothing about the Weyl tensor injected directly into $G^{\mu\nu}$. Taken on its own, the value of $G^{\mu\nu}$ at a point in space–time can only tell us part of the story about gravitational effects.

The Ricci tensor is formed from the Riemann tensor by contracting the first and third indices of $R^{\alpha}_{\beta\gamma\delta}$:

$$R_{\mu\nu} = \sum_{\rho} R^{\rho}_{\mu\rho\nu}$$

$$R^{\alpha\beta} = \sum_{\mu,\nu} g^{\alpha\mu} g^{\beta\nu} R_{\mu\nu}.$$

The scalar curvature for space–time (it means about the same as it did in our discussion of elliptic geometry) is obtained this way:

$$R = \sum_{\nu} R^{\nu}_{\nu} = \sum_{\mu,\nu} g^{\nu\mu} R_{\mu\nu} = \sum_{\mu,\nu,\rho} g^{\nu\mu} R^{\rho}_{\mu\rho\nu}$$

The notation may seem a little confusing. The way to tell the Riemann and Ricci tensors apart, and to distinguish them from the scalar curvature, is to count the number of indices.

Note that the metric tensor $g^{\nu\mu}$ will include things like those $(1 - \frac{2GM}{rc^2})$ factors that were floating around in the Schwarzschild metric. The Einstein tensor is built from the Ricci tensor, the metric tensor, and the scalar curvature:

$$G^{\mu\nu} = R^{\mu\nu} - \frac{1}{2} g^{\mu\nu} R$$

---

[2]Again, see a text on differential geometry or else the Wikipedia article: http://en.wikipedia.org/wiki/Weyl_tensor.

It is a statement about the geometry of space–time. If its 16 components are known everywhere in space–time, the behavior of the four-velocities of "weightless" particles—particles in free fall—can be calculated as they move through space as *long as we provide additional information about the distribution of mass in the universe.*

The Schwarzschild metric

In empty space, even near a black hole where the spatial geometry deviates wildly from Euclidean, all elements of the stress–energy tensor are zero. At first, you might think that the Einstein Equations force the induced space–time curvature to be zero since they reduce to $G^{\mu v} = 0$ when $T^{\mu v}$ is zero. But that's not actually the case. Since

$$G^{\mu v} = R^{\mu v} - \frac{1}{2}g^{\mu v}R$$

a properly chosen metric $g^{\mu v}$ will permit the Einstein tensor to be zero by allowing a non-zero scalar curvature to cancel a non-zero Ricci tensor.

Karl Schwarzschild found that the metric for space–time in which a compact mass M is at rest at the origin is

$$g_{\mu v} = \begin{bmatrix} \left(1 - \dfrac{2GM}{rc^2}\right) & 0 & 0 & 0 \\ 0 & -\left(1 - \dfrac{2GM}{rc^2}\right)^{-1} & 0 & 0 \\ 0 & 0 & -r^2 & 0 \\ 0 & 0 & 0 & -r^2\sin^2\theta \end{bmatrix}$$

With this metric, the square of the distance between points in space at r, θ, ϕ and $r + dr,\ \theta + d\theta,\ \phi + d\phi$ is

$$dL^2 = -g_{11}dr^2 - g_{22}d\theta^2 - g_{33}d\varphi^2$$

$$= \frac{dr^2}{1 - \frac{2GM}{rc^2}} + r^2 d\theta^2 + r^2\sin^2\theta d\varphi^2$$

The distance dL has the conventional meaning: an observer at (coordinate radius) r, θ, ϕ using a tape measure to determine the distance to the nearby point $r + dr, \theta + d\theta, \phi + d\phi$ will measure dL.

The square of the amount of time that passes on the face of a clock as it moves from r, θ, ϕ to $r + dr, \theta + d\theta, \phi + d\phi$ in an amount of time dt (where dt is the time difference determined by a distant observer stationed at a point in space where the curvature is negligible) is

$$d\tau^2 = g_{00}dt^2 + \frac{1}{c^2}[g_{11}dr^2 + g_{22}d\theta^2 + g_{33}d\varphi^2]$$

$$= \left(1 - \frac{2GM}{rc^2}\right)dt^2 - \frac{dL^2}{c^2}$$

Here's a reminder: the next few paragraphs are just a repeat of the material from the last unit.

What the coordinates actually mean

Here's an engineering approach to establishing a coordinate system around a mass M at the origin: let's build a large number of gigantic hoops to create a coordinate grid. We proceed as follows.

Creating the spatial coordinate grid

1. Construct a large number of rigid circular hoops in a region of the universe that is so far from massive objects that its geometry is Euclidean. Build them so that, while under construction, they are parallel to the x, y plane and centered on the z-axis. Fabricate the jth hoop so that its circumference is $C_j = 2\pi j$.

   Recall that I defined the *coordinate radius* in terms of the circumference: $r = C/2\pi$. This way, the jth hoop has coordinate radius $r_j = C_j/2\pi = j$ meters. In this part of the universe, where the curvature is negligible, the measure-it-with-a-tape-measure radius is also j meters.
2. Paint fiducial (measurement) markings on each hoop at $0°$, $1°, \ldots 359°$.
3. Install a clock near each of the fiducial marks on each hoop. Synchronize all of the clocks.
4. Transport the hoops along the z-axis so that they eventually lie in the x, y plane, centered on the origin.

   These hoops define the r, ϕ coordinate grid at $\theta = 90°$ and are to be used by all observers at rest with respect to the origin. If a meteor wipes out a clock on the $\phi = 15°$ mark on a ring with coordinate radius $12\,\text{m}$, we'll say that the clock-destruction event happened at $r = 12$, $\theta = 90°$, and $\phi = 15°$.

5. Build another set of circular hoops by repeating Steps 1 through 4, but this time construct them parallel to the x, z plane and centered on the y axis, at a great distance from the origin.
   Once they are moved into place, they define an r, θ coordinate grid at $\phi = 0°$ and are to be used by all observers at rest with respect to the origin.
6. Repeat step 5, but install the hoops at $\phi = 1°$. Keep installing sets of hoops at one degree ϕ increments up to $\phi = 179°$.

   These hoops define the r, θ, ϕ coordinate grid and are to be used by all observers at rest with respect to the origin. We can interpolate between grid points—we have a fiducial mark every meter in coordinate radius and every degree in both θ and ϕ.

   The presence (or absence) of a mass at the center of our collection of concentric rings does not change any of the rings' circumferences and, consequently, does not alter any of their coordinate radii.

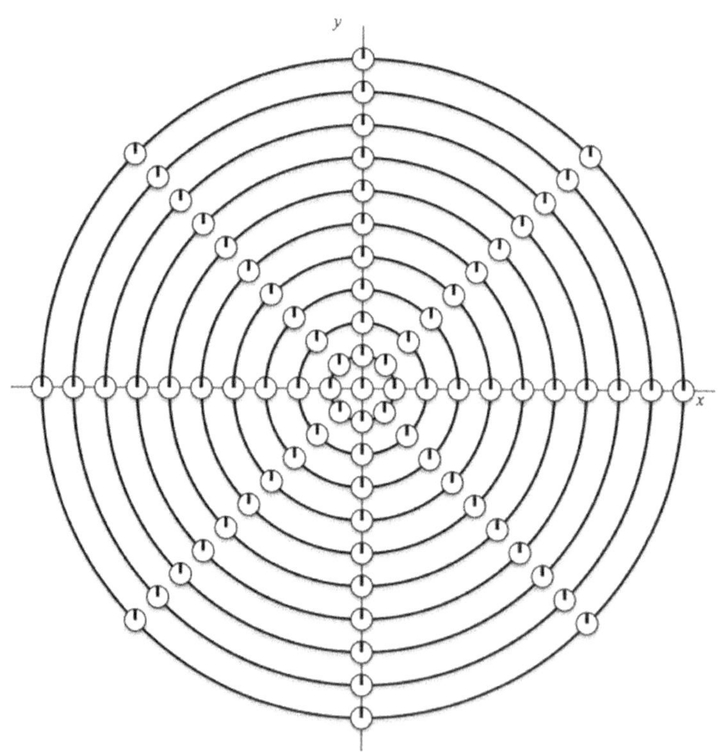

Defining the time interval dt

Imagine that the clocks we installed on our rings were manufactured to tick once per second. An observer at rest next to one of the clocks will see that it continues to tick once per second. This will be the case even if the clock and observer are close to a very massive object.

Here's a way to think about this: if the observer and clock are suddenly detached from the ring, they will begin to free fall toward the mass. For a very short time after entering free fall, they will be moving at negligible speed and will not feel the effects of any (non-gravitational) net force. It will seem to the observer that they live in a force-free inertial frame and that the clock behaves as designed by the manufacturer. The observer will conclude that the (proper) time interval between ticks is $\Delta\tau = 1$ second.

Each time it ticks, the clock sends a laser pulse toward the distant engineer who supervised the construction of all those rings. The transit time from the laser to the engineer might be substantial, but each pulse will take the same amount of time to make the trip.

The time interval between arriving pulses measured by the distant engineer is dt.

What we mean by a measurement of the space–time interval between two events

Our distant engineer maps out the transit time of a light beam from each clock to his/her position by sending a laser pulse toward each clock and measuring how long it takes for the reflection off the clock to return to his/her position. The one-way transit time is half this, of course.

The distant engineer observes that a clock at r, θ, ϕ is zapped by a meteor. Later, the engineer sees the explosion when the clock at $r + dr$, $\theta + d\theta$, $\phi + d\phi$ is struck by a second meteor. After correcting for the different transit times of light from the two explosions, the engineer concludes that the time interval between the explosions is dt, though an observer next to the clocks would measure the time interval to be $\Delta\tau$, which could be different from dt.

The distant engineer is unable to perform a direct measurement of the distance between the two clocks, but they see which hoops had supported the now-destroyed time pieces and can use the dr, $d\theta$, $d\phi$ information gleaned in this fashion to calculate dL using the Schwarzschild metric.

The speed of light

The observers next to the now-destroyed clocks will always measure the speed of a light beam that passes their positions to be 299,792,458 meters per second.

Exercise 9.1: The Schwarzschild radius

Two of the elements of the Schwarzschild metric include factors of $(1 - 2GM/rc^2)$, either in the numerator or the denominator. Not surprisingly, the quantity $2GM/c^2$ has dimensions of length; we refer to it as the Schwarzschild radius for an object of mass M. Recall that Newton's gravitational constant G is 6.674×10^{-11} m^3/kg/s^2.

Pull out your calculators to determine the following Schwarzschild radii. For the water-filled sphere, assume that the density of water is (initially!!) constant and 10^3 kg/m^3. (The radius of the sphere puts its surface midway between the orbits of Mars and Jupiter.)

Here's what I get:

Object	Mass (kg)	Schwarzschild radius
Small cantaloupe	1 kg	1.485×10^{-23} m
Proton	1.673×10^{-27} kg	2.485×10^{-54} m
Earth	5.972×10^{24} kg	8.87 mm
Sun	1.989×10^{30} kg	2.954 km
Water-filled sphere of radius 4.9×10^{11} m	4.93×10^{38} kg	7.32×10^{11} m (uh oh! It has collapsed into a black hole!)

Exercise 9.2: Time

The distant engineer watches a clock at coordinate radius $r = 10,000$ m and sees that the clock emits a flash of light each time it ticks. At the origin is a neutron star whose mass is 1.4 solar masses. (The distant engineer is much, much farther from the origin than 10 km.) Another observer standing next to the clock at coordinate radius $r = 10,000$ m determines that the interval between ticks is $d\tau = 1$ second. What is the time dt between flashes as seen by the distant engineer?

Solution

$$d\tau = \sqrt{1 - \frac{2GM}{rc^2}}\, dt$$

$$\frac{2GM}{c^2} = 4.136\,\text{km}$$

so $$\sqrt{1 - \frac{2GM}{rc^2}} = \sqrt{1 - \frac{4.136\,\text{km}}{10\,\text{km}}} = 0.76579$$

$$\text{so } d\tau = 1\,\text{second} = 0.76579\,dt$$

$$dt = 1.31\,\text{seconds}$$

Exercise 9.3: *dL*

The distant engineer in the above problem watches the observer at $r = 10,000\,\text{m}$ extend a tape measure radially inwards toward the surface of the neutron star. How much must that observer extend the tape measure to reach the ring with coordinate radius $9,999\,\text{m}$?

Solution

Same idea as earlier: we have the change in coordinate radius $dr = 1\,\text{m}$, but we need to calculate ds, the change in "string length," or in this case, tape measure radial extension:

$$ds = \frac{dr}{\sqrt{1 - \frac{2GM}{rc^2}}} = \frac{1}{0.76579} = 1.31\,\text{m}.$$

Exercise 9.4: *c*

The distant engineer calculates the speed of light for a radially directed beam that passes between two clocks deep in the gravitational field of that neutron star at the origin. He/she sees that the beam travels from r, θ, ϕ to $r + dr, \theta, \phi$ (a distance he/she calculates to be dL by making use of the Schwarzschild metric) in an amount of time dt.

Keep in mind that an observer next to the clock at r, θ, ϕ will calculate the speed of the light beam as $dL/d\tau = 299{,}792{,}458\,\text{m/s}$.

Calculate dL and the apparent speed of the light beam, dL/dt, according to the distant engineer.

If the distant observer had instead calculated the speed of the light beam as dr/dt, what value would she/he have obtained?

Solution

$$dL = \frac{dr}{\sqrt{1 - \frac{2GM}{rc^2}}} = 1.31 dr \quad \text{and} \quad dt = \frac{d\tau}{\sqrt{1 - \frac{2GM}{rc^2}}} = 1.31 d\tau \quad \text{so}$$

$$\frac{dL}{dt} = \sqrt{1 - \frac{2GM}{rc^2}} \frac{dL}{d\tau} = 0.76579 c$$

$$\frac{dr}{dt} = \frac{\sqrt{1 - \frac{2GM}{rc^2}} dL}{dt} = \frac{\sqrt{1 - \frac{2GM}{rc^2}} dL}{\frac{d\tau}{\sqrt{1 - \frac{2GM}{rc^2}}}}$$

$$= \left(1 - \frac{2GM}{rc^2}\right) \frac{dL}{d\tau} = (0.76579)^2 \, c = 0.58643 \, c$$

The Kerr metric

You certainly noted that the Schwarzschild metric, written in spherical coordinates, cares mostly about r and that the only place any angles enter is in addressing the difference between the polar and azimuthal angles associated with ordinary spherical coordinates in Euclidean geometry. But we don't expect the effects on space–time's geometry to be so symmetric near a rotating black hole!

Recall that parallel transport stuff: everything needed to specify all that there is about a space's geometry comes from working up the Riemann tensor.

Also, recall that the Einstein Equations are $G^{\mu\nu} = \frac{8\pi G}{c^4} T^{\mu\nu}$. Here,

$$G^{\mu\nu} = R^{\mu\nu} - \frac{1}{2} g^{\mu\nu} R, \quad G^{\mu\nu} = R^{\mu\nu} - \frac{1}{2} g^{\mu\nu} R$$

So, the Einstein Equations connect space–time curvature with the distribution of stuff in the universe.

If "stuff" includes a rotating black hole, there'll be effects due to its rotation that change the stress–energy tensor from its simpler form that applies to a non-rotating black hole. Karl Schwarzschild derived the metric for a non-rotating black hole in 1915, just months after Einstein published his first work on General Relativity, but the analysis of a rotating compact object was more challenging. Roy Kerr, a mathematician from New Zealand, finally cracked the problem 48 years later. The "Kerr metric" for an object of mass M rotating around the z-axis with total angular momentum J is as follows:

First, some definitions:

$$r_s \equiv \frac{2GM}{c^2} \qquad \alpha \equiv \frac{J}{Mc} \qquad \rho^2 \equiv r^2 + \alpha^2 \cos^2\theta \qquad \Delta \equiv r^2 - r_s r + \alpha^2$$

Here's the Lorentz-invariant proper time between events:

$$c^2 d\tau^2 = \left(1 - \frac{r_s r}{\rho^2}\right) c^2 dt^2 - \frac{\rho^2}{\Delta} dr^2 - \rho^2 d\theta^2$$

$$- \left(r^2 + \alpha^2 + \frac{r_s r \alpha^2}{\rho^2}\sin^2\theta\right)\sin^2\theta d\phi^2 + \frac{2r_s r \alpha \sin^2\theta}{\rho^2} cdtd\phi$$

Note that when J is zero, this reduces to the Schwarzschild value: the above becomes

$$c^2 d\tau^2 = \left(1 - \frac{r_s r}{r^2}\right) c^2 dt^2 - \frac{r^2}{r\left(r - r_s\right)} dr^2 - r^2 d\theta^2 - \left(r^2\right)\sin^2\theta d\phi^2$$

$$= \left(1 - \frac{r_s}{r}\right) c^2 dt^2 - \frac{1}{\left(1 - \frac{r_s}{r}\right)} dr^2 - r^2 d\theta^2 - r^2\sin^2\theta d\phi^2$$

So, here it is! First, Schwarzschild

$$g_{\mu v} = \begin{bmatrix} \left(1 - \dfrac{r_S}{r}\right) & 0 & 0 & 0 \\ 0 & -\left(1 - \dfrac{r_S}{r}\right)^{-1} & 0 & 0 \\ 0 & 0 & -r^2 & 0 \\ 0 & 0 & 0 & -r^2\sin^2\theta \end{bmatrix}$$

and now, Kerr. Note the off-diagonal terms!

$$g_{\mu v} = \begin{bmatrix} \left(1 - \dfrac{r_s r}{\rho^2}\right) & 0 & 0 & \dfrac{2cr_s r\alpha\sin^2\theta}{\rho^2} \\ 0 & -\dfrac{\rho^2}{\Delta} & 0 & 0 \\ 0 & 0 & -\rho^2 & 0 \\ \dfrac{2cr_s r\alpha\sin^2\theta}{\rho^2} & 0 & 0 & -\left(r^2 + \alpha^2 + \dfrac{r_s r\alpha^2}{\rho^2}\sin^2\theta\right) \\ & & & \times \sin^2\theta \end{bmatrix}$$

Here's the Kerr metric for something dropped above the equator. In this case, $\theta = 90°$, $d\theta = 0$. J is constant, so α is too. The metric elements, plugged into the proper time expression, give

$$c^2 d\tau^2 \Rightarrow \left(1 - \frac{r_s}{r}\right) c^2 dt^2 - \frac{1}{1 - \frac{r_s}{r} + \frac{\alpha^2}{r^2}} dr^2$$

$$- \left(r^2 + \alpha^2 \left[1 + \frac{r_s}{r}\right]\right) d\phi^2 + \frac{2r_s\alpha}{r} c\, dt\, d\phi$$

$$= \left[\left(1 - \frac{r_s}{r}\right) c^2 - \frac{1}{1 - \frac{r_s}{r} + \frac{a^2}{r^2}} \left(\frac{dr}{dt}\right)^2 - \left(1 + \left(\frac{\alpha}{r}\right)^2 \left[1 + \frac{r_s}{r}\right]\right)\right.$$

$$\left. \times \left(\frac{r\, d\phi}{dt}\right)^2 + \frac{2r_s\alpha}{r} c\frac{d\phi}{dt}\right] dt^2$$

Look at the last term on the right side of the equation. What happens as the object begins to fall from rest, so that $r \to r_S$? Rewriting slightly, for $\theta = 90°$, the last term is

$$c^2 d\tau^2 = \cdots + \frac{2J}{M}\frac{r_s}{r}\frac{d\phi}{dt} dt^2$$

Clocks in free fall follow trajectories that maximize the rate of passage of proper time. Picking up a non-zero $d\phi/dt$ of the same sign as the angular momentum will increase the change in $d\tau$ per unit change in dt: the falling object's velocity picks up a component in the ϕ direction!

Things in free fall always move to maximize the proper time that passes. At first, a dropped object falls straight down, and $d\tau$ is reduced relative to dt by a factor of

$$(1 - r_s r/\rho^2)$$

But picking up a small positive $d\phi$ (when J is in the positive z-direction) will add a positive contribution to $d\tau$ because of that strange $d\phi dt$ term. This will overcome the negative term that is proportional to $d\phi^2$.

As a result, the object's trajectory begins to deviate from falling straight down, picking up a velocity component parallel to the equatorial rotation of the massive object at the origin.

This is called "frame dragging."

LIGO

Here's another cool GR thing. LIGO is the "Laser Interferometer, Gravitational-Wave Observatory." The original device comprised enormous, independently functioning facilities in Hanford, WA, and Livingston, LA. You can read about it in all sorts of places.[3]

LIGO was built to detect the small changes in the metric—the geometry of space–time—that were expect to propagate outwards from the merger of black holes.

Following is a picture of the Livingston site, from the Caltech LIGO web pages.[4]

On January 4, 2017, both detectors saw indications of the merger of a binary block hole system, in which a pair of black holes that had been in an ever-tightening orbit around their center of mass, coalesced.

The black holes were thought to have masses of 36 and 29 solar masses. It is likely that they were formed by the collapse of stars formed about 100 million years after the Big Bang.

---

[3]For example, https://www.ligo.caltech.edu/page/what-is-ligo.
[4]*Ibid.*

The system radiated away as gravitational waves about three solar masses of energy, doing so in about a fifth of a second. The peak instantaneous radiated power was approximately 3.6×10^{49} W. (Recall that a Watt is a Joule per second.)

Here's a plot of the data from both devices, along with a model of the expected time-dependent signal.[5] Take note of how well the model fits the data.

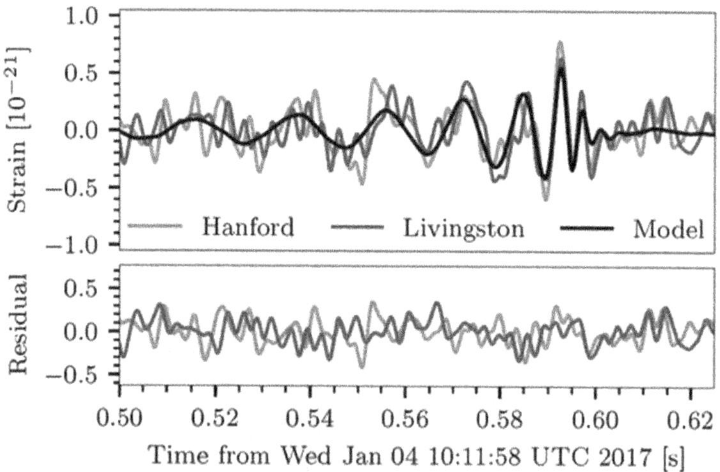

That's an incomprehensibly large amount of radiated power. But here's a way to think about it: our Sun puts out about 3.8×10^{26} W, so the LIGO black holes were about as "bright" as 10^{23} suns.

There are thought to be 10^{22} to 10^{24} stars in the *entire* universe.

The peak instantaneous power radiated as gravitational waves by LIGO's black holes was about as big as the instantaneous (electromagnetic) power output of the ENTIRE UNIVERSE. That's insane.

Oh, here's how to "visualize" the large number 10^{24}. Take 10^{24} tiny cubes, each 1/8 inch (3 mm) on a side. Pack them tightly and stack them into a giant cube 200 miles on a side. (That's the Space Shuttle's orbital altitude.)

The number of stars in the entire universe is about the same as the number of little cubes.

Here's a map:

---

[5] *Ibid.*

So, what is it that is actually being radiated?

The metric for in-spiraling black holes is time-dependent and much more complicated than the Schwarzschild or Kerr metrics.

At a point in space–time, the metric indicates how squirrelly the geometry right there, at that spot, at that instant in time has become. The disruption of the metric—the disruption of the geometry of space–time—propagates outwards, away from the in-spiraling black holes.

That propagating disruption is what comprises a gravitational wave. It stretches/compresses distances and messes around with time intervals. The effects are miniscule but—now, thanks to LIGO— measurable.

Summary

We discussed

- the differences in the approach of Newton and Einstein;
- the stress–energy tensor;

- parallel transport and the Riemann curvature tensor;
- the Einstein tensor;
- the Schwarzschild metric;
- what the coordinates actually mean;
- the Schwarzschild radius and the speed of light near a black hole;
- the Kerr metric and frame dragging.

Problem 1: Magic gas

A large, spherically symmetric cloud of very cold gas with magical properties floats in intergalactic space. The "number density"—the number of gas molecules per cubic meter—is constant inside the cloud, which has radius R.

For reasons having to do with pedagogical necessity, the individual gas atoms are nearly massless, so gravitational effects can be ignored UNTIL SUDDENLY something happens and all the gas atoms simultaneously come to have masses rather like those of nitrogen molecules. The cloud acquires a (constant) density of $1\,\mathrm{kg/m^3}$, fairly close to the density of the Earth's atmosphere in Denver, Colorado.

(a) For what value of R is the Schwarzschild radius of the cloud equal to the radius R of the cloud?

Solution

$$r_{Schwarzschild} = \frac{2GM}{c^2} = \frac{2G}{c^2}\frac{4}{3}\pi R^3 \rho$$

$$= \frac{(2)6.674 \times 10^{-11}}{(3 \times 10^8)^2}\frac{4}{3}\pi R^3 = 6.212 \times 10^{-27} R^3.$$

When $R = r_{Schwarzschild}$, we have

$$R^2 = (6.212 \times 10^{-27})^{-1}$$

$$\text{so } R = 1.269 \times 10^{13}\ \mathrm{m}$$

By way of comparison, the orbit of Pluto varies in Sun-Pluto distance from about 4×10^{12} to 7×10^{12} m.

(b) Assume the cloud's radius R is, in fact, the same as the value you found for part (a). For the moment, pretend that you can calculate the gravitational force on a test particle a distance R from the center of the cloud in an entirely Newtonian fashion, assigning the cloud's mass to the origin and using the usual Newtonian expression for the gravitational field from a point mass at the origin.

How large is the gravitational acceleration at R compared to the Earth's surface gravitational acceleration $g = 9.81\,\mathrm{m/sec^2}$?

Solution

$$M = \frac{4}{3}\pi R^3 \rho = \frac{4}{3}\pi \left(1.269 \times 10^{13}\right)^3 1$$

$$= 8.55 \times 10^{39}\,\mathrm{kg}$$

so gravitational acceleration is

$$\frac{GM}{R^2} = \frac{(6.674 \times 10^{-11})(8.55 \times 10^{39})}{(1.269 \times 10^{13})^2} = 3.545 \times 10^3$$

$$= 361.4\,g$$

Problem 2: Crouton's hoops

Infamous sociopath Lawrence Crouton purchases large quantities of an incredibly strong alloy from the nefarious Rumpdock family and forms it into perfectly circular hoops of measured circumferences $60{,}000\,\mathrm{m}$ and $(60{,}000 + 0.02\pi)\,\mathrm{m}$. Crouton uses a large handful of Mrs. Rumpdock's daughter Hikalope's hair twisties to suspend the smaller hoop inside the larger, so the hoops are concentric and lie in a plane.

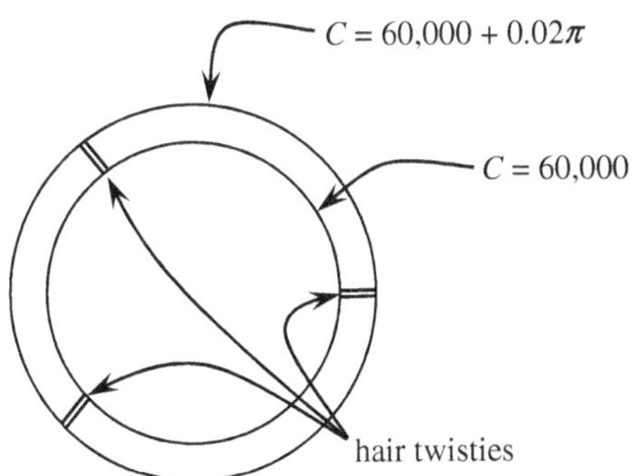

(a) Ignoring any effects associated with space–time curvature and gravity, how long are the hair twisties, as measured by an observer next to a twisty? [**Solution:** $(C_2 - C_1)/2\pi = .01$ m]

(b) While Crouton is napping, Mrs. Rumpdock pops a one-solar mass neutron star into the exact center of the assembly of rings. So perfectly centered is the neutron star that the rings stay where they are. Naturally, they are made of steel so strong that the hoops' circumferences are unchanged. How long are the hair twisties now? [**Solution:** $.01 \times (1 - r_s/r)^{-1/2} = .0123$ m]

Problem 3: GPS tunings

In this problem, assume that the Earth is a perfect *non-rotating* sphere of radius $R_\oplus = 6{,}371$ km and uniform density, with mass $M_\oplus = 5.97 \times 10^{24}$ kg. Recall that Newton's gravitational constant is $G = 6.673 \times 10^{-11}$ m^3 kg^{-1} s^{-2} and that $c = 299{,}792{,}458$ m $\cdot$ s^{-1}. (Also, recall that you'll need to convert masses to kilograms and distances to meters!)

The Global Positioning System employs a constellation of satellites in circular orbits with radii of 26,600 km. (This places them at an altitude of 20,229 km.) By determining the arrival time of signals from several GPS satellites, a receiver can calculate its location relative to the positions of the satellites. The signals broadcast by the satellites include orbit data and the time according to their onboard clocks.

(a) With what speed does a GPS satellite travel as it orbits the Earth? (Calculate this in the style of introductory physics, using purely Newtonian arguments.) What is the period of its orbit?

Solution

$$\frac{GM}{r^2} = \frac{v^2}{r} \quad \text{so}$$

$$v = \sqrt{\frac{GM}{r}} = 3.86996 \times 10^3 \text{ meters per second}$$

(b) What is the Schwarzschild radius for an object of mass $M_\oplus$?

Solution

$$r_s = \frac{2GM}{c^2}$$

$$\text{so } r_s = 8.853 \times 10^{-3}\,\text{m}$$

An observer on the surface of the (non-rotating) Earth notes that the clocks carried by the GPS satellites "tick" at a different rate than they would if they were at rest on the Earth's surface.

(c) Including *only* the change in rate associated with time dilation, how much will a satellite clock deviate from an identical ground-based stationary clock at the end of one day? You'll need to make sure your calculator/software/whatever is working to sufficient precision to yield an accurate answer, if you don't do this with clever binomial approximations.

Solution

Time at the surface of the stationary Earth: 86,400 seconds.
Satellite time:

$$86{,}400\sqrt{1 - \frac{v^2}{c^2}} \approx 86{,}400 - 43{,}200\left(\frac{3.86996 \times 10^3}{3 \times 10^8}\right)^2$$

$$= 86{,}400 - 7.18876 \times 10^{-6}$$

Deviation from ground based (stationary) clocks is 7.18876 microseconds after one day, with the satellite clocks running behind.

(d) Neglecting time dilation, but including the General Relativistic effect of the different r values for a ground-based clock and a satellite-based clock, how much will the satellite clock deviate from the ground-based clock at the end of one day?

Solution

Clock on the Earth's surface: in one day, this much time passes on it:

$$86{,}400\sqrt{1 - \frac{2GM}{rc^2}} = 86{,}400(1 - 1.38957 \times 10^{-9})^{1/2}$$

$$\approx 86{,}400 - 0.600 \times 10^{-4}\,\text{seconds}$$

so the daily loss due to general relativity is about 60 microseconds.

On the satellite, the general relativistic loss is less. Just plug-and-chug to find that it is only 14.3778 microseconds per day. As a result, the Earth clock loses, each day, (60–14.38) microseconds per day, or about 45.6 microseconds per day.

(e) Including both effects calculated in parts (c) and (d), how much will the satellite clock deviate from the ground-based clock at the end of one day?

Solution

The two effects go in the opposite direction: GR makes the Earth clock run slower and SR makes the satellite clock run slower. The net effect is that the Earth clock loses $45.6 - 7.2 = 38.4$ microseconds per day relative to the satellite clock.

Note how large the accumulated error is! Since the speed of light is about 300 meters per microsecond, navigational inaccuracies caused by a failure to include relativistic corrections would be substantial.

Problem 4: Mrs. Rumpdock tries to sell cough drops to space aliens

Inspired by a commercial for cough drops in which Venusians blow through strange wind instruments that resemble sewer pipes, Mrs. Rumpdock builds the unusual structure shown in the figure out of a semicircular (nickel-plated) tube, a laser, and a photodetector.

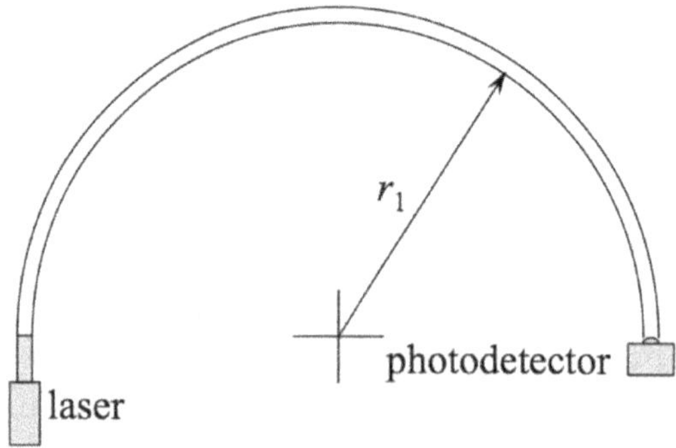

A pulse of light fired by the laser enters the open end of a semicircle of radius r_1. It skitters around the inside of the pipe so that it travels in a perfect semicircle (always with speed c, according to observers at the hoop), then strikes the photodetector at the right.

(a) In the absence of gravitational effects, how long does it take the light to travel from the laser to the photodetector, according to observers at a distance r_2 away from the origin? (Assume $r_2 > r_1$.)

Solution

Keep in mind that circumferences are not changed by the introduction of a mass at the origin and that observers always measure the speed of light as a light beam passes by their positions as c. So, this one's easy! In the absence of gravitational effects, all observers everywhere will agree that the light beam travels a distance πr_1, so $T = \pi r_1/c$.

(b) A compact object of mass M (whose Schwarzschild radius is smaller than r_1) is placed at the center of things. The tubes are not deformed by the presence of the gravitating object. How long does it take for the light beam to travel from the laser to the photodetector now, according to observers at r_1? According to observers at r_2?

Solution

Now, we have clocks at r_1 running slowly; the same is true for clocks at r_2. Imagine that technicians operating the laser and photodetector, who hold synchronized clocks at coordinate radius r_1, send simultaneous signals to observers at r_2 and also to observers at $r \to \infty$. These distant observers can use these calibration signals to figure out the difference in propagation times in order to correct for this when analyzing their observations of the laser/photodetector system.

Observers at r_1 will always see the beam traveling with speed c, and since the circumference doesn't change with the introduction of the mass, they will find it still takes the light beam time $t_1 = \pi r_1/c$ to travel the half-circle. Observers at infinity will see that the r_1 clocks are running slow by a factor of $(1 - r_{sch}/r_1)^{1/2}$, but they will still see the path length as unchanged from πr_1. Consequently, they conclude that the light beam takes longer:

$$t_\infty = \frac{t_1}{\sqrt{1 - \frac{2GM}{r_1 c^2}}} = \frac{\pi r_1}{c\sqrt{1 - \frac{2GM}{r_1 c^2}}}$$

The clocks at r_2 tick faster than those at r_1 but slower than those at infinity, so they see the time for the light beam's trip as

$$t_2 = t_\infty \sqrt{1 - \frac{2GM}{r_2 c^2}} = \frac{\pi r_1}{c} \sqrt{\frac{1 - \frac{2GM}{r_2 c^2}}{1 - \frac{2GM}{r_1 c^2}}}$$

Châteauneuf du Pape, France

Unit 10

Motion in Curved Space–Time

Extremum principles

Fermat's principle of least time

Toward the end of introductory electricity and magnetism, you probably discussed geometrical optics—the propagation of light through lenses and prisms, in circumstances that allow you to neglect the wavelength of light. One important tool would have been Snell's law, which describes the change in angle when a light ray crosses an interface between media with different indices of refraction. (A medium's index of refraction n is the ratio between the speed of light in vacuum and the speed with which light travels inside the medium. It can be greater than unity thanks to the complicated stimulus and re-radiation of light that takes place as the incident light travels through transparent media.)

Snell's law is best illustrated with an equation and a diagram: $n_1 \sin \theta_1 = n_2 \sin \theta_2$. The light ray enters medium 1 at point A and exits medium 2 at point B. The angles are defined with respect to the normal to the interface between the media.

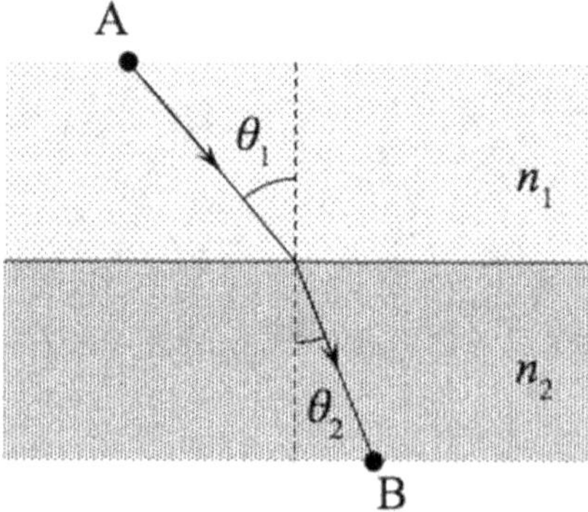

Snell's law is equivalent to Fermat's Principle of Least Time: if we know that the beam passed through points A and B, then the beam will follow a path which minimizes its transit time between A and B. It's not hard to show that the point on the interface through which the beam travels (based on Fermat's Principle) is also the point which satisfies Snell's law.

Fermat's Principle gives us a different (though equivalent) way to think about the propagation of light through complicated geometries of transparent objects.

Principle of least action

In intermediate mechanics, you will learn about Lagrangians and the Principle of Least Action. This is another extremum technique, with a certain similarity to Fermat's Principle. It is a curious and deceptively simple principle and serves as the underpinning for much of advanced classical mechanics and relativistic quantum field theory.

In intro mechanics, you analyzed the motion of conservative systems using conservation of energy, where the energy is defined as the sum of the kinetic and potential energies: $E = K + U$.

The Lagrangian is the *difference* between the kinetic and potential energies: $L \equiv K - U$. It is not a constant of motion; as an object falls, its potential energy will decrease as its kinetic energy increases, so its Lagrangian will increase with time.

Imagine that the (falling) object under the influence of conservative forces passes through point A at time t_A and through point B at time t_B. The classical action S is defined as the time integral of the Lagrangian from t_A to t_B:

$$S \equiv \int_{t_A}^{t_B} L \, dt$$

Hamilton's Principle is that the object will move in a way that minimizes its action. Any trial variation from the correct trajectory (that you can derive from integration of $F = ma$ while holding the points A and B and times t_A and t_B constant) will yield a larger action than the true trajectory. In intermediate mechanics, you will see (have already seen) a derivation of the equivalence of the Principle of Least Action and Newton's second law.

Principle of maximal aging

This is Taylor and Wheeler's term[1] for one of the kinematic consequences of special and general relativity. An object not subject to forces will move along a world line that maximizes the time that passes on its internal clock. This is true in both the flat space–time of special relativity and the curved space–time of general relativity. Gravitational effects arise from space–time curvature, so gravity is not considered a force.

Taylor has a nice derivation of it on his website.[2] Here's a sketch of his presentation, in which he shows that (for the non-relativistic

---

[1]Edwin F. Taylor and John Archibald Wheeler, *Exploring Black Holes: Introduction to General Relativity First Edition*, Addison Wesley Longman, (2000).

[2]See the work of Edwin F. Taylor, "Deriving the Nonrelativistic Principle of Least Action from the Schwarzschild Metric and the Principle of Maximal Aging," http://www.eftaylor.com/pub/GRtoPLA.pdf (2004).

case) it reduces to the classical principle of least action. His derivation assumes a single mass M at the origin so that the Schwarzschild metric is appropriate.

Consider the behavior of a free-falling clock at $\theta = 90°$ that moves through a short space–time interval, during which an amount of time $d\tau$ passes on the clock. A distant observer sees that the space–time interval corresponds to changes in coordinates of dt, dr, and $d\phi$. This observer knows that the time interval measured by the falling clock is

$$d\tau = \sqrt{\left(1 - \frac{2GM}{c^2 r}\right)(dt)^2 - \left(1 - \frac{2GM}{c^2 r}\right)^{-1}\frac{(dr)^2}{c^2} - \frac{r^2(d\phi)^2}{c^2}}$$

Let's consider something falling non-relativistically near the Earth's surface so that $r \gg 2GM/c^2$. Use binomial approximations to rewrite the terms under the square root after factoring out a dt:

$$d\tau = dt\sqrt{\left(1 - \frac{2GM}{c^2 r}\right) - \left(1 - \frac{2GM}{c^2 r}\right)^{-1}\frac{1}{c^2}\left(\frac{dr}{dt}\right)^2 - \frac{1}{c^2}\left(\frac{rd\phi}{dt}\right)^2}$$

$$\approx dt\sqrt{1 - \left(\frac{2GM}{c^2 r}\right) - \frac{1}{c^2}\left(\frac{dr}{dt}\right)^2 - \left(\frac{2GM}{c^2 r}\right)\frac{1}{c^2}\left(\frac{dr}{dt}\right)^2 - \frac{1}{c^2}\left(\frac{rd\phi}{dt}\right)^2}$$

Define dr/dt to be v_r and $rd\phi/dt$ to be v_ϕ. With one more binomial approximation, we have

$$d\tau \approx dt\left[1 - \frac{GM}{c^2 r} - \frac{v_r^2}{2c^2} - \frac{v_\phi^2}{2c^2} - \frac{GM}{c^2 r}\frac{v_r^2}{c^2}\right]$$

The last term is small enough to neglect.

Multiply both sides by the rest energy of the clock to obtain

$$mc^2 d\tau \approx dt\left[mc^2 - \frac{GMm}{r} - \frac{mv_r^2}{2} - \frac{mv_\phi^2}{2}\right]$$

$$= \left[mc^2 - \frac{GMm}{r} - \frac{1}{2}mv^2\right]dt$$

Take note of what we've found: the last two terms in the square bracket on the ride side are the potential energy and the negative of the kinetic energy.

Integrate the motion of the clock between two points with finite separation: the left side of the previous equation gives us

$$\int_{\tau_{begin}}^{\tau_{end}} mc^2 d\tau = mc^2 \Delta\tau$$

while the right side yields

$$\int_{t_{begin}}^{t_{end}} \left[mc^2 - \frac{GMm}{r} - \frac{1}{2}mv^2 \right] dt$$

$$= mc^2 \Delta t - \int_{t_{begin}}^{t_{end}} \left[\frac{1}{2}mv^2 + \frac{GMm}{r} \right] dt$$

$$= \text{constant} - \int_{t_{begin}}^{t_{end}} [K - U] dt$$

$$= \text{constant} - \int_{t_{begin}}^{t_{end}} L \, dt$$

We have, then

$$mc^2 \Delta\tau = \text{constant} - \int_{t_{begin}}^{t_{end}} L \, dt$$

Motion that maximizes the proper time that passes on the clock corresponds to motion that makes the time integral of the Lagrangian—this is the classical action—as small as possible. To summarize, physically realized motion that satisfies the (non-relativistic) principle of least action will also maximize the proper time that passes on the falling clock.

Behavior of a falling object near the Earth's surface

Nothing I've said so far about general relativity actually shows that it predicts that an object released near the surface of the Earth will accelerate as it falls! Let's use the Principle of Maximal Aging to show that this is the case.

The setup

Consider an object that is released from rest 2 m above the surface of the Earth. We'll refer to the midpoint of its trajectory—1 m above the ground—as r, where the center of the earth is at $r = 0$. The beginning and end of the trajectory are at $r + \Delta$ and $r - \Delta$, where $\Delta = 1\,\mathrm{m}$.

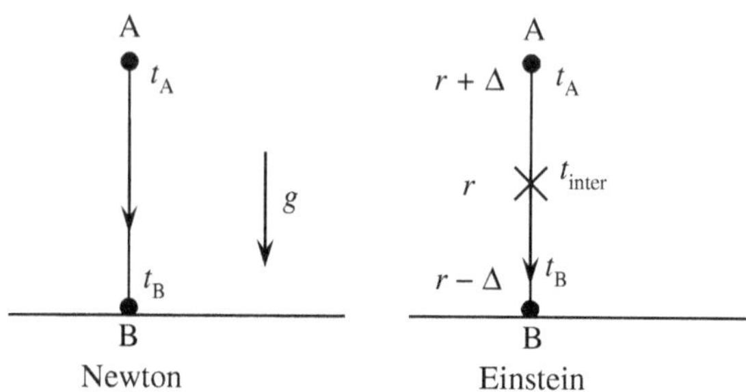

Let me be excruciatingly clear about what I mean by r and Δ. We commissioned our Distant Engineer from the last unit to make three large circular hoops. The smallest has a circumference that allows it to lie on a great circle at the surface of the Earth. We place it so that our dropped object will land on it. The largest ring has a circumference that is 4π meters greater than the smallest. We place it so that it lies in the same plane as the smallest ring and is concentric with it. The middle ring has a circumference that is 2π meters larger than the smallest. It too is placed concentric with the smallest ring and lies in the same plane. The middle ring has coordinate radius r; the coordinate radii of the inner and outer rings differ by $\Delta = 1$ m from that of the middle ring.

As we drop the object from the largest ring, we fire a laser pulse from the largest ring toward a very distant observer. As the object falls past the middle ring, a laser on the middle ring sends a signal toward the distant observer. As the object crashes into the smallest ring, a third laser on the ground fires off a pulse. The distant observer takes note of the arrival times of the three pulses, corrects for their slightly different propagation times (the middle and inner rings are farther away than the outer ring), and records the times (and time differences) of the launching of the pulses. The observer refers to these times as t_A, t_{inter}, and t_B. For convenience, let's assume that the observer declares that t_A is zero. I will refer to the difference between t_A and t_B as T.

Let's calculate separately the proper times for the upper and lower halves of the trajectory. I'll refer to the proper times for the upper and lower halves of the trip as τ_A and τ_B, and their sum as τ. From the Principle of Maximal Aging, we know that calculating τ with the right values of t_A and t_B, but the wrong value of t_{inter} will yield a smaller total proper time. Using the wrong t_{inter} corresponds to working with an unphysical trajectory.

Once we have an algebraic expression for the total proper time, we can differentiate it with respect to t_{inter} to check that we've been given the correct t_{inter}. The correct value of t_{inter} maximizes the proper time, and the derivative will evaluate to zero. We will keep the differences in coordinate radius for the upper and lower halves relative to the middle ring the same (that's Δ) and hold t_A and t_B fixed. We will only allow t_{inter} to vary.

If space–time curvature really is able to make the falling object accelerate, we should find a t_{inter} value that is **very** close to the Newtonian prediction.

For the upper half of the trajectory, I will approximate the metric using the coordinate radius of its midpoint, namely $r + \Delta/2$. For the bottom half, I'll use $r - \Delta/2$.

We have

$$\tau_A^2 = (dt_{upper\ half})^2\, g_{00} - \left(\frac{dr_{upper\ half}}{c} \right)^2 g_{11}$$

$$= (t_{inter} - 0)^2 \left(1 - \frac{2GM}{\left(r + \frac{\Delta}{2}\right) c^2} \right) - \frac{\Delta^2/c^2}{\left(1 - \frac{2GM}{\left(r+\frac{\Delta}{2}\right)c^2} \right)}$$

and

$$\tau_B^2 = (T - t_{inter})^2 \left(1 - \frac{2GM}{\left(r - \frac{\Delta}{2}\right) c^2} \right) - \frac{\Delta^2/c^2}{\left(1 - \frac{2GM}{\left(r-\frac{\Delta}{2}\right)c^2} \right)}$$

The total proper time between points A and B is $\tau = \sqrt{\tau_A^2} + \sqrt{\tau_B^2}$.

Since Newtonian gravity characterizes the behavior of the falling object extremely well, we can use it to estimate what we should find for t_{inter}. Doing the intro physics calculation, I find

$$t_{inter} \approx \sqrt{\frac{\Delta}{g}} = 0.451 \, \text{sec}$$

$$t_B \approx \sqrt{\frac{2\Delta}{g}} = \sqrt{2} t_{inter} = 0.638 \, \text{sec}.$$

Zero-gravity result

In the absence of gravity (this corresponds to setting the Earth's mass M to zero), the general relativistic calculation ought to tell us that t_{inter} is half the total time, namely $T/2$. Let's check that this earthless intermediate time really does maximize the total proper time. (Keep in mind that we are requiring the object to make it from point A to point B in time T, so we might have needed to start it with a non-zero initial velocity.)

We have, in the absence of gravity (after using a binomial approximation to eliminate the square root),

$$\tau_A = \sqrt{(t_{\text{inter}} - 0)^2 - \Delta^2/c^2}$$

$$\approx t_{\text{inter}} - \frac{\Delta^2}{2t_{\text{inter}}c^2}$$

and

$$\tau_B = \sqrt{(T - t_{\text{inter}})^2 - \Delta^2/c^2}$$

$$\approx T - t_{\text{inter}} - \frac{\Delta^2}{2(T - t_{\text{inter}})c^2}$$

so that

$$\tau = \tau_A + \tau_B \approx T - \frac{\Delta^2}{2c^2}\left[\frac{1}{t_{\text{inter}}} + \frac{1}{T - t_{\text{inter}}}\right]$$

$$= T - \frac{\Delta^2}{2c^2}\left[\frac{T}{t_{\text{inter}}(T - t_{\text{inter}})}\right]$$

$$= T - T\frac{\Delta^2}{2c^2}\left[\frac{1}{t_{\text{inter}}T - t_{\text{inter}}^2}\right]$$

We maximize this by choosing t_{inter} to **minimize** the second term in the last equation, which is negative. To do so, we **maximize** the (positive) denominator in the square brackets. Take its derivative, and see what makes it zero:

$$\frac{d(t_{\text{inter}}T - t_{\text{inter}}^2)}{dt_{\text{inter}}} = T - 2t_{\text{inter}}$$

which is zero for $t_{\text{inter}} = T/2$. So far so good.

Constant-curvature result

If the curvature were non-zero but constant (in an elliptical geometry, for example), we would have

$$\tau_A = \sqrt{g_{00}(t_{\text{inter}} - 0)^2 + g_{11}\Delta^2/c^2}$$

$$\approx t_{\text{inter}}\sqrt{g_{00}} + \frac{g_{11}\Delta^2}{2t_{\text{inter}}c^2\sqrt{g_{00}}}$$

and

$$\tau_B = \sqrt{g_{00}(T - t_{\text{inter}})^2 g_{11}\Delta^2/c^2}$$

$$\approx (T - t_{\text{inter}})\sqrt{g_{00}} + \frac{g_{11}\Delta^2}{2(T - t_{\text{inter}})c^2\sqrt{g_{00}}}$$

(see the definitions of τ_A, τ_B above) so that

$$\tau = \tau_A + \tau_B \approx T\sqrt{g_{00}} - \frac{g_{11}\Delta^2}{2c^2\sqrt{g_{00}}}\left[\frac{1}{t_{\text{inter}}} + \frac{1}{T - t_{\text{inter}}}\right]$$

$$= T\sqrt{g_{00}} - T\frac{g_{11}\Delta^2}{2c^2\sqrt{g_{00}}}\left[\frac{1}{t_{\text{inter}}T - t_{\text{inter}}^2}\right]$$

Once again, $t_{\text{inter}} = T/2$ maximizes the proper time.

So far so good. We see that turning off gravity causes the object to move with constant speed: it takes the same amount of time to traverse the first and second halves of its path.

Take note of how small the t_{inter}-dependent term in our proper time expression is. We have (neglecting any subtleties involving g_{00} and g_{11})

$$\tau = T - T\frac{\Delta^2}{2c^2}\left[\frac{1}{t_{\text{inter}}T - t_{\text{inter}}^2}\right]$$

The first term (T) is about 0.6 seconds, but $\Delta^2/(2c^2 t_{\text{inter}})$ is roughly 10^{-17}. That tiny second term has an inordinate amount of influence! A maximum is a maximum, and it is the second term that guarantees motion with constant speed in the absence of gravity, even though it is minuscule compared to the constant term T.

Free-fall near the surface of the Earth

Let's now include the Earth's gravity by setting the Earth's mass M to 5.97219×10^{24} kg. I'll use 6,371 km as the Earth's radius and set the time T to the Newtonian estimate of 0.6382 seconds.

It is a major nuisance to do the calculation of t_{inter} analytically. (I slogged through it when teaching general relativity in an undergraduate course some years ago.) You can set it up like this:

$$0 = \left. \frac{d(\tau_A + \tau_B)}{dt_{\text{inter}}} \right|_{\text{correct } t_{\text{inter}}}$$

$$= \frac{d}{dt_{\text{inter}}} \left\{ \sqrt{(t_{\text{inter}} - 0)^2 \left(1 - \frac{2GM}{\left(r + \frac{\Delta}{2}\right)c^2}\right) - \frac{\Delta^2/c^2}{\left(1 - \frac{2GM}{\left(r + \frac{\Delta}{2}\right)c^2}\right)}} + \sqrt{(T - t_{\text{inter}})^2 \left(1 - \frac{2GM}{\left(r - \frac{\Delta}{2}\right)c^2}\right) - \frac{\Delta^2/c^2}{\left(1 - \frac{2GM}{\left(r - \frac{\Delta}{2}\right)c^2}\right)}} \right\} \Bigg|_{\text{correct } t_{\text{inter}}}$$

Take the derivatives, then solve for the t_{inter} value that gives zero.

It is less taxing to write a Python script to hunt for the value that maximizes the elapsed proper time instead of cranking through the differentiations. (If you are inclined to try this yourself, you'll want to find an extended precision package to import into Python. I used the mpmath 0.19 library, which can be found at https://pypi.python. org/pypi/mpmath/).

I find that the best t_{inter} value is 0.43767 seconds, reasonably close to the Newtonian expectation of 0.451 seconds. Keep in mind that I've made approximations—using the curvatures at the midpoints of each half-trajectory to represent the changing curvature—so I can't really expect an exact result.

The maximum proper time is slightly less than 0.63823. In the following figure, I plot the difference between the maximum proper time and the value obtained with other values of t_{inter}. Note that I have multiplied the difference in proper times by a factor of 10^{18} to make it display sensibly!

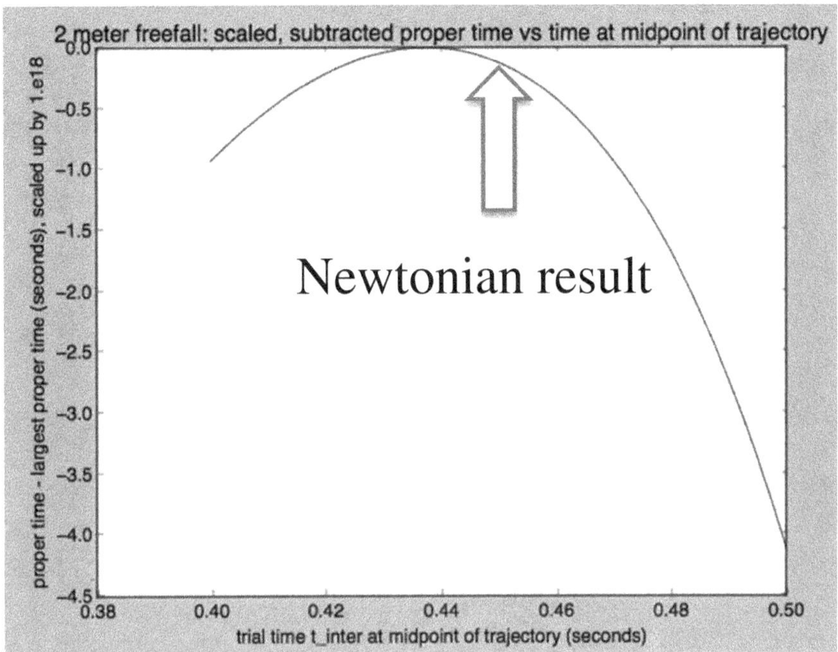

We see that space–time curvature, combined with the Principle of Maximal Aging, was responsible for the apple that bonked Isaac Newton[3] on the head!

Energy in free fall

The Principle of Maximal Aging guarantees that an object in free fall near the Earth will accelerate as it falls. But I showed this by searching for an intermediate time during its downwards trajectory that maximized the total proper time rather than actually doing the differentiations. Cranking out the math would have revealed a useful conservation law that is the general relativistic analog of conservation of total mechanical energy for a falling object in a Newtonian universe.

---

[3]The story is apocryphal, I expect. But see this account: https://www.new scientist.com/article/2170052-newtons-apple-the-real-story/.

Let's say that a very distant observer watches a clock as it falls straight down toward a neutron star of mass M at the origin. As the clock passes through coordinate radius r, it is seen (by the distant observer) to require an amount of time dt to undergo a change in coordinate radius of dr. As the clock moves through dr, the time it displays on its face changes by (the proper time) $d\tau$.

The differentiations that I skipped would have allowed me to conclude that

$$\frac{dt}{d\tau}\left(1 - \frac{2GM}{rc^2}\right)$$

is a constant of motion for the free-falling object.

Let me multiply this by mc^2 (m is the rest mass of the falling clock, of course) to define something I will call "Energy," which is a constant of motion:

$$E \equiv mc^2\frac{dt}{d\tau}\left(1 - \frac{2GM}{rc^2}\right)$$

As the clock falls, it runs more and more slowly for two reasons: it is moving faster (thus time dilation) and it is deeper in the neutron star's gravitational field.

We already know from the Schwarzschild metric that

$$d\tau^2 = \left(1 - \frac{2GM}{rc^2}\right)dt^2 - \frac{1}{\left(1 - \frac{2GM}{rc^2}\right)}\frac{dr^2}{c^2}$$

or

$$\left(\frac{d\tau}{dt}\right)^2 = \left(1 - \frac{2GM}{rc^2}\right) - \frac{1}{\left(1 - \frac{2GM}{rc^2}\right)}\frac{1}{c^2}\left(\frac{dr}{dt}\right)^2$$

In a following problem, you will combine the expression for the energy E with the above equation to solve for the falling object's apparent speed (as viewed by the distant observer) dr/dt.

More generally

We can analyze motion that is not perfectly radial in a similar fashion to what I did earlier for an object falling near the Earth.

Take a look at the right-hand figure in the following diagram for trajectories near a neutron star. By holding the intermediate point fixed in space but varying the time that a mass passes through it, we can maximize the proper time to find that best estimate for the mass's position as a function of time. I'm not going to pursue that here.

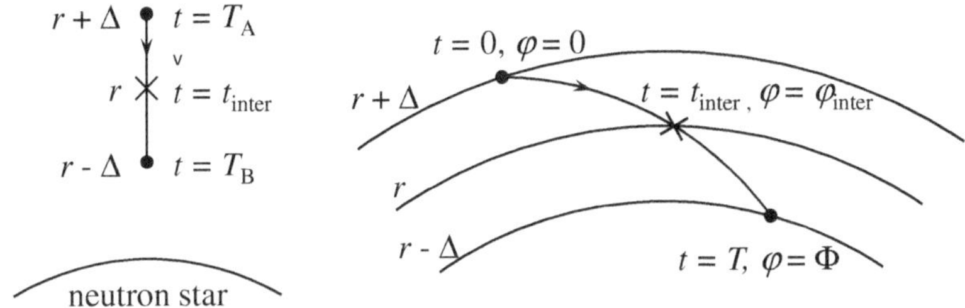

Propagation of light toward a massive object

You're going to derive this one yourselves, with some help from me.

Imagine that a laser held at fixed coordinate radius r_2 fires a pulse directly toward the center of a neutron star. An observer at coordinate radius r_1 measures the speed of the light pulse as it passes through his/her laboratory.

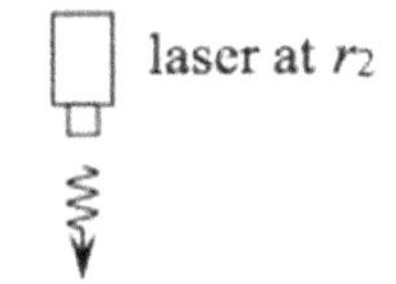

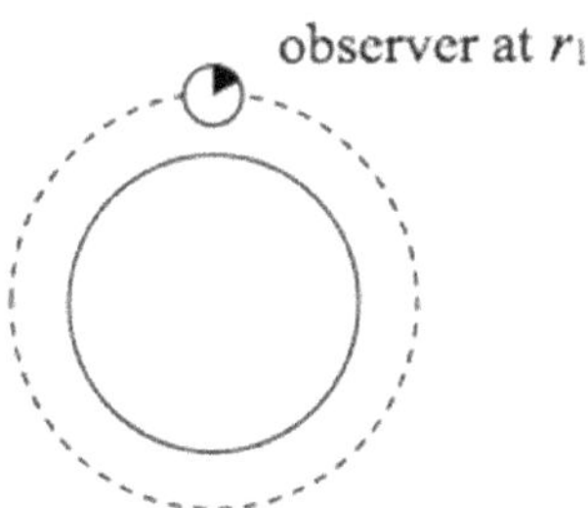

Let's assume that the observer (and laboratory) are at rest but go into free fall a few nanoseconds before the measurement so that, for all practical purposes, the lab is an inertial frame. As a result, the measured speed of the light beam will be c since all measurements of the speed of light performed in (locally) inertial frames must yield this value.

The observer at r_1 holds a ruler of length ds oriented radially and measures a time dt_1 for the beam to travel the length of the ruler. Naturally, they find that $c = ds/dt_1$ so that $dt_1 = ds/c$.

A second observer is so far from the neutron star that they reside in a region that is effectively free of curvature. This distant observer relies on the concentric rings built by the Distant Engineer to ascertain the r, θ, ϕ coordinates of events in the vicinity of the neutron star. This observer, having previously determined the propagation time of signals from all of the rings to his/her position, watches the falling light beam travel the length of the r_1 observer's ruler in a time dt.

Exercise 10.1: Time ratio

Determine the ratio dt/dt_1.

Solution

The clock at r_1 runs slow by a factor of

$$\sqrt{1 - \frac{2GM}{r_1 c^2}}$$

so

$$\frac{dt}{dt_1} = \frac{1}{\sqrt{1 - \frac{2GM}{r_1 c^2}}}$$

Exercise 10.2: Ruler from afar

The distant observer sees that the difference in the coordinate radii of the two ends of the ruler is dr_1. The r_1 observer knows that the measure-it-with-a-tape-measure length of the ruler is ds.

Determine the ratio dr_1/ds.

Solution

The connection between ruler length ds and the difference in coordinate radius of the two ends of the ruler dr is

$$ds = \frac{dr}{\sqrt{1 - \frac{2GM}{r c^2}}}$$

so for the ruler

$$ds = \frac{dr_1}{\sqrt{1 - \frac{2GM}{r_1 c^2}}} \quad \text{giving us} \quad \frac{dr_1}{ds} = \sqrt{1 - \frac{2GM}{r_1 c^2}}$$

Exercise 10.3: Infinitesimal

Use your results from the previous two exercises to express dt in terms of $r_1, dr_1, c, G,$ and M.

Solution

Collect what we know:

$$\frac{ds}{dt_1} = c$$

$$\frac{dt}{dt_1} = \frac{1}{\sqrt{1 - \frac{2GM}{r_1 c^2}}}$$

so

$$dt = \frac{dt_1}{\sqrt{1 - \frac{2GM}{r_1 c^2}}}$$

$$= \frac{ds/c}{\sqrt{1 - \frac{2GM}{r_1 c^2}}}$$

$$ds = \frac{dr_1}{\sqrt{1 - \frac{2GM}{r_1 c^2}}}$$

so

$$dt = \frac{\dfrac{dr_1}{\sqrt{1 - \frac{2GM}{r_1 c^2}}} \Big/ c}{\sqrt{1 - \frac{2GM}{r_1 c^2}}} = \frac{dr_1}{c\left(1 - \frac{2GM}{r_1 c^2}\right)}$$

Exercise 10.4: Calculus time

The distant observer measures the amount of time for the light to descend from r_2 to r_1 to be T. We can calculate T for ourselves as an integral:

$$T(r_2 \to r_1) = \int_{t \text{ at } r_2}^{t \text{ at } r_1} dt$$

If all went well during your completion of the first three exercises, you should be able to rewrite the integral this way:

$$T(r_2 \to r_1) = \int_{t \text{ at } r_2}^{t \text{ at } r_1} dt = -\int_{r_2}^{r_1} \frac{dr/c}{\left(1 - \frac{2GM}{rc^2}\right)} = -\frac{1}{c}\int_{r_2}^{r_1} \frac{r\, dr}{\left(r - \frac{2GM}{c^2}\right)}$$

This is easy to evaluate if you define $u = r - \frac{2GM}{c^2}$ so that $r = u + \frac{2GM}{c^2}$ and $du = dr$.

Use this to evaluate the integral to prove that

$$T(r_2 \to r_1) = \frac{r_2 - r_1}{c} - \frac{2GM}{c^3} \ln\left(r - \frac{2GM}{c^2}\right)\Big|_{r_2}^{r_1}$$

$$= \frac{1}{c}\left[r_2 - r_1 + r_{\text{Sch}} \ln\left(\frac{r_2 - r_{\text{Sch}}}{r_1 - r_{\text{Sch}}}\right)\right]$$

where r_{Sch} is the neutron star's Schwarzschild radius.

Solution

$$T(r_2 \to r_1) = -\frac{1}{c}\int_{r_2}^{r_1} \frac{r\,dr}{\left(r - \frac{2GM}{c^2}\right)}$$

Let $u \equiv \left(r - \frac{2GM}{c^2}\right)$ and note that $du = dr$. Rewrite the integral:

$$T = -\frac{1}{c}\int_{r_2}^{r_1} \frac{\left(u + \frac{2GM}{c^2}\right)du}{u} = -\frac{1}{c}\int_{r_2}^{r_1} du - \frac{2GM}{c^3}\int_{r_2}^{r_1} \frac{du}{u}$$

$$= \frac{r_2 - r_1}{c} + \frac{2GM}{c^3} \ln\left(\frac{u_2}{u_1}\right)$$

$$= \frac{1}{c}\left[(r_2 - r_1) + r_{\text{Sch}} \ln\left(\frac{r_2 - r_{\text{Sch}}}{r_1 - r_{\text{Sch}}}\right)\right]$$

Exercise 10.5: In the limit

What happens to the time for the laser beam to fall to r_1 as r_1 approaches r_{Sch}?

Solution

As r_1 approaches r_{Sch}, the logarithm diverges, so the time approaches infinity.

Propagation delay of a light beam

If we put a mirror at r_1 so that the high observer could measure the round-trip time from r_2 to r_1 and back to r_2, we'd find that the total elapsed time was twice the value calculated above.

For a 1.4 solar mass neutron star, the coefficient of the logarithm is r_{Sch}/c, which is about 13.8 microseconds. For the Sun, it's about 9.8 microseconds.

Here's a plot of the extra round-trip time induced by gravitational effects for a laser shot fired from 150 million kilometers (about the same as the radius of the Earth's orbit around the Sun) toward a mirror above a 1.4 solar mass neutron star. The horizontal axis gives the r value for the mirror; the minimum r is 10 km.

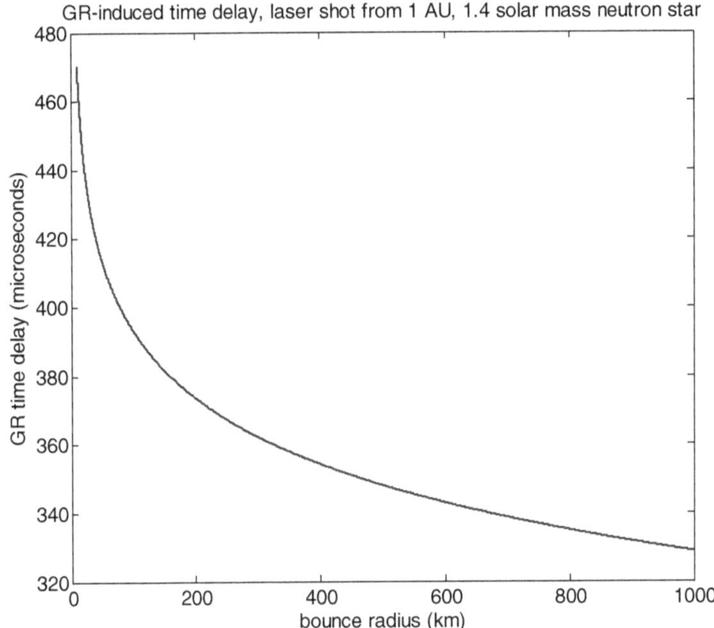

Now, this time delay is as measured by an observer who is very far from the neutron star. If our laboratory clocks are "only" 150 million kilometers away, they'll be running slow by a factor of $\sqrt{1 - 2GM/rc^2} = \sqrt{1 - r_{\text{Sch}}/r}$ (which is slightly smaller than unity), relative to the clocks of the very distant observer.

Imagine we fired a laser beam (or a radar signal!) at Mercury from the Earth and measured the round-trip time. The orbital radii of the two planets are approximately 58 million and 150 million kilometers.

Running the numbers gives a gravitation-induced round-trip delay of about 19 microseconds, compared to a total signal travel time of about 10 minutes.

Embedding diagram

You have probably seen diagrams like the one that follows,[4] but chances are that nobody has ever explained to you what the vertical axis really represents.

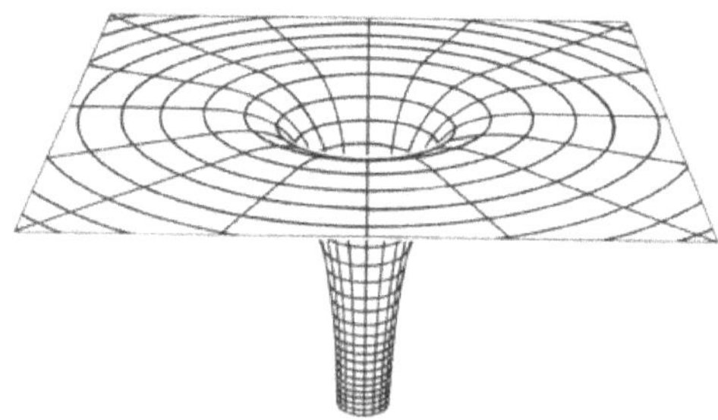

The horizontal position indicates where you are in a plane that cuts through the center of a massive object. But the vertical axis does *not* represent the gravitational potential at points in the plane.

It's called an "embedding diagram" and is too abstract to be meaningful to all those viewers watching PBS science programs. I'll bet almost none of your physics professors know what the vertical axis represents, either. It's a cool-looking thing (and becomes flat as the gravitational field weakens) but not a good pedagogical tool for the uninitiated.

So, now, I'll show you something most of your professors don't know.

Let's look at a slice through the embedding diagram for a 1.4 solar mass neutron star. Have the slice cut through the figure along the plane $y = 0$. All you'll see is a pair of curves, placed symmetrically on either side of the vertical axis.

---

[4]From http://www.khadley.com/courses/Astronomy/ph_206/topics/ns_blackholes/images/slide64.png

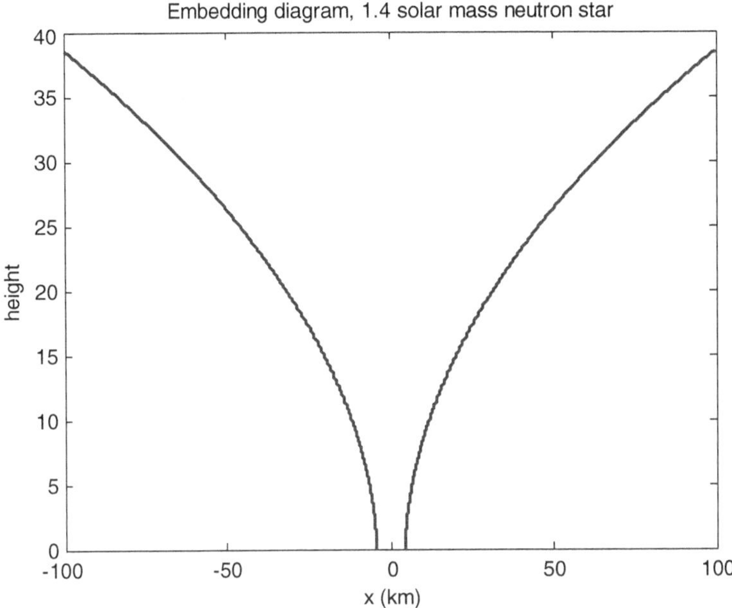

The horizontal coordinate is the distance away from the center of the neutron star, in terms of r rather than s: no surprise there. Here's a graphical representation of how the vertical coordinate is generated.

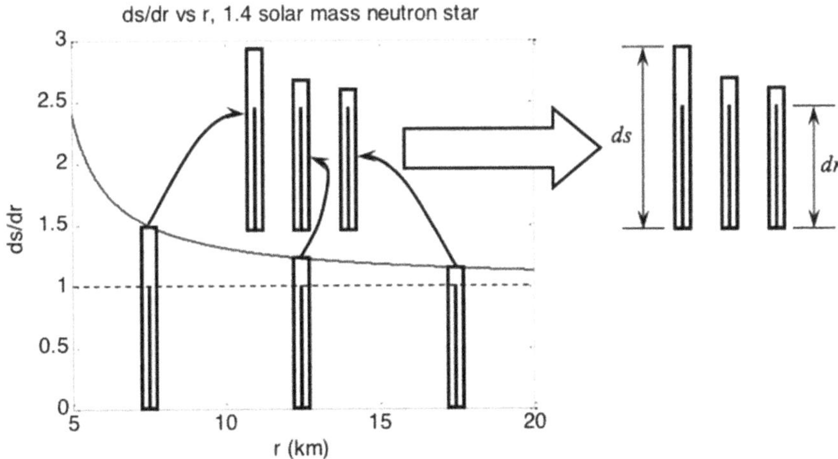

For each value of r, pick a value of dr you'll use to calculate ds at that value of r. In the above figure, the vertical straight lines represent dr values of 5 km that are assigned to r values of 7.5 km, 12.5 km, and 17.5 km. The heights of the narrow rectangles represent ds for each of the dr values. The first one, corresponding to $r = 7.5$ km, is about 1.5 times larger than its accompanying dr value since ds/dr is about 1.5 at $r = 7.5$ km.

Next, rotate each narrow rectangle so that its projection on the r-axis has the same length as dr.

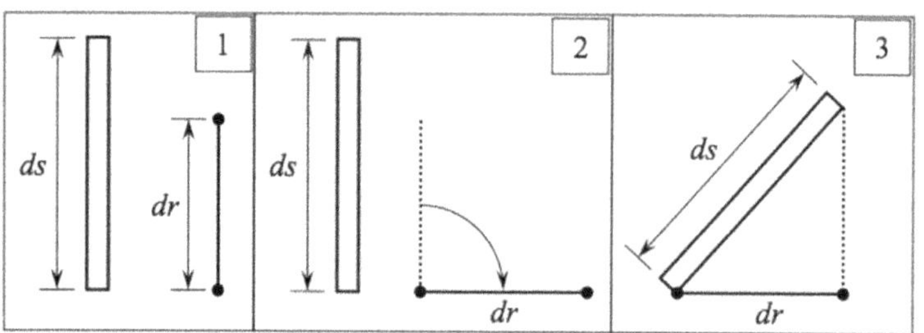

Then daisy-chain the (rotated) ds rectangles together like this:

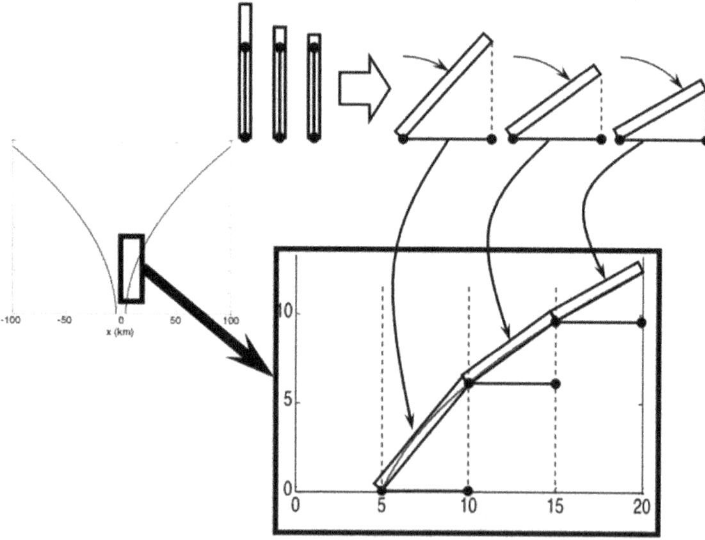

That's it! In the limit that dr goes to zero, the chain of narrow rectangles lies on the curve that's a slice through the embedding diagram.

Finish by spinning the figure around the vertical axis to generate the surface that is so often used in popular representations of space–time curvature.

A little more on the formalism

This is, in part, a reminder of what we've already discussed: "indexology" and the Einstein summation convention.

Four-vectors

Let's define a four-component object (a "contravariant four-vector") that includes both space and time information this way: (ct, x, y, z). (Note that this is "the other" convention from one in which ct is the last rather than the first component.)

For the sake of using indices to refer to the components, it's convenient to define $x_0 \equiv ct$, $x_1 \equiv x$, $x_2 \equiv y$, and $x_3 \equiv z$.

When I want to write equations concerning the components of contravariant four-vectors, I'll use lower-case Greek superscripts, e.g. x^μ. In fact, I'll usually write x^μ to refer to the entire four-vector. So, we have

$$x^\mu = (ct, x, y, z) = (x_0, x_1, x_2, x_3)$$

Since we'll want to make it easy to calculate quantities like

$$\tau^2 \equiv t^2 - (\Delta x^2 + \Delta y^2 + \Delta z^2)/c^2 = (x_0^2 - x_1^2 - x_2^2 - x_3^2)/c^2$$

it is helpful to define a *covariant* four-vector this way: $x_\mu = (ct, -x, -y, -z) = (x_0, -x_1, -x_2, -x_3)$. Note the sign flip on the last three components.

This way, we can write $\tau^2 = \frac{1}{c^2} \sum_{\mu=0}^{\mu=3} x_\mu x^\mu$.

We can write the Lorentz transformations for relative motion in the x-direction this way:

$$\begin{bmatrix} ct' \\ x' \\ y' \\ z' \end{bmatrix} = \begin{bmatrix} \gamma & -\gamma\beta & 0 & 0 \\ -\gamma\beta & \gamma & 0 & 0 \\ 0 & 0 & 1 & 0 \\ 0 & 0 & 0 & 1 \end{bmatrix} \begin{bmatrix} ct \\ x \\ y \\ z \end{bmatrix}$$

or

$$
\begin{bmatrix} x'_0 \\ x'_1 \\ x'_2 \\ x'_3 \end{bmatrix}
=
\begin{bmatrix} \gamma & -\gamma\beta & 0 & 0 \\ -\gamma\beta & \gamma & 0 & 0 \\ 0 & 0 & 1 & 0 \\ 0 & 0 & 0 & 1 \end{bmatrix}
\begin{bmatrix} x_0 \\ x_1 \\ x_2 \\ x_3 \end{bmatrix}
$$

Defining

$$
\underset{\sim}{\Lambda} =
\begin{bmatrix} \gamma & -\gamma\beta & 0 & 0 \\ -\gamma\beta & \gamma & 0 & 0 \\ 0 & 0 & 1 & 0 \\ 0 & 0 & 0 & 1 \end{bmatrix}
$$

and referring to its components as Λ^{μ}_{v} allows us to write a Lorentz transformation like this:

$$
x'^{\mu} = \sum_{v=0}^{v=3} \Lambda^{\mu}_{v} x^{v}
$$

Note the placement of indices:

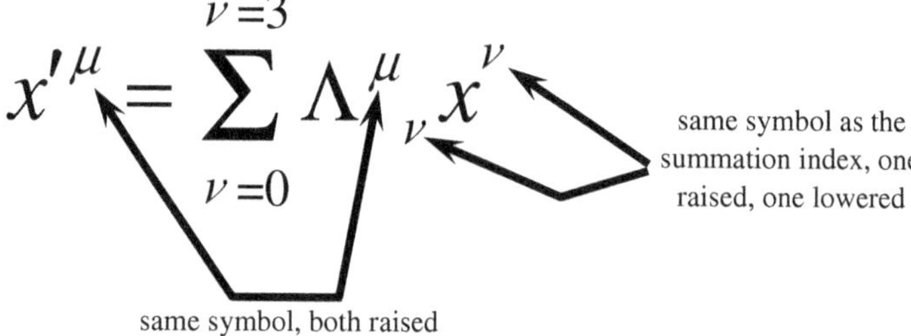

The metric tensor

We define the *metric* tensor in flat space–time this way:

$$
\underset{\sim}{g} \equiv
\begin{bmatrix} 1 & 0 & 0 & 0 \\ 0 & -1 & 0 & 0 \\ 0 & 0 & -1 & 0 \\ 0 & 0 & 0 & -1 \end{bmatrix}
$$

There are two versions of the metric—tensor-covariant and contravariant—which are identical in special relativity, except for the placement of indices. (This is *not* necessarily the case in general relativity in a curved space–time.) We refer to their components as $g_{\mu\nu}$ (covariant) and $g^{\mu\nu}$ (contravariant). Both versions look like the identity matrix except for the sign flip on three of the diagonal components.

The identity matrix is $\delta_{\mu\nu} = \delta^{\mu\nu} = \begin{cases} 1 & \mu = \nu \\ 0 & \mu \neq \nu \end{cases}$

We can use the metric tensor to turn a covariant four-vector into a contravariant four-vector and vice versa. For x, a covariant four-vector (lowered index: x_μ), we have

$$\underset{\sim}{g} x^\mu \equiv \begin{bmatrix} 1 & 0 & 0 & 0 \\ 0 & -1 & 0 & 0 \\ 0 & 0 & -1 & 0 \\ 0 & 0 & 0 & -1 \end{bmatrix} \begin{bmatrix} x_0 \\ x_1 \\ x_2 \\ x_3 \end{bmatrix} = \begin{bmatrix} x_0 \\ -x_1 \\ -x_2 \\ -x_3 \end{bmatrix} = x_\mu$$

More compactly, we can write

$$x_\mu = \sum_{\nu=0}^{\nu=3} g_{\mu\nu} x^\nu$$

Not surprisingly, it is also true that

$$x^\mu = \sum_{\nu=0}^{\nu=3} g^{\mu\nu} x_\nu$$

All this means is that we've done some algebra to flip the sign on the space components. There's no physics in this, just mathematical convenience.

Note that another convention for the metric is to define the *ct* component as the one that changes sign. The two are equivalent; you'll just need to keep in mind which convention you'll use for ALL of your calculations.

We can rewrite the Lorentz invariant s^2 using the metric tensor:

$$s^2 = -\sum_\mu x^\mu x_\mu = -\sum_\mu \left[x^\mu \sum_\nu g_{\mu\nu} x^\nu \right] = -\sum_\mu \sum_\nu g_{\mu\nu} x^\mu x^\nu$$

If you're unsure that it's really OK to manipulate the sums this way, consider writing everything out explicitly as a series of terms to confirm that it's fine.

You may have noted that a sum over an index involves products of terms with one raised (contravariant) and one lowered (covariant) index but not a pair of indices that are both covariant or both contravariant. That is going to be true pretty much *always*. In special (and general) relativity, if you ever encounter a term like

$$\sum_\mu A_\mu B_\mu$$

it'll be a sign that you've made a mistake somewhere.

The metric tensor is not just a convenient algebraic entity: it tells us something very deep about the connection between space and time. When an observer changes frames of reference, the quantity that remains unchanged is calculated with a relative minus between the space and time portions of what otherwise looks like a dot product of two vectors.

Einstein summation convention

I'm not going to use Einstein's summation convention very much, but you'll see it all the time in relativistic equations, and it's pretty much the industry standard, so you'll become familiar with it soon enough.

It's simple: whenever you see a repeated index in a product of terms in an expression, there's an implied sum over that index. (One of the indices will be contravariant, one will be covariant: one upstairs, one downstairs.) For example,

$$x_\mu = \sum_{v=0}^{v=3} g_{\mu v} x^v$$

becomes

$$x_\mu = g_{\mu v} x^v$$

and

$$s^2 = -\sum_\mu x_\mu x^\mu = -\sum_\mu \sum_v g^{\mu v} x_\mu x_v$$

becomes

$$s^2 = -x_\mu x^\mu = -g^{\mu v} x_\mu x_v$$

But it is confusing when you first encounter it, so I won't make all that much use of the Einstein convention.

More four-vectors, including the four-vector potential of electrodynamics

We can construct four-vectors by combining Lorentz scalars with other four-vectors.

One example is the four-velocity. Imagine that a particle travels a distance Δx in time Δt so that the (time dilation-influenced) amount of time that passes on the clock it carries is $\Delta \tau = \Delta t / \gamma$. ($\Delta \tau$ is a Lorentz scalar.) We define the four-velocity this way:

$$u^\mu \equiv \frac{\Delta x^\mu}{\Delta \tau} = \frac{(c\Delta t, \Delta \vec{x})}{\Delta \tau} = \gamma \frac{(c\Delta t, \Delta \vec{x})}{\Delta t}$$

$$= \gamma \left(c, \frac{\Delta \vec{x}}{\Delta t} \right) = \gamma(c, \vec{v})$$

What I've done is taken a known four-vector—x^μ, or more precisely, the difference between two space–time four-vectors—and divided each component by a Lorentz scalar.

Another example is the four-momentum, calculated as the product of the rest mass and the four-velocity:

$$p^\mu \equiv mu^\mu = \gamma m(c, \vec{v}) = (E/c, \vec{p})$$

since $\vec{p} = \gamma m \vec{v}$ and $E = \gamma mc^2$.

Still another is the four-current in electrodynamics, which is the product of the rest charge density and four-velocity:

$$J^\mu \equiv \rho_0 u^\mu = \gamma \rho_0(c, \vec{v}) = (\rho c, \vec{J}).$$

Note that $\vec{J}$ is a current, not an angular momentum.

It will be true for any four-vector K^μ that it transforms under changes of reference frame like this:

$$K'^\mu = \sum_{v=0}^{v=3} \Lambda^\mu_v K^v$$

This resembles the transformation rule for three-vectors under rotations: $\vec{A}' = \underset{\sim}{R} \vec{A}$ or $A'_j = \sum_{k=1}^{k=3} R_{jk} A_k$.

Of course, the conserved quantity under rotations is the three-vector's length:

$$A^2 = \sum_{j=1}^{j=3} A_j A_j = \sum_{j=1}^{j=3} A_j \sum_{i=1}^{i=3} \delta_{ij} A_i = \sum_{i=1}^{i=3} \sum_{j=1}^{j=3} \delta_{ij} A_i A_j$$

while the conserved quantity under Lorentz transformations is

$$\sum_{\mu} K_\mu K^\mu = \sum_{\mu} \sum_{v} g^{\mu v} K_\mu K_v = \sum_{\mu} \sum_{v} g_{\mu v} K^\mu K^v$$

The four-vector potential in electrodynamics is a combination of the scalar and vector potentials: we'll discuss both in a later unit, but here's an initial mention of it. For a point charge q at position $\vec{r}\,'$ moving with velocity $\vec{v}$, the scalar and vector potentials V and $\vec{A}$ are

$$V(\vec{r}) = \frac{1}{4\pi\varepsilon_0} \frac{q}{|\vec{r} - \vec{r}\,'|} \frac{\mu_0 c^2}{4\pi} \frac{q}{|\vec{r} - \vec{r}\,'|}$$

and

$$\vec{A}(\vec{r}) = \frac{\vec{v}}{c^2} V(\vec{r}) = \frac{\mu_0}{4\pi} \frac{q\vec{v}}{|\vec{r} - \vec{r}\,'|}.$$

I'm not going to go into the details of retarded position: $\vec{r}\,'$ is where the charge *was*, not where it is *now*.

Note that the units of $\vec{A}(\vec{r})$ are the same as those for $V(\vec{r})/c$.

The four-vector potential is $A^\mu = (V(\vec{r})/c, \vec{A}(\vec{r}))$. You know (or will learn in a bit) that $\vec{B} = \vec{\nabla} \times \vec{A}$, so that $B_x = \partial A_z/\partial y - \partial A_y/\partial z = \partial A^3/\partial y - \partial A^2/\partial z$ and so forth.

Do you remember from other courses that $\vec{E} = -\vec{\nabla} V - \partial \vec{A}/\partial t$? This means that

$$E_x = -\partial V/\partial x - \partial A^1/\partial t = -\partial c A^0/\partial x - \partial c A^1/\partial(ct)$$

Let's define a slick kind of notation for derivatives:

$$\partial_0 \equiv \frac{\partial}{\partial ct} = \frac{\partial}{\partial x^0}, \quad \partial_1 \equiv \frac{\partial}{\partial x} = \frac{\partial}{\partial x^1}, \quad \partial_2 \equiv \frac{\partial}{\partial y} = \frac{\partial}{\partial x^2} \quad \text{and}$$

$$\partial_3 \equiv \frac{\partial}{\partial z} = \frac{\partial}{\partial x^3}$$

More compactly,

$$\partial_\mu \equiv \frac{\partial}{\partial x^\mu} = \left(\frac{\partial}{\partial x^0}, \vec{\nabla} \right)$$

Note that the sub- and superscripts are *not* exponents.

Naturally, we can define a contravariant version of this in the usual way:

$$\partial^\mu = \sum_{v=0}^{v=3} g^{\mu v}\partial_v \quad \text{or} \quad \partial^\mu = \left(\frac{\partial}{\partial x^0}, -\vec{\nabla}\right)$$

We can rewrite the connection between the four-potential and the electromagnetic field like this:

$$B_x = \partial A^3/\partial y - \partial A^2/\partial z \;\Rightarrow\; -B_x = \partial^2 A^3 - \partial^3 A^2$$

and

$$E_x = -c\partial A^0/\partial x - c\partial A^1/\partial(ct) \;\Rightarrow\; -E_x/c = \partial^0 A^1 - \partial^1 A^0$$

There are similar equations for the other components.

Lorentz tensors

You might recall that objects that transform as rank-two tensors under spatial rotations behave this way: $\underset{\sim}{T'} = \underset{\sim}{R}\,\underset{\sim}{T}\,\underset{\sim}{R}^T$ or

$$T'_{ij} = \sum_{p=1}^{p=3}\sum_{q=1}^{q=3} R_{ip}R_{jq}T_{pq}$$

(The "T" superscript on the last R indicates the transpose.)

Similarly, objects that transform as tensors under Lorentz transformations behave like this: $\underset{\sim}{F'} = \underset{\sim}{\Lambda}\,\underset{\sim}{F}\,\underset{\sim}{\Lambda}^T$ or

$$F'^{\mu v} = \sum_{\rho=0}^{\rho=3}\sum_{\sigma=0}^{\sigma=3} \Lambda^\mu_\rho \Lambda^v_\sigma F^{\rho\sigma}.$$

One way to construct a Lorentz tensor is to take the outer product of a pair of four-vectors: $\underset{\sim}{T} = A \otimes B$. In terms of components, $T^{\mu v} = A^\mu B^v$. If we wanted to construct an *antisymmetric* Lorentz tensor, we could make it this way:

$$T^{\mu v} = A^\mu B^v - A^v B^\mu$$

The expressions for the electric and magnetic fields in terms of derivatives of the four-potential are suggestive of an outer product that has been "antisymmetrized."

The 4×4 object whose elements are $\partial^\mu A^\nu - \partial^\nu A^\mu$ is called the electromagnetic field strength tensor:

$$F^{\mu\nu} \equiv \partial^\mu A^\nu - \partial^\nu A^\mu = \begin{bmatrix} 0 & -E_x/c & -E_y/c & -E_z/c \\ +E_x/c & 0 & -B_z & +B_y \\ +E_y/c & +B_z & 0 & -B_x \\ +E_z/c & -B_y & +B_x & 0 \end{bmatrix}$$

It does, in fact, transform like a Lorentz tensor.

We can construct the "dual" of $F^{\mu\nu}$ by substituting $-\vec{E}/c$ for $\vec{B}$ and $\vec{B}$ for $+\vec{E}/c$

$$G^{\mu\nu} \equiv \begin{bmatrix} 0 & -B_x & -B_y & -B_z \\ +B_x & 0 & +E_z/c & -E_y/c \\ +B_y & -E_z/c & 0 & +E_x/c \\ +B_z & +E_y/c & -E_x/c & 0 \end{bmatrix}$$

$G^{\mu\nu}$ is also a tensor.

Tensors with two raised indices are contravariant; tensors with two lowered indices are covariant; tensors with one raised and one lowered index are "mixed."

Multiplication of a tensor by the metric tensor can be used to raise or lower one index or both indices:

$$T^\nu_\mu = \sum_\rho g_{\mu\rho} T^{\rho\nu} \quad \text{and}$$

$$T_{\mu\nu} = \sum_\sigma g_{\nu\sigma} T^\sigma_\mu = \sum_\sigma \left[g_{\nu\sigma} \sum_\rho g_{\mu\rho} T^{\rho\sigma} \right]$$

(Einstein summation convention: $T^\nu_\mu = g_{\mu\rho} T^{\rho\nu}$ and $T_{\mu\nu} = g_{\nu\sigma} T^\sigma_\mu = g_{\mu\rho} g_{\nu\sigma} T^{\rho\sigma}$.)

Laws of nature

We want equations that are candidates for consideration as "laws of nature" to have the same form in all inertial frames. A statement like $\vec{E} = 0$ (true outside an electrically neutral current-carrying wire at rest, in the absence of externally supplied electric fields) won't be true in a frame in which the wire is seen to be moving. So, that's not going to cut it as a law of nature.

As long as we build our equations from objects that behave well under Lorentz transformations—four-vectors, Lorentz scalars, and Lorentz tensors—we'll do fine. An equation of this sort that is true in one frame of reference will necessarily be true in any other frame of reference.

A handy fact (you can prove it if you're up to it) is that ∂^μ behaves like a contravariant four-vector while ∂_μ transforms like a covariant four-vector. So, we can build up equations involving derivatives that will keep their form as we shift to other reference frames.

Here's an example of a physical law, cast this way. The conservation of (electric) charge requires that the rate of change of charge in a volume V be entirely determined by the rate of flow of charge through the surface bounding the volume into or out from the volume. In equations,

$$\frac{d}{dt}\left[\int_V \rho\, dV\right] = -\oint_S \rho\vec{v}\cdot d\vec{A} = -\oint_S \vec{J}\cdot d\vec{A} = -\int_V \vec{\nabla}\cdot\vec{J}\, dV$$

where the last step comes from the divergence theorem. Since

$$\frac{d}{dt}\left[\int_V \rho\, dV\right] = \int_V \frac{d\rho}{dt}\, dV$$

we conclude that $\int_V \frac{d\rho}{dt} dV = -\int_V \vec{\nabla}\cdot\vec{J}\, dV$ so that integrands in the volume integrals are equal:

$$\frac{d\rho}{dt} = -\vec{\nabla}\cdot\vec{J} \quad\text{or}\quad \frac{d\rho}{dt} + \vec{\nabla}\cdot\vec{J} = 0$$

We can rewrite this in terms of the four-current and ∂_μ:

$$\vec{\nabla}\cdot\vec{J} + \frac{d\rho}{dt} = \vec{\nabla}\cdot\vec{J} + \frac{dc\rho}{dct} = \sum_{\mu=1}^{\mu=4} \partial_\mu J^\mu = 0$$

If I were using the Einstein summation convention, I'd have written the equation for conservation of charge even more compactly as $\partial_\mu J^\mu = 0$. This is an equation that keeps its form in all inertial frames. (I mean to look at the rate of change of ρ when $x, y,$ and z

are held fixed, so the difference between partial and total derivatives is unimportant here.)

Here's another example. You'll see in a later unit that the Maxwell equations are

$$\vec{\nabla} \cdot \vec{E} = \frac{\rho}{\varepsilon_0} \qquad \vec{\nabla} \times \vec{B} - \frac{1}{c^2}\frac{\partial \vec{E}}{\partial t} = \frac{1}{\varepsilon_0 c^2}\vec{J}$$

$$\vec{\nabla} \times \vec{E} + \frac{\partial \vec{B}}{\partial t} = 0 \quad \vec{\nabla} \cdot \vec{B} = 0$$

Note that each curl equation is really shorthand for the three equations describing the Cartesian components of the vectors so calculated.

It's easy to show (just crank it out, term by term) that we can write the first two Maxwell equations as the single equation:

$$\sum_{\mu} \partial_\mu F^{\mu v} = J^v/\varepsilon_0 \quad \text{(Einstein: } \partial_\mu F^{\mu v} = J^v/\varepsilon_0\text{)}$$

It's also true that the other two equations are the same as the single equation:

$$\sum_{\mu} \partial_\mu G^{\mu v} = 0 \quad \text{(Einstein : } \partial_\mu G^{\mu v} = 0\text{)}$$

The Maxwell equations do not change form as an observer moves from one inertial frame to another. They can be stated concisely as

$$\sum_{\mu} \partial_\mu F^{\mu v} = J^v/\varepsilon_0 \quad \sum_{\mu} \partial_\mu G^{\mu v} = 0$$

Observers in another frame will find the corresponding equations to be

$$\sum_{\mu} \partial'_\mu F'^{\mu v} = J'^v/\varepsilon_0 \quad \sum_{\mu} \partial'_\mu G'^{\mu v} = 0$$

Some general principles concerning Lorentz scalars, four-vectors, and tensors

1. *Lorentz scalars*

   - The sum, difference, product, or quotient of two Lorentz scalars is a Lorentz scalar.

- The scalar product of any two four-vectors is a Lorentz scalar: $\sum_\mu A_\mu B^\mu$ is a scalar. (Einstein: $A_\mu B^\mu$)
- The "contraction" of both indices of any two tensors is a Lorentz scalar:

$$b = \sum_{\mu,\nu} S^{\mu\nu} T_{\mu\nu} = \sum_{\mu,\nu} \left[S^{\mu\nu} \sum_{\rho,\sigma} g_{\mu\rho} g_{\nu\sigma} T^{\rho\sigma} \right]$$

$$\text{(Einstein: } b = S^{\mu\nu} T_{\mu\nu} = S^{\mu\nu} g_{\mu\rho} g_{\nu\sigma} T^{\rho\sigma})$$

2. *Four-vectors*

- The product of a Lorentz scalar and a four-vector is a four-vector.
- The sum or difference of two four-vectors is a four-vector.
- The contraction of a four-vector and one index of a tensor is a

$$\text{four-vector: } K^\nu = \sum_\mu A_\mu S^{\mu\nu} = \sum_\mu \left[\sum_\rho g_{\mu\rho} A^\rho \right] S^{\mu\nu}$$

$$\text{(Einstein: } K^\nu = A_\mu S^{\mu\nu} = g_{\mu\rho} A^\rho S^{\mu\nu})$$

3. *Tensors*

- The product of a Lorentz scalar and a tensor is a tensor.
- The sum or difference of two tensors is a tensor.
- The outer product of two four-vectors is a tensor: $T^{\mu\nu} = A^\mu B^\nu$.
- A product of two tensors obtained by the contraction of one index is a tensor:

$$A^{\mu\nu} = \sum_\rho [B^\mu_\rho C^{\rho\nu}] = \sum_\rho \left\{ \left[\sum_\sigma g_{\rho\sigma} B^{\mu\sigma} \right] C^{\rho\nu} \right\}$$

$$\text{(Einstein: } A^{\mu\nu} = B^\mu_\rho C^{\rho\nu} = g_{\rho\sigma} B^{\mu\sigma} C^{\rho\nu})$$

4. *General covariance of "good" equations*

- An equation written as a relationship among Lorentz scalars, four-vectors, and Lorentz tensors that is known to be true in one inertial frame will be true in all inertial frames.

Where's the physics?

There's actually not very much physics in this: the speed of light is constant, and so space–time intervals transform from frame to frame according to the Lorentz transformations. The rest is just slick, powerful machinery that enables rapid calculations, as well as easy recognition of a system's essential features. Try to keep that in mind: there's much less "there" there than meets the eye. It's all about how space and time transform when shifting frames of reference and is all contained in the Lorentz transformations.

Summary

We discussed:

- extremum principles: Fermat's for optics, principle of least action for classical mechanics, and the principle of maximal aging for general relativity;
- behavior of objects in free fall;
- energy in free fall;
- propagation of light toward a massive object;
- the embedding diagram;
- a review of four-vectors, and so forth;
- a preview of the Maxwell Equations

Problem 1: Energy in free fall

Recall that the energy for a falling clock of mass m is

$$E \equiv mc^2 \frac{dt}{d\tau} \left(1 - \frac{2GM}{rc^2} \right)$$

(a) Imagine that the clock is so far away from the neutron star that the $2GM/rc^2$ term is negligible. But it could be moving—perhaps the owner of the clock launched it toward the neutron star using a fabulously powerful catapult.

Prove that my definition of energy reduces to the familiar form for the total energy of a (possibly fast-moving) mass in a gravitation-free region. (By gravitation-free, I mean that you should take r to be so large that the term proportional to M is negligible.)

Solution

In a gravitation-free region, the $2GM/rc^2$ term is absent, so

$$E = mc^2 \frac{dt}{d\tau}$$

An observer watching the clock move will see it ticking slowly by a factor of γ: $dt_{\text{observer}} = \gamma d\tau$. As a result,

$$E = mc^2 \frac{dt}{d\tau} = mc^2 \gamma \frac{d\tau}{d\tau} = mc^2 \gamma$$

which is the special relativistic result. Since there's no gravity, there's no potential energy.

(b) An object of mass m is (momentarily) at rest at coordinate radius r very far from a neutron star of mass M. Prove (with the help of a binomial approximation) that the energy of the mass agrees with your (non-relativistic) expectation from introductory physics, except for the inclusion of the object's rest energy, mc^2.

Solution

Now, we have

$$E \equiv mc^2 \frac{dt}{d\tau} \left(1 - \frac{2GM}{rc^2} \right)$$

Keep in mind that dt is the time that passes according to the distant observer, during which $d\tau$ passes on the face of the clock at coordinate radius r. Since the clock is seen to run slow by a factor of

$$\sqrt{1 - \frac{2GM}{rc^2}} \quad \text{we have} \quad \frac{dt}{d\tau} = \frac{1}{\sqrt{1 - \frac{2GM}{rc^2}}} \quad \text{so that}$$

$$E = mc^2 \sqrt{1 - \frac{2GM}{rc^2}}.$$

For large r we have

$$E \approx mc^2 \left(1 - \frac{1}{2} \frac{2GM}{rc^2} \right) = mc^2 - \frac{GMm}{r}$$

That last term is just the Newtonian (non-relativistic) potential energy.

Problem 2: Free fall, take two

A clock of mass m falls straight down toward a black hole of mass M at the origin. When it was enormously far from the black hole, the clock was moving very slowly so that its (conserved) energy is

$$E = mc^2 \frac{dt}{d\tau} \left(1 - \frac{2GM}{rc^2} \right) \approx mc^2.$$

Naturally, as the clock passes through coordinate radius r, it is seen (by the distant observer) to require an amount of time dt to undergo a change in coordinate radius of dr. And as the clock moves through dr, the time it displays on its face changes by (the proper time) $d\tau$.

(a) Using the definition of energy, above, derive in a line or two of algebra an expression for $d\tau$ in terms of dt, r, M, and various physical constants.

Solution

We have

$$E = mc^2 \frac{dt}{d\tau} \left(1 - \frac{2GM}{rc^2} \right) \approx mc^2 \text{ is constant as it falls, so}$$

$$\cancel{mc^2} \frac{dt}{d\tau} \left(1 - \frac{2GM}{rc^2} \right) \approx 1\cancel{mc^2} \text{ or}$$

$$dt \left(1 - \frac{2GM}{rc^2} \right) = d\tau$$

(b) Recall that the Schwarzschild metric guarantees

$$d\tau = \left[\left(1 - \frac{2GM}{rc^2} \right) dt^2 - \frac{1}{\left(1 - \frac{2GM}{rc^2} \right)} \frac{dr^2}{c^2} \right]^{\frac{1}{2}}$$

Take your expression for $d\tau$ that you derived in part (a), divide it by dt, and then set it equal to my expression for $d\tau/dt$ that comes from the Schwarzschild metric. From that equation, solve for dr/dt as a function of r and various constants.

Solution

Algebra! From above, we have

$$\frac{d\tau}{dt} = \left(1 - \frac{2GM}{rc^2} \right) \quad \text{from before, and also}$$

$$\frac{d\tau}{dt} = \sqrt{\left(1 - \frac{2GM}{rc^2}\right) - \frac{1}{\left(1 - \frac{2GM}{rc^2}\right)}\frac{1}{c^2}\left(\frac{dr}{dt}\right)^2}$$

so that

$$\left(1 - \frac{2GM}{rc^2}\right) = \sqrt{\left(1 - \frac{2GM}{rc^2}\right) - \frac{1}{\left(1 - \frac{2GM}{rc^2}\right)}\frac{1}{c^2}\left(\frac{dr}{dt}\right)^2}$$

$$\left(1 - \frac{2GM}{rc^2}\right)^2 = \left(1 - \frac{2GM}{rc^2}\right) - \frac{1}{\left(1 - \frac{2GM}{rc^2}\right)}\frac{1}{c^2}\left(\frac{dr}{dt}\right)^2.$$

Do some more algebra to find

$$\frac{dr}{dt} = -c\left(1 - \frac{2GM}{rc^2}\right)\sqrt{\frac{2GM}{rc^2}}.$$

I've chosen the negative sign in taking a square root since I know the object falls toward the origin.

(c) At what coordinate radius r does the falling clock obtain its maximum value of dr/dt?

Solution

This is such a surprise: you might think the falling object speeds up until it reaches the Schwarzschild radius, then disappears. But it doesn't! Take the derivative of dr/dt and find where it is zero:

$$\frac{d}{dt}\left(\frac{dr}{dt}\right) = \frac{d}{dt}\left(-c\left(1 - \frac{2GM}{rc^2}\right)\sqrt{\frac{2GM}{rc^2}}\right) = 0.$$

Simplify a little by discarding multiplicative constants that don't matter:

$$\frac{d}{dt}\left(\left(1 - \frac{2GM}{rc^2}\right)\sqrt{\frac{1}{r}}\right) = 0$$

$$-\frac{1}{2}r^{-3/2}\frac{dr}{dt} + \frac{3GM}{c^2}r^{-5/2}\frac{dr}{dt} = 0 \quad \text{or}$$

$$\frac{3GM}{c^2} = \frac{1}{2}r \quad \text{or} \quad r = 3 \times \frac{2GM}{c^2} = 3r_{\text{Sch}}$$

The falling object hits its maximum apparent speed at three times the Schwarzschild radius, then begins to slow down, as seen by the distant observer!

(d) As the clock approaches $r = 2GM/c^2$, what does dr/dt approach?

Solution

Just plug in to our dr/dt expression:

$$\lim_{r \to \frac{2GM}{c^2}} \left(\frac{dr}{dt} \right) = \lim_{r \to \frac{2GM}{c^2}} -\left[c \left(1 - \frac{2GM}{rc^2} \right) \sqrt{\frac{2GM}{rc^2}} \right] = 0$$

Problem 3: Mrs. Rumpdock's sunburn

Approximately one microsecond after the Big Bang, the early universe was filled with a hot plasma of photons, electrons, baryons (three-quark objects like protons and neutrons), and neutrinos. About 379,000 years after that, the universe had cooled to about 3,000 K and protons and electrons were able to form neutral hydrogen atoms. The universe rapidly became transparent to the blackbody radiation filling space. Now, billions of years after the Big Bang, the "color temperature" of the blackbody radiation has dropped to 2.7 K, corresponding to a typical wavelength of 1 millimeter.

This is a problem for the Rumpdocks, who have decided to vacation near the Schwarzschild horizon of a 10^9 solar mass black hole at the center of NGC 3115, without having remembered to pack any sunscreen. ("NGC" is the *New General Catalogue of Nebulae and Clusters of Stars*.) The Silver Conveyance, an interplanetary sports car purchased with ill-gotten gains, is in a parking orbit, and the flux of cosmic microwave background photons falling on it has up-shifted to a wavelength of 300 nanometers, harder than that used by tanning salons.

Perhaps the easiest way to approach this problem is to recall that the Rumpdocks will see photons striking them traveling at exactly the speed of light and that the wavelength they measure will be inversely proportional to the frequency they measure. Since their clocks run slowly, they will determine a different frequency from that of a 1 millimeter wavelength photon far from the black hole.

Calculate r for their parking orbit. (Ignore special relativistic effects such as their orbital motion around the black hole.)

Solution

The shift in wavelength is by a factor of $10^{-3}/300 \times 10^{-9} = 3333.33$, so the Randocks see the photons' frequency upshifted by this amount.

That means their clocks are running slow by this same factor:

$$\sqrt{1 - \frac{2GM}{rc^2}} = \frac{1}{3333.33} \quad \text{or}$$

$$1 - \frac{r_{\text{Sch}}}{r} = 9 \times 10^{-8}$$

$$r - r_{\text{Sch}} = 9 \times 10^{-8} r$$

$$r(1 - 9 \times 10^{-8}) = r_{\text{Sch}}$$

$$r \approx \left(1 + 9 \times 10^{-8}\right) r_{\text{Sch}}$$

A billion solar mass black hole has $r_{\text{Sch}} = 2.953 \times 10^{12}\,$m, so the Rumpdocks are in an orbit that is $2.658 \times 10^5\,$m further out.

Problem 4: Hikalope's honeymoon

Fearing she has, once again, been abandoned by her cruel mother, Hikalope sets out for NGC 3115, hoping to regain the love of Mr. and Mrs. Rumpdock by flying straight into the black hole, waving to her parents as she flashes past at high speed, then popping out the other side. Hikalope expects to get home in time to clear away the beer cans from her parents' pre-vacation bash and convinces fiancé Lawrence to accompany her. They take Crouton's afterburner-retrofitted pickup truck.

Mr. and Mrs. Rumpdock, realizing that they should have brought some UV-B-rated sunscreen, have moved to a higher parking orbit, at twice the black hole's Schwarzschild radius.

Crouton and Hikalope arrive at NGC 3115; Crouton temporarily pauses the truck at a point above the black hole where the coordinate radius r is

$$r_0 = \frac{20GM}{c^2} = 10 r_{\text{sch}}$$

and then lets the truck fall straight down. Hikalope salutes her sunburned mother as she flashes past.

(a) Assuming that the combined mass of Hikalope, Crouton, and the truck is m, what is the total energy of the two individuals plus the truck?

Solution

Start with the conserved energy and work from there. The conserved quantity during the fall is $E = mc^2 \frac{dt}{d\tau} \left(1 - \frac{2GM}{rc^2}\right) \approx mc^2$ is constant as it falls, so $\frac{dt}{d\tau}\left(1 - \frac{2GM}{rc^2}\right)$ is also constant.

Since Hikalope pauses at $r = 10\, r_{\text{Sch}}$, we'll have nothing but gravitation causing dt to be bigger than $d\tau$:

$$dt = \frac{d\tau}{\sqrt{1 - \frac{2GM}{rc^2}}} \quad \text{when } r = 10 r_{\text{Sch}}$$

As a result, at this r we'll have

$$E = mc^2 \frac{dt}{d\tau}\left(1 - \frac{2GM}{rc^2}\right) = mc^2 \frac{1}{\sqrt{1 - \frac{2GM}{rc^2}}}\left(1 - \frac{2GM}{rc^2}\right)$$

$$= mc^2 \sqrt{1 - \frac{2GM}{rc^2}} = mc^2\sqrt{9/10}$$

Note that this is less than mc^2.

(b) Calculate $dt/d\tau$ for Hikalope and her fiancé as they shoot past Mrs. Rumpdock at twice the Schwarzschild radius of the black hole. Note that $d\tau/dt = (dt/d\tau)^{-1}$: you'll make use of this shortly.

Solution

At a different r, we'll have

$$E = mc^2 \frac{dt}{d\tau}\left(1 - \frac{r_{\text{Sch}}}{r}\right) = mc^2\sqrt{9/10}$$

so that

$$\frac{dt}{d\tau}\left(1 - \frac{1}{2}\right) = \sqrt{9/10} \quad \text{or} \quad \frac{dt}{d\tau} = 2\sqrt{9/10} = \sqrt{18/5}.$$

(c) I showed earlier that

$$\left(\frac{d\tau}{dt}\right)^2 = \left(1 - \frac{2GM}{rc^2}\right) - \frac{1}{\left(1 - \frac{2GM}{rc^2}\right)}\frac{1}{c^2}\left(\frac{dr}{dt}\right)^2$$

Use this to calculate Hikalope's dr/dt as she passes Mrs. Rumpdock.

Solution

At twice the Schwarzschild radius, we have

$$\left(\frac{d\tau}{dt}\right) = \sqrt{\frac{5}{18}}$$

so

$$\left(\frac{d\tau}{dt}\right)^2 = \frac{5}{18} = \left(1 - \frac{r_{\text{Sch}}}{2r_{\text{Sch}}}\right) - \frac{1}{\left(1 - \frac{r_{\text{Sch}}}{2r_{\text{Sch}}}\right)}\frac{1}{c^2}\left(\frac{dr}{dt}\right)^2$$

so

$$\frac{5}{18} = \frac{1}{2} - \frac{2}{c^2}\left(\frac{dr}{dt}\right)^2$$

so that

$$\frac{c^2}{9} = \left(\frac{dr}{dt}\right)^2$$

$$\frac{dr}{dt} = \frac{c}{3}$$

(d) Mrs. Rumpdock builds a device to measure Hikalope's speed as she falls past the two-Schwarzschild-radii position. She does this by constructing a vertical ruler of length ds and measures the time required for the pickup truck to fall a distance ds. What is the ratio of ds and dr for the ruler?

Solution

$$ds = \frac{dr}{\sqrt{1 - \frac{2GM}{rc^2}}} = \sqrt{2}dr$$

(e) The small time interval dt in an expression like dr/dt is that measured by distant observers. If we define dt_{Rumpdock} as the time Mrs. Rumpdock measures for Hikalope to fall past her ruler, what is the ratio dt/dt_{Rumpdock}?

Solution

$$\frac{dt}{dt_{\text{Rumpdock}}} = \frac{1}{\sqrt{1 - \frac{r_{\text{Sch}}}{2r_{\text{Sch}}}}} = \sqrt{2}$$

(f) Use your answers to the previous parts of this problem to calculate Hikalope's speed as measured by Mrs. Rumpdock, namely ds/dt_{Rumpdock}.

Solution

$$\frac{ds}{dt_{\text{Rumpdock}}} = \frac{\sqrt{2}dr}{dt/\sqrt{2}} = 2\frac{dr}{dt} = \frac{2}{3}c$$

A little more material for you

What's the difference in total energy between a mass m at rest at infinity and the same mass at rest (so $dr = 0$) at coordinate radius r? (Use the $d\tau/dt$ result from a few pages back.)

$$E_\infty = mc^2 \frac{dt}{d\tau}$$

$$= \gamma mc^2$$

$$= mc^2$$

$$E_r = mc^2 \frac{dt}{d\tau}\left(1 - \frac{2GM}{rc^2}\right)$$

$$= mc^2 \left(1 - \frac{2GM}{rc^2}\right)^{-\frac{1}{2}}\left(1 - \frac{2GM}{rc^2}\right)$$

$$= mc^2 \sqrt{1 - \frac{2GM}{r^2}}$$

$$\boxed{E_\infty - E_r = mc^2 \left(1 - \sqrt{1 - \frac{2GM}{rc^2}}\right)}$$

Here's another. What is dr/dt for something falling from $\sim$ rest at infinity toward a mass M?

Conserved energy:

$$E = mc^2$$

$$= mc^2 \frac{dt}{dt}\left(1 - \frac{2GM}{rc^2}\right)$$

$$1 = \frac{dt}{d\tau}\left(1 - \frac{2GM}{rc^2}\right)$$

$$d\tau = dt\left(1 - \frac{2GM}{rc^2}\right)$$

Schwarzschild metric:

$$d\tau^2 = \left(1 - \frac{2GM}{rc^2}\right) dt^2 - \left(1 - \frac{2GM}{rc^2}\right)^{-1} \left(\frac{dr}{c}\right)^2$$

$$= \left[\left(1 - \frac{2GM}{rc^2}\right) - \left(1 - \frac{2GM}{rc^2}\right)^{-1} \frac{1}{c^2} \left(\frac{dr}{dt}\right)^2\right] dt^2$$

Use our $d\tau$ and $d\tau^2$ expressions from before:

$$d\tau^2 = \left[\left(1 - \frac{2GM}{rc^2}\right) - \left(1 - \frac{2GM}{rc^2}\right)^{-1} \frac{1}{c^2} \left(\frac{dr}{dt}\right)^2\right] dt^2 \quad \text{(last slide)}$$

$$d\tau = dt \left(1 - \frac{2GM}{rc^2}\right) \quad \text{(last slide)}$$

$$dt^2 \left(1 - \frac{2GM}{rc^2}\right)^2 = \left[\left(1 - \frac{2GM}{rc^2}\right) - \left(1 - \frac{2GM}{rc^2}\right)^{-1} \frac{1}{c^2} \left(\frac{dr}{dt}\right)^2\right] dt^2$$

(square the $d\tau$ equation)

$$\left(1 - \frac{2GM}{rc^2}\right)^2 = \left[\left(1 - \frac{2GM}{rc^2}\right) - \left(1 - \frac{2GM}{rc^2}\right)^{-1} \frac{1}{c^2} \left(\frac{dr}{dt}\right)^2\right]$$

(divide out the dt^2)

$$\frac{1}{c^2} \left(\frac{dr}{dt}\right)^2 = \left(1 - \frac{2GM}{rc^2}\right)^2 - \left(1 - \frac{2GM}{rc^2}\right)^3$$

(rearrange terms, etc.)

$$\frac{dr}{dt} = c \left(1 - \frac{2GM}{rc^2}\right) \sqrt{1 - \left(1 - \frac{2GM}{rc^2}\right)} = c \left(1 - \frac{2GM}{rc^2}\right) \sqrt{\frac{2GM}{rc^2}}$$

And here's a graph!

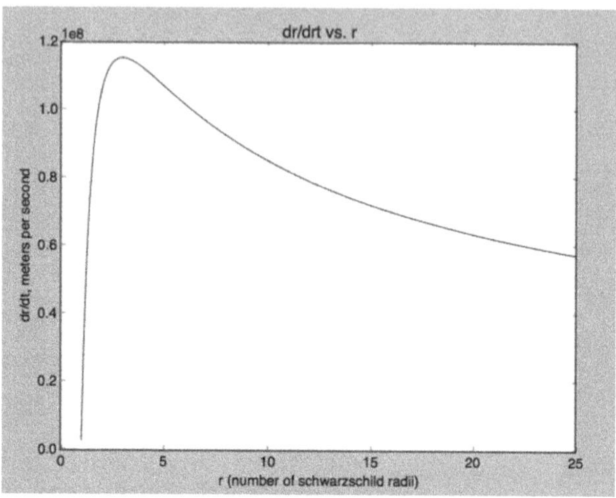

And that's it for General Relativity!

Le Pont du Gard, Vers-Pont-du-Gard, France

Unit 11

Fields, Fluids, Line Integrals, and Curl

Say what?

My own introduction to vector calculus came about during freshman E&M, when the professor taught us about the connection between changing electric and magnetic fields. There were other topics in the course for which we used vector calculus—the Biot-Savart law, Ampère's law, stuff like that—but it was a long time ago.

The subject of electricity and magnetism is exceptionally abstract, with the early weeks of introductory E&M moving from a discussion of the Coulomb force to the mathematical construct of a field, and then a potential (distinct from potential energy), then to flux and Gauss's law, and more. I thought it would make sense to partially embed the material on vector calculus into a more intuitive context, using mathematics to quantify the behavior of systems that you already understand, at least conceptually.

Here's what I mean. Consider the following examples:

- You are motoring along in a boat on the surface of a still lake. When you dip your hand into the water, the water pushes on your hand in the opposite direction from the boat's velocity. You know that the faster the boat travels, the greater the force on your hand.
- You are using a leaf blower to dislodge the autumn leaves from a roof gutter on a neighbor's house. You know that you should aim the air stream parallel to the gutter to have the greatest effect on the soggy leaf mass.
- You submerge a dry sponge in a sink filled with water. After the sponge absorbs as much water as it can, you lift it out of the sink. You know that the sponge will be heavier than it was before.
- You are swimming underwater and blunder into a region where the water temperature is uncomfortably hot. If you know the position dependence of the water's temperature, you will swim in the direction in which the water's temperature drops most rapidly.

You already know these things.

There'll be electrodynamics too, of course, once I've taken you through some of the necessary mathematical preliminaries.

The building blocks

We make use of the following ingredients. Some of them—the velocity-dependence of viscous forces, for example—are only approximately true, but we'll pretend that they are exact.

Viscous drag forces

An object moving with velocity $\vec{v}$ through a stationary fluid (for example, air or water) will suffer a force that is opposite in direction to its velocity and proportional to its speed:

$$\vec{F}_{drag} = -\mu\vec{v}$$

(The same is true if the fluid flows past a stationary object.) The positive quantity μ depends on the properties of both the fluid and the object.

Conservation laws

Some things don't disappear; they just go elsewhere. If a damp sponge becomes lighter as it dries, it is because water molecules are leaving it through evaporation. If the amount of stuff inside a defined volume is constant, there must be as much flow into the volume as out of it.

Partial derivatives

The partial derivative of a function of several variables tells us how the function changes if we hold all but one of the variables fixed. The formal definition is this:

$$\frac{\partial f(x, y, z, w, \ldots)}{\partial x}$$

$$= \lim_{\substack{\Delta x \to 0 \\ y,z,w,\ldots,\text{constant}}} \frac{f(x + \Delta x, y, z, w, \ldots) - f(x, y, z, w, \ldots)}{\Delta x}$$

You saw a problem some time ago in which I asked you to take the partial derivative of a function of x, y, and z. Partials are easy to evaluate: you just treat the to-be-held-constant variables as if they were ordinary constants like π or e and take a derivative in the usual way.

Line integrals

Always *always always* keep in mind that an integral is just a big sum. An integral sign is drawn as a vertically stretched "S" for this reason.[1] A definite integral is the continuum limit of a discrete sum,

---

[1] https://en.wikipedia.org/wiki/Integral_symbol.

as you learned in first-semester calculus:

$$f(x_i)\Delta x + f(x_i + \Delta x)\Delta x + f(x_i + 2\Delta x)\Delta x + \cdots f(x_i + N\Delta x)\Delta x$$

$$= \sum_{n=0}^{n=N} f(x_i + n\Delta x)\Delta x \xrightarrow[N\to\infty,\Delta x\to 0]{} \int_{x=x_i}^{x=x_f} f(x)dx$$

Sometimes we will want to keep track of the direction between a vector in the integrand and the direction of the integration path. A good example from introductory mechanics is the calculation of the work done by a force acting on an object that moves along a complicated path:

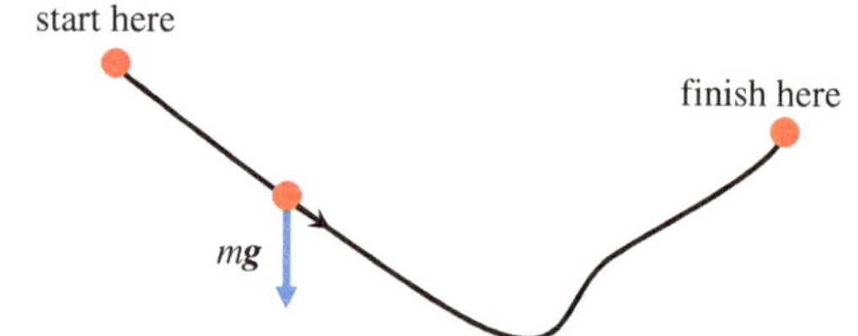

Work done by gravity near the Earth's surface

Break the path into short segments

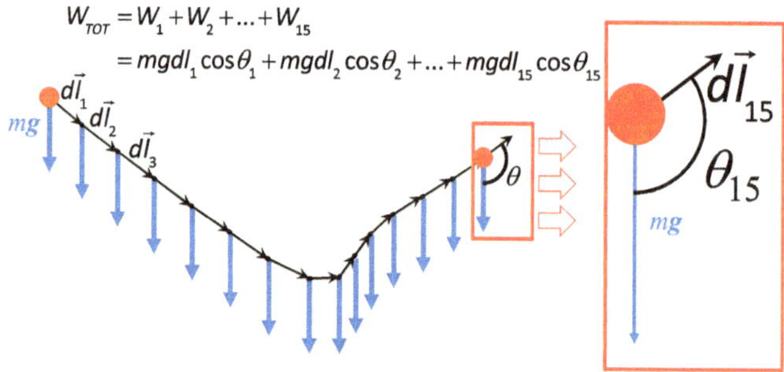

In the above example, θ is a function of position as we move along the path:

$$W_{\text{tot}} = (-mg\hat{y})\cdot \vec{dl_1} + (-mg\hat{y})\cdot \vec{dl_2} + \cdots + (-mg\hat{y})\cdot \vec{dl_N}$$

$$= -mg \sum_{n=0}^{n=N} \cos\theta_i dl \xrightarrow[N\to\infty,dl\to 0]{} \int_{x=x_i}^{x=x_f} \cos[\theta(l)]dl$$

This can be messy, so we may need to be clever about how we set up the problem.

And now to work. Let's start with some line integrals.

Exercise 11.1: Heavy bead on a wire

Recall the work-kinetic energy theorem:

$$\Delta K_{1\to 2} = \int_{\vec{x}_1}^{\vec{x}_2} dK = \int_{\vec{x}_1}^{\vec{x}_2} \vec{F}(x, y, z) \cdot d\vec{l}$$

Imagine that a heavy bead of mass m is free to slide along a frictionless wire that runs along the x-axis. At $t = 0$, the bead is at rest at the origin.

The entire system is immersed in a fluid that flows with position-dependent velocity:

$$\vec{u}(x, y, z) = u_x(x, y, z)\hat{x} + u_y(x, y, z)\hat{y} + u_z(x, y, z)\hat{z}$$

where $\hat{x}, \hat{y}$, and $\hat{z}$ are unit vectors. The unit vectors just serve to attach directions to things and are, themselves, dimensionless: $u_x(x, y, z)\hat{x}$, etc. have dimensions of velocity since u_x, u_y, u_z have units of distance divided by time.

Assume that the fluid in which the bead and wire are immersed moves with velocity:

$$\vec{u}(x, y, z) = a_0[(x^2 - y^2 + 2z^2)\hat{x} + (xy + yz + zx)\hat{y} + (x^2 + y^2 - 2xy)\hat{z}]$$

so that the viscous force on the initially stationary bead is $\vec{F}(x, y, z) = \mu\vec{u}(x, y, z)$.

Also, assume that the fluid velocity is so fast (or μ is so small, or m so large) that the total force, even after the bead begins to move, is well approximated by $\vec{F} = \mu\vec{u}(x, y, z)$. (The exact expression for the force involves the relative velocity of the bead and fluid, namely $\mu(\vec{u}(x, y, z) - \vec{v})$.)

How fast is the bead moving when it reaches the point $x = b$?

Solution

We have $\vec{F} = \mu\vec{u}$. Since the object is constrained by the wire to move only in the x-direction, we have

$$F_x = \mu u_x = \mu a_0(x^2 - y^2 + 2z^2)$$

We move along the x-axis, so $y = z = 0$, yielding $F = \mu a_0 x^2$. As a result,

$$\Delta K = \int_0^b \vec{F} \cdot d\vec{x} = \int_0^b \mu a_0 x^2 dx$$

$$= \frac{\mu a_0 b^3}{3} = \frac{1}{2} m v^2$$

$$v = \sqrt{\frac{2\mu a_0 b^3}{3}}$$

Exercise 11.2: Same bead, different fluid velocity field

The pumps driving the fluid are adjusted so that the fluid's velocity is entirely in the x, y plane:

$$\vec{u}(x, y, z) = u_x(x, y, z)\hat{x} + u_y(x, y, z)\hat{y}$$

The velocity components are chosen so that u_x and u_y satisfy the following equations:

$$u_x^2 + u_y^2 = u_0^2 \qquad \frac{u_x}{\sqrt{u_x^2 + u_y^2}} = e^{-ax}$$

where u_0 and a are positive constants.

The bead starts from rest at the origin. Once again, assume that the force on the bead is well approximated by $\vec{F} = \mu\vec{u}(x, y, z)$. How fast is it moving when it reaches the point $x = D$?

(**Hint:** how would you express the cosine of the angle between the fluid's velocity and the x-axis in terms of the fluid's x, y velocity components?)

Solution

We need to play with the algebra a little since we want to get the cosine term sorted out: the motion is along x but the viscous force is not:

$$u_x^2 + u_y^2 = u_0^2$$

$$\frac{u_x}{\sqrt{u_x^2 + u_y^2}} = e^{-ax} = \frac{u_x}{u_0}$$

so

$$u_x = u_0 e^{-ax}$$

$$F_x = \mu u_x = \mu u_0 e^{-ax}$$

$$\Delta K = \int_0^b \mu u_0 e^{-ax}\, dx = -\frac{\mu u_0}{a} e^{-ax}\Big|_0^b$$

$$\frac{1}{2}mv^2 = \frac{\mu u_0}{a}(1 - e^{-ab})$$

$$v = \sqrt{\frac{2\mu u_0}{am}(1 - e^{-ab})}$$

Exercise 11.3: Same bead but now electrified

We drain the tank holding the fluid, bend the wire into a rectangular loop, and add positive charge Q to the bead. The dimensions of the loop are as shown in the figure; the corners are slightly rounded so that the bead can execute a complete circuit without getting stuck. One corner of the loop is at the origin, and the loop lies entirely in the y, z plane.

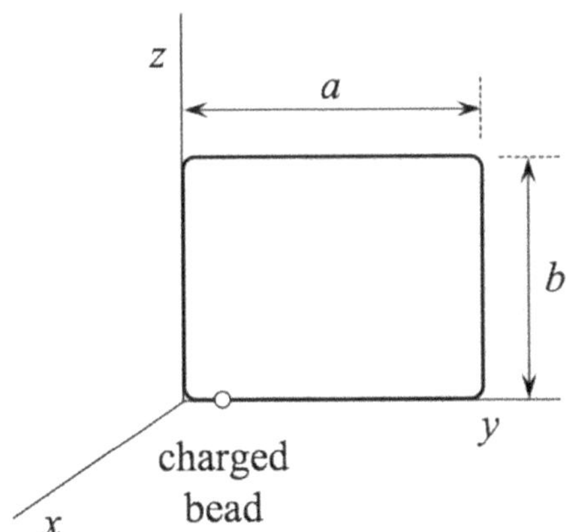

Through obscure and clever techniques, we create an electric field

$$\vec{E}(x, y, z) = E_0 e^{-\gamma z}\hat{y}$$

in the vicinity of the loop, with $\gamma > 0$. (We ignore any possible magnetic effects at this time.) Note that this field points in the y-direction, but its strength varies with z.

We release the bead from rest at the origin. How fast is it moving when it completes one full circuit of the loop?

Solution

Work is only done along the bottom and top legs of the loop. And for fixed z, the electric field is constant! We have

$$\vec{F} = QE_0\hat{y} \quad \text{at } z = 0$$

$$\vec{F} = QE_0 e^{-\gamma b}\hat{y} \quad \text{at } z = b$$

$$\Delta K = QE_0 a - QE_0 e^{-\gamma b} a = \frac{1}{2}mv^2$$

$$v = \sqrt{\frac{2QE_0 a(1 - e^{-\gamma b})}{m}}$$

Exercise 11.4: Lots of beads, lots of (little) loops

We attach a generator and battery to the loop in the previous exercise. Through mysterious means, we transfer all of the kinetic energy gained by the bead in one full turn around the loop in the battery.

We build another device of many tiny wire loops, each equipped with an identical bead of the same mass and charge as the bead in the one-loop system. The new device is shown in the following diagram.

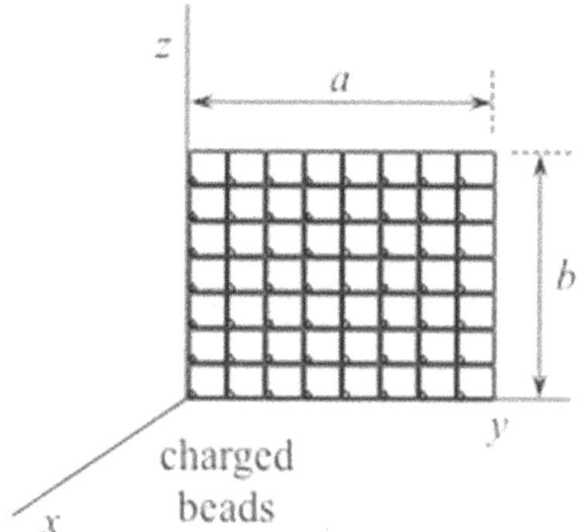

We release each bead from rest at the lower left corner of its tiny wire loop, then collect the energy imparted to each bead by the electric field upon completion of a single turn around its tiny loop. How does the energy we collect compare to the previous case? Ignore the Coulomb force exerted by each bead on the other beads.

Solution

Think about the work done along the adjacent sides of two tiny loops, as shown in the following. Note that the direction of travel of a bead along a side of a loop that is in contact with another loop is opposite to that of the other loop's bead on that side.

The work done by the field on these two beads along the adjacent sides sums to zero.

Only exterior sides do not exhibit this nearest-neighbor cancellation. As a result, the total work done on the system is the same as it was for the single, larger loop.

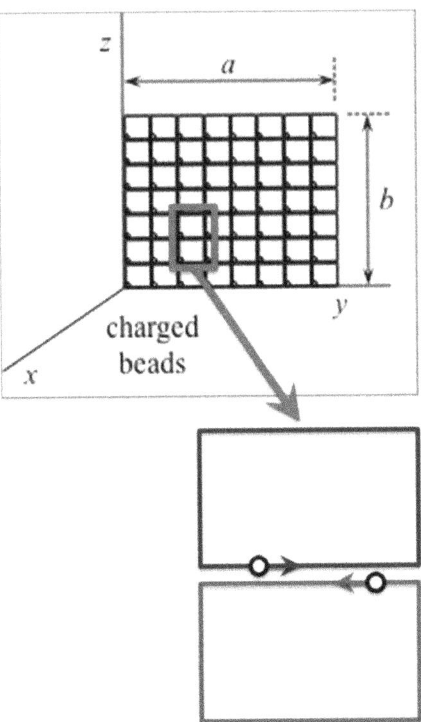

Exercise 11.5: One bead, one tiny loop, and some differential calculus

Consider the behavior of one bead that moves around a tiny loop with sides of length Δy and Δz and centered at y, z. (See the following figure.) Starting with the bottom side and moving around the loop counterclockwise, the centers of the sides are at

$$(0, y, z - \Delta z/2), (0, y + \Delta y/2, z), (0, y, z + \Delta z/2), (0, y - \Delta y/2, z)$$

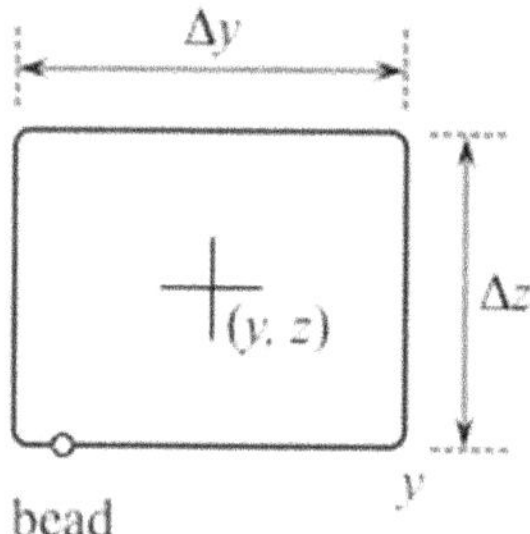

We can approximate the electric field that acts on the particle as it travels along one side of the loop as the field at the center of that side. Only the component of the field parallel to that side will be able to accelerate the particle. The fields of interest are

$$E_y(x = 0, y, z - \Delta z/2), \quad E_z(x = 0, y + \Delta y/2, z),$$
$$E_y(x = 0, y, z + \Delta z/2), \quad E_z(x = 0, y - \Delta y/2, z)$$

(In this particular case, but not in general, E_z is zero.)

Write an expression for the work done by the electric field as the particle makes one circuit of the tiny loop, then (assuming that the sides of the loop are infinitesimal) rewrite this in terms of partial derivatives of the electric field and the infinitesimals Δy and Δz. Do this for the general case, in which E_z might not be zero.

Solution

The work is

$$dW = QE_y(x = 0, y, z - \Delta z/2)\Delta y + QE_z(x = 0, y + \Delta y/2, z)\Delta z$$
$$- QE_y(x = 0, y, z + \Delta z/2)\Delta y - QE_z(x = 0, y - \Delta y/2, z)\Delta z$$

Note that

$$E_z(x = 0, y + \Delta y/2, z) - E_z(x = 0, y - \Delta y/2, z) = \left. \frac{\partial E_z}{\partial y} \Delta y \right|_{(x=0,y,z)}$$

and

$$E_y(x = 0, y, z + \Delta z/2) - E_y(x = 0, y, z - \Delta z/2) = \left. \frac{\partial E_y}{\partial z} \Delta z \right|_{(x=0,y,z)}$$

Therefore,

$$dW = \left. Q \frac{\partial E_z}{\partial y} \Delta y \Delta z \right|_{(x=0,y,z)} - \left. Q \frac{\partial E_y}{\partial z} \Delta y \Delta z \right|_{(x=0,y,z)}$$

Curl of a vector function

We can use the right-hand rule to assign a direction to the motion of the bead. The field I've specified drives it in a counterclockwise direction in the y, z plane, to which we assign the positive x-direction.

We can also define a direction to the area patch with sides Δy and Δz; it is most convenient to pick a direction perpendicular to the surface of the patch. We'll need more information to decide if it is more useful to choose the positive or negative x-directions, so for now, let's pick the same sign as we've attached to the motion of the bead, namely in the positive direction. Attaching vector signs to everything, we have $d\vec{A} = dy dz \hat{x}$.

In Cartesian coordinates, we define the curl of a vector function (in this case, the electric field) this way:

$$\vec{\nabla} \times \vec{E}(x, y, z) = \left(\frac{\partial E_y}{\partial x} - \frac{\partial E_x}{\partial y} \right) \hat{z} + \left(\frac{\partial E_z}{\partial y} - \frac{\partial E_y}{\partial z} \right) \hat{x}$$
$$+ \left(\frac{\partial E_x}{\partial z} - \frac{\partial E_z}{\partial x} \right) \hat{y}$$

This—at least the x component—should look familiar.

Exercise 11.6: Surface integral

For the electric field $\vec{E}(x, y, z) = E_0 e^{-\gamma z}\hat{y}$, evaluate the surface integral

$$Q \int_{z=0}^{z=b} \int_{y=0}^{y=a} \vec{\nabla} \times \vec{E} \cdot \hat{x}\, dy\, dz$$

in the plane $x = 0$. (Note that $\hat{x}dydz$ is the same thing as $d\vec{A}$.) Look at your calculus texts if you are unsure how to do a multiple integral.

Solution

$$Q \int_{z=0}^{z=b} \int_{y=0}^{y=a} \vec{\nabla} \times \vec{E}(x, y, z) \cdot \hat{x}dydz$$

$$= Q \int_{z=0}^{z=b} \int_{y=0}^{y=a} \left(\frac{\partial E_z}{\partial y} - \frac{\partial E_y}{\partial z} \right) dydz$$

$$= Q \int_{z=0}^{z=b} \int_{y=0}^{y=a} \left(-\frac{\partial E_0 e^{-\gamma z}}{\partial z} \right) dydz$$

$$= Q \int_{z=0}^{z=b} \int_{y=0}^{y=a} -(-\gamma)E_0 e^{-\gamma z}dydz$$

$$= Q\gamma a \int_{z=0}^{z=b} E_0 e^{-\gamma z}dydz = \frac{Q\gamma a E_0}{\gamma} \left(1 - e^{-\gamma b} \right)$$

$$= QaE_0(1 - e^{-\gamma b})$$

Exercise 11.7: A special case

How does your value for the surface integral, above, compare to the line integral around the original $a \times b$ rectangular loop? This is the calculus version of Exercise 11.4.

Solution

Note that this is the same as the line integral around the loop that you calculated in an earlier problem!

Stokes' theorem

More generally, it is always true that—for ***any*** vector field—

$$\int_{\text{surface}} \vec{\nabla} \times \vec{J} \cdot d\vec{A} = \oint_{\text{perimeter}} \vec{J} \cdot d\vec{l}$$

In words, the surface integral of the curl of a vector field is the same as the line integral around the perimeter of the surface. This is called *Stokes' theorem* (sometimes the *Kelvin-Stokes Theorem*), named for Irish-born British mathematician George Stokes.

A practical example: Vector potential of an infinite line charge

By now, I expect you've learned in an introductory course that the magnitude of the magnetic field created by a current I flowing in a very long, straight wire is

$$B = \frac{\mu_0 I}{2\pi r}.$$

The field lines form concentric circles.

We can calculate the current arising from a flow of (positive!!) charge: if there are λ coulombs of positive charge per meter, the current I will be λv where v is the magnitude of the charges' average velocity.

You've also learned from intro physics that the scalar potential of a line charge λ (when we set the zero of the potential to be at $r = 1$) is

$$V(r) = \frac{-\lambda}{2\pi\varepsilon_0}\ln(r).$$

The *vector potential* of a (non-relativistic) moving point charge is the same as the scalar potential caused by the charge, multiplied by its velocity and a few constants. That's all it is: the vector potential is the scalar potential with a direction attached and its magnitude adjusted with some constants and the speed of the moving charge.

For an infinite, straight filament of charge in motion, its vector potential (when written in terms of the current I, where the vector version of the current is in the direction of the current flow) is

$$\vec{A}(r) = \frac{-\vec{I}\mu_0}{2\pi}\ln(r)$$

Replacing I by λv (and recalling the intro physics fact that $\mu_0\varepsilon_0 = 1/c^2$) gives

$$\vec{A}(r) = \frac{-\lambda\vec{v}\mu_0}{2\pi}\ln(r)$$

$$\vec{A}(r) = \mu_0 \varepsilon_0 V(r)\vec{v}$$

$$= \frac{V(r)}{c}\frac{\vec{v}}{c}.$$

The raison d'être of the vector potential is that its curl is the magnetic field.

You can see the importance of relativity in magnetism: note the v/c factor in the vector potential. The ratio of the scalar potential and the speed of light arises because of our archaic choice of units in electrodynamics. In an electromagnetic wave, the electric and magnetic fields carry the same energy, but a wave with a magnetic field amplitude of 1 Tesla will have an electric field amplitude of 299,792,458 volts per meter. That's a familiar number!

There's a subtle point that I've brushed under the rug. The potentials depend on the "retarded positions" of the source charges: if the instantaneous position of a charge is $\vec{r}(t)$, we need to include its contribution to the potential at the point $\vec{r}'$ based on where it *was*. Instead of determining the potential as $V(\vec{r}', \vec{r}(t))$, we must calculate it as $V(\vec{r}', \vec{r}(t_{\rm ret}))$, where the "retarded time" $t_{\rm ret}$ is

$$t_{\rm ret} = t - \frac{|\vec{r}' - \vec{r}(t_{\rm ret})|}{c}$$

The idea is fairly intuitive, but sometimes actually solving for $t_{\rm ret}$ can be complicated.

For the time being, we are going to ignore retardation effects, though they are important, and are what actually give rise to electromagnetic radiation.

Exercise 11.8: The curl of a current's vector potential

Imagine that a current I flows along the y-axis. In the y, z plane, the resulting vector potential is

$$\vec{A}(x = 0, y, z) = \frac{-I\mu_0}{2\pi}\ln(z)\hat{y}.$$

Calculate the curl of the vector potential in the y, z plane, then verify that it is the same as the magnetic field. By this, I mean, show that

$$\vec{\nabla} \times \vec{A}(x = 0, y, z) = \frac{I\mu_0}{2\pi z}\hat{x}$$

$$= \vec{B}(x = 0, y, z).$$

Solution

$$\vec{\nabla} \times \vec{A}(x, y, z)$$

$$= \left(\frac{\partial A_y}{\partial x} - \frac{\partial A_x}{\partial y} \right) \hat{z} + \left(\frac{\partial A_z}{\partial y} - \frac{\partial A_y}{\partial z} \right) \hat{x} + \left(\frac{\partial A_x}{\partial z} - \frac{\partial A_z}{\partial x} \right) \hat{y}$$

$$= \frac{\partial A_y}{\partial x} \hat{z} - \frac{\partial A_y}{\partial z} \hat{x} = -\frac{\partial A_y}{\partial z} \hat{x} = -\frac{\partial \frac{-I\mu_0}{2\pi} \ln(z)}{\partial z} \hat{x}$$

$$= \frac{I\mu_0}{2\pi z} \hat{x}$$

Summary

We discussed

- derivatives and partial derivatives;
- line integrals;
- curl;
- surface integrals;
- Stokes' theorem;
- vector potential.

Problem 1: Do young children still play with jacks?

A child's toy consists of six small spheres, each of mass m, attached to rods of length a. The rods are joined together so that two rods lie parallel to the x-axis, two parallel to the y-axis, and two parallel to the z-axis, as shown in the figure. The center of the toy is held fixed at the point (x_0, y_0, z_0), but the toy is able to rotate freely about its center without any constraints on the direction of its angular velocity.

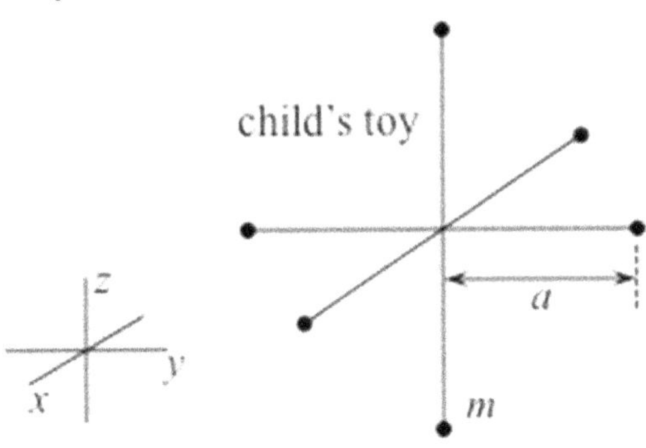

The toy is initially prevented from rotating and is immersed in a fluid flowing with velocity

$$\vec{v}(x, y, z) = v_x(x, y, z)\hat{x} + v_y(x, y, z)\hat{y} + v_z(x, y, z)\hat{z}$$

The force exerted by the fluid on a small sphere at the end of a rod, before the toy is allowed to rotate, is $\mu\vec{v}$.

(a) Write an expression for the torque exerted by the fluid on the toy. Keep in mind that torque is a vector, so your expression must yield a vector, not a scalar.

Solution

The support system will cancel any force on a sphere parallel to the rod holding it. So, let's crank this out:

local sphere coordinate	$\vec{r} \times \vec{F}$
$(a, 0, 0)$	$\mu a[v_y(x_0 + a, y_0, z_0)\hat{z} - v_z(x_0 + a, y_0, z_0)\hat{y}]$
$(-a, 0, 0)$	$-\mu a[v_y(x_0 - a, y_0, z_0)\hat{z} - v_z(x_0 - a, y_0, z_0)\hat{y}]$
$(0, a, 0)$	$\mu a[v_z(x_0, y_0 + a, z_0)\hat{x} - v_x(x_0, y_0 + a, z_0)\hat{z}]$
$(0, -a, 0)$	$-\mu a[v_z(x_0, y_0 - a, z_0)\hat{x} - v_x(x_0, y_0 - a, z_0)\hat{z}]$
$(0, 0, a)$	$\mu a[v_x(x_0, y_0, z_0 + a)\hat{y} - v_y(x_0, y_0, z_0 + a)\hat{x}]$
$(0, 0, -a)$	$-\mu a[v_x(x_0, y_0, z_0 - a)\hat{y} - v_y(x_0, y_0, z_0 - a)\hat{x}]$

As a result,

$$\begin{aligned}
\vec{\tau} = \vec{r} \times \vec{F} = {}&\mu a\hat{z}[v_y(x_0 + a, y_0, z_0) - v_y(x_0 - a, y_0, z_0) \\
&+ v_x(x_0, y_0 - a, z_0) - v_x(x_0, y_0 + a, z_0)] + \mu a\hat{x}[v_z(x_0, y_0 + a, z_0) \\
&- v_z(x_0, y_0 - a, z_0) + v_y(x_0, y_0, z_0 - a) - v_y(x_0, y_0, z_0 + a)] \\
&+ \mu a\hat{y}[v_z(x_0 - a, y_0, z_0) - v_z(x_0 + a, y_0, z_0) + v_x(x_0, y_0, z_0 + a) \\
&- v_x(x_0, y_0, z_0 - a)]
\end{aligned}$$

(b) In the approximation that a is very small compared to the distance over which the fluid velocity changes appreciably, it is possible to approximate your answer to (a) using the curl of the velocity. Do so:

$$v_y\left(x_0 + a, y_0, z_0\right) - v_y\left(x_0 - a, y_0, z_0\right) \approx 2a\frac{\partial V_y}{\partial x}, \quad \text{etc.},$$

so

$$\vec{\tau} \approx 2\mu a^2 \left\{ \hat{z}\left[\frac{\partial V_y}{\partial x} - \frac{\partial V_x}{\partial y}\right] + \hat{x}\left[\frac{\partial V_z}{\partial y} - \frac{\partial V_y}{\partial z}\right] + \hat{y}\left[\frac{\partial V_x}{\partial z} - \frac{\partial V_z}{\partial x}\right] \right\}$$

$$= 2\mu a^2 \nabla \times \vec{v}$$

Problem 2: Point charge magnetic field

The scalar potential of a (non-relativistic) point charge Q is $Q/(4\pi\varepsilon_0 r)$, where r is the distance from the charge to the observer. Imagine that at $t = 0$ a charge is passing through the origin, traveling along the y-axis. Neglecting retardation, calculate the magnitude and direction of the magnetic field it produces in the plane defined by $x = 0$.

Solution

$$V = \frac{Q}{4\pi\varepsilon_0 r}$$

so that

$$\vec{A} = \frac{Q}{4\pi\varepsilon_0 rc^2} \quad \vec{v} = \vec{A} = \frac{Q}{4\pi\varepsilon_0 rc^2}v\hat{y}.$$

The $x = 0$ plane means that

$$r = \sqrt{y^2 + z^2}$$

so that

$$\vec{A} = \frac{Qv\hat{y}}{4\pi\varepsilon_0 c^2}\frac{1}{\sqrt{y^2 + z^2}}$$

and

$$\vec{\nabla} \times \vec{A} = \frac{Qv}{4\pi\varepsilon_0 c^2} \left[\left(\frac{\partial A_y}{\partial x} - \frac{\partial A_x}{\partial y} \right) \hat{z} + \left(\frac{\partial A_z}{\partial y} - \frac{\partial A_y}{\partial z} \right) \hat{x} \right.$$
$$\left. + \left(\frac{\partial A_x}{\partial z} - \frac{\partial A_z}{\partial x} \right) \hat{y} \right]$$
$$= \frac{Qv}{4\pi\varepsilon_0 c^2} \left[(0-0)\hat{z} + \left(0 - \frac{\partial \left(y^2 + z^2 \right)^{-1/2}}{\partial z} \right) \hat{x} + (0-0)\hat{y} \right]$$
$$= \frac{Qv}{4\pi\varepsilon_{c_0} c^2} \frac{z}{r^3} \hat{x}$$

Recall that

$$\varepsilon_0 c^2 = 1/\mu_0 \text{ so}$$

$$\vec{\nabla} \times \vec{A} = \frac{Qv\mu_0}{4\pi r^2} \frac{z}{r} \hat{x}$$

Think about the geometry a little: $z/r = \sin\theta$, so we can write

$$\vec{B} = \vec{\nabla} \times \vec{A} = \frac{Qv\mu_0}{4\pi r^2} \sin\theta \hat{x}$$

Problem 3: Mrs. Rumpdock's green slime

The manufacturer of a particular brand of shampoo used to advertise the merits of its product by dropping "pearls" into bottles of shampoo. The viscosity of the gooey liquid caused the beads to fall slowly, reaching terminal velocity in a small fraction of a second.

Hoping to capitalize on the manufacturer's large advertising budget, Mrs. Rumpdock formulates a copycat product of her own from vodka, moldy bread (for color), and cornstarch. Before filming a commercial, she notes that the cornstarch will not remain suspended in the vile fluid for more than a few seconds before sinking to the bottom of the production tank. A hooligan to whom she had sold a chemical engineering degree advises her to stir the green glop in the tank while shooting the commercial. She does so.

The bottom of the cylindrical tank is at the origin; the enormous tank contains fake shampoo up to a height of $z = 10$ m. After some experimentation with electric charge (spray painted onto the surface of her "pearls") and strong electric fields, Rumpdock is able to cause the bead to travel vertically downwards along the z-axis. But she notes that viscous forces cause the bead to spin wildly during its downward journey.

The stirring mechanism induces a position-dependent velocity to the fluid in the tank, with

$$\vec{v}_{\text{fluid}}(x, y, z) = v_0 \sin(\pi z)\hat{x} + v_0 \cos(\pi z)\hat{y}$$

where $v_0 > 0$.

Naturally, the **torque** exerted by the fluid on the bead will increase with the bead's diameter, as well as the parameter v_0. But the torque will also depend on the curl of the velocity field, very much as it did in the first problem! You should assume that viscous forces immediately stop the bead from spinning when it enters regions in which the curl is zero.

Determine the direction of the bead's rotation axis as a function of z.

Solution

The direction of the rotation axis will be the same as the direction of the torque, which is the same as the direction of the curl of the fluid's velocity field. (Many components of the curl are zero!)

$$\vec{\nabla} \times \vec{v} = \left[\left(\frac{\partial v_y}{\partial x} - \frac{\partial v_x}{\partial y} \right) \hat{z} + \left(\frac{\partial v_z}{\partial y} - \frac{\partial v_y}{\partial z} \right) \hat{x} + \left(\frac{\partial v_x}{\partial z} - \frac{\partial v_z}{\partial x} \right) \hat{y} \right]$$

$$= [(0 - 0)\hat{z} + (0 + v_0\pi \sin(\pi z))\hat{x} + (v_0\pi \cos(\pi z) - 0)\hat{y}]$$

so the axis is parallel to $\sin(\pi z)\hat{x} + \cos(\pi z)\hat{y}$. Identifying θ as πz, we can write this as $\sin(\theta)\hat{x} + \cos(\theta)\hat{y}$, with the angle of the rotation axis as relative to the positive y-axis.

Problem 4: Hikalope's rocket sled

In order to test a purported treatment for detached retinas, Mrs. Rumpdock instructs her daughter Hikalope to assemble a replica of the rocket sled used by Col. John Stapp ("the fastest man on Earth") for his groundbreaking (and perilous) studies of pilot survivability. Rumpdock provides her with plans after opening a $10 million life insurance policy on her dimwitted adult offspring.

First, some history. In Stapp's work during the 1950s, his rocket sled (with Stapp on board) would routinely exceed speeds of 400 miles per hour, then stop suddenly when scoops mounted below the sled dipped into a trough of water placed between the rails at the end of the track. In his fastest run, Stapp hit 632 mph and suffered a deceleration of 46 g. It was dangerous work, and Stapp was frequently injured.

In the following photo,[2] Stapp is in the middle of a run at Edwards Air Force Base. There are, of course, YouTube videos.

In Rumpdock's plans for Hikalope's sled, the rocket booster and braking mechanism are combined into a single system in which the rocket thrust is vectored—directed in a controllable fashion—so that the thrust angle varies from 0 to $-\pi$ radians as the sled travels down the track. Naturally, while $|\theta| < \pi/2$, the sled speeds up, but while $\pi/2 < |\theta| < 3\pi/2$, the sled slows down. See the following figure, drawn at an instant when $\theta \sim -\pi/6$ (about $-30°$).

---

[2]See https://en.wikipedia.org/wiki/Rocket_sled, visited February 14, 2025.

The gimbal (pivot mechanism) joining the rocket motor to the sled is programmed to change the thrust angle as the sled's position changes so that $\theta(x) = -\pi x/1200$ radians. The combined mass of sled and doomed offspring is $1{,}000\,\text{kg}$, while the sled's rocket motor has a thrust of $2{,}000\,\text{N}$.

The sled, with Hikalope strapped to it, begins its first (and final) run at $x = 0$. Unfortunately, Rumpdock has installed a brick wall built across the rails at $x = 1000\,\text{m}$.

How fast is Hikalope traveling when she hits the wall?

Solution

$$\Delta K = \int_0^{1000} \vec{F} \cdot \vec{dx} = \int_0^{1000} 2000 \cos\left(\frac{\pi x}{1200}\right) dx$$

$$= -2000 \left(\frac{1200}{\pi}\right) \sin\left(\frac{\pi x}{1200}\right)\Bigg|_0^{1000} = \frac{1.2 \times 10^6}{\pi}$$

$$= \frac{1}{2}mv^2 \quad \text{so} \quad v = \sqrt{\frac{2.4 \times 10^6}{\pi}}$$

$$= 27.6\,\text{meters per second} = 61.74\,\text{mph}.$$

Grand Canyon, United States

Unit 12

Gradient, Divergence, Surface Integrals, the Divergence Theorem, and Gauss's Law

A reminder

See the last problem in Unit 2 for a reminder about partial derivatives. To jog your memory, the definition of a partial derivative of a function of several variables is quite similar to the definition of the derivative of a function of one variable. It is this:

$$\frac{\partial f(x, y, z, w, \ldots)}{\partial x}$$
$$= \lim_{\substack{\Delta x \to 0 \\ y, z, w, \ldots \text{constant}}} \frac{f(x + \Delta x, y, z, w, \ldots) - f(x, y, z, w, \ldots)}{\Delta x}$$

275

The machinery is easy to use. Just treat all "the other" variables as if they were constants and take the derivative of the function in the usual way.

The gradient

The gradient of a scalar function is

$$\vec{\nabla} f(x, y, z) \equiv \frac{\partial f}{\partial x}\hat{x} + \frac{\partial f}{\partial y}\hat{y} + \frac{\partial f}{\partial z}\hat{z}$$

It is a vector function, of course, and points in the direction in which f changes most rapidly. You show this in the following exercise.

Exercise 12.1: The meaning of the gradient

Imagine that in a small region around the point (x_0, y_0, z_0) the temperature T is well approximated as

$$T(x, y, z) \approx T_0 + ax + by$$

Under an infinitesimal displacement $\vec{\delta} = \delta \cos\theta \hat{x} + \delta \sin\theta \hat{y}$ away from (x_0, y_0, z_0), the value of the temperature changes slightly. Calculate the gradient of T at (x_0, y_0, z_0), then show that

(a) the change in T is approximately $dT = \vec{\nabla} T \cdot \vec{\delta}$.

Solution

$$T = T_0 + ax + by \quad \text{so} \quad \vec{\nabla} T = \left(\hat{x}\frac{\partial}{dx} + \hat{y}\frac{\partial}{dy} + \hat{z}\frac{\partial}{dz} \right) T$$

$$= a\hat{x} + b\hat{y}$$

Imagine moving first along x, then along y. The change when moving by

$$\delta_x \hat{x} \text{ is } \delta_x \frac{\partial T}{\partial x}\bigg|_{x=x_0, y=y_0, z=z_0} = a\delta_x$$

The change when moving through

$$\delta_y \hat{y} \text{ is } \delta_y \frac{\partial T}{\partial y}\bigg|_{x=x_0, y=y_0, z=z_0} = b\delta_y$$

The net change, then, is

$$a\delta_x + b\delta_y = (\vec{\nabla}T) \cdot \vec{\delta} = \Delta T$$

(b) Also, show that choosing θ so that $\tan\theta = b/a$ maximizes the change in T. (This shouldn't be a surprise in the light of the previous result.)

Solution

For $\delta_x = \delta\cos\theta$ and $\delta_y = \delta\sin\theta$ we have $\Delta T = \delta(a\cos\theta + b\sin\theta)$. Maximizing ΔT with respect to θ : $\frac{d\Delta T}{d\theta} = 0$ or $\frac{d(a\cos\theta+b\sin\theta)}{d\theta} = -a\sin\theta + b\cos\theta = 0$. This is satisfied for $\frac{\sin\theta}{\cos\theta} = \tan\theta = \frac{b}{a}$.

Exercise 12.2: Now in three dimensions

It is also true in three dimensions, of course, that $df = \vec{\nabla}f \cdot \vec{\delta}$ for f any scalar function of position.

Imagine that you are swimming underwater in a lake with known temperature distribution:

$$T(x, y, z) \approx T_0 + \kappa\sin(\alpha x)\cos(\beta y)e^{\gamma z}.$$

Upon arriving at the point (x_0, y_0, z_0), you realize that the water has become uncomfortably warm and that you risk the same fate as that befalling a lobster at Mrs. Rumpdock's Clam Shack.

Your maximum underwater swimming speed is v, and you set off in a direction that drops the temperature to which you are exposed most quickly. How rapidly does the temperature you are experiencing change when you first begin to flee the hot spot?

Solution

The fastest direction (and rate) of change of temperature is determined by taking the gradient of the temperature field:

$$\vec{\nabla}T = \kappa\alpha\cos(\alpha x)\cos(\beta y)e^{\gamma z}\hat{x} - \kappa\beta\sin(\alpha x)\sin(\beta y)e^{\gamma z}\hat{y}$$
$$+ \kappa\gamma\sin(\alpha x)\cos(\beta y)e^{\gamma z}\hat{z}$$

If we swim with speed v in the opposite direction to that of the gradient, we'll have a temperature change of

$$-v\kappa e^{\gamma z}(\alpha\cos(\alpha x)\cos(\beta y) - \beta\sin(\alpha x)\sin(\beta y) + \gamma\sin(\alpha x)\cos(\beta y))$$

degrees per second: the change in temperature when we move a distance $\vec{ds}$ is $\vec{\nabla T} \cdot \vec{ds}$ and the rate of change will be $\frac{\vec{\nabla T} \cdot \vec{ds}}{dt}$ degrees per second.

Exercise 12.3: Force and potential energy

You probably remember from intro physics that the work done by a force $\vec{F}$ acting through a small displacement $\vec{\delta}$ is $dW = \vec{F} \cdot \vec{\delta}$.

Imagine that you apply a position-dependent force $\vec{F}$ to an object, lifting it in opposition to a conservative force such as gravity. The force you apply is just barely sufficient to counter the conservative force so that the object moves slowly, with negligible kinetic energy. The work you do causes the object's potential energy to change, of course: $dW = dU$.

Explain in a couple of sentences (which you should write down on paper) why the connection between the conservative force acting on the object and the potential energy is $\vec{F}_{conservative} = -\vec{\nabla}U$. Your explanation should include the reason for the appearance of the negative sign.

Solution

We have $dW = \vec{F} \cdot \vec{\delta} = dU$. From before, we know that the gradient of a scalar function dotted with a small displacement is equal to the change in the function. As a result, $\vec{F} \cdot \vec{\delta} = dU = \vec{\nabla}U \cdot \vec{\delta}$, so the force *you* exert must be the same as $\vec{\nabla}U$. But since you are applying a force that exactly cancels the conservative force, we have $\vec{F}_{cons} = -\vec{F} = -\overrightarrow{\nabla U}$. Therefore $\vec{F}_{cons} = -\overrightarrow{\nabla U}$.

Flux, flow, and divergence

A wet sponge in a dry environment will lose water to evaporation: the rate at which water molecules flow out of the sponge will exceed the rate at which water from the air flows into the sponge.

If the difference in the outflow and inflow rates is $J\,\mathrm{kg/m^2/s}$, we can calculate the total rate of water loss, in kg/s, by multiplying the total surface area of the sponge by J.

Exercise 12.4: Flux

A very small rectangular wire frame of side lengths dx and dy is placed in a tank of fluid of (position-dependent) density $\rho(x, y, z)\,\mathrm{kg/m^3}$. The frame is centered at the point (x_0, y_0, z_0) and lies parallel to the x, y plane. The velocity of the fluid flow at the point (x, y, z) is

$$\vec{v}(x, y, z) = v_x(x, y, z)\hat{x} + v_y(x, y, z)\hat{y} + v_z(x, y, z)\hat{z}$$

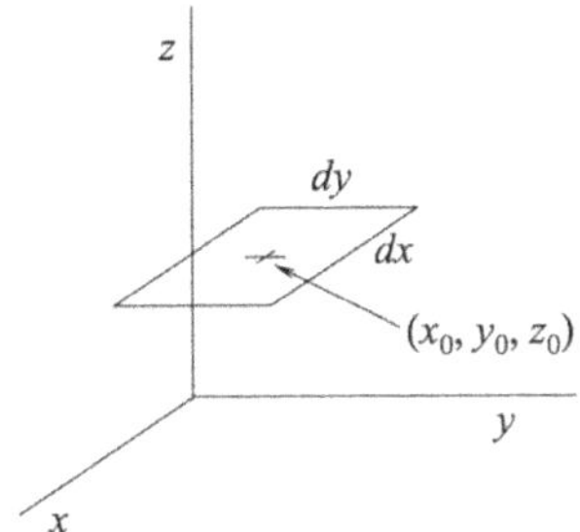

The wire frame is so small that it is safe to approximate the fluid's density and velocity as the density and velocity at its center.

What is the fluid's flow rate, in kg/s, through the wire frame?

Solution

Only the z-component of velocity will drive the fluid through the plane of the wire loop. In time dt, all molecules within $dz = v_z dt$ will pass through the same plane as that in which the loop lies. As a result, the mass of all the molecules passing through the loop in time dt will be $dm = \rho\, dx\, dy\, dz = \rho\, dx\, dy\, v_z dt$. So, the flow rate is $dm/dt = \rho v_z dx dy$.

Exercise 12.5: Now as a vector

For convenience, we can assign a direction to the wire frame and refer to its area (which is $dA = dx dy$) as a vector. The most useful assignment of direction is perpendicular to the plane of the wire frame, as we did last unit. Re-express your answer to the previous exercise in terms of $\rho, \vec{v}$, and $d\vec{A}$.

Solution

Define a direction attached to an area patch this way: $d\vec{A} = \hat{z}\, dx dy$.

We can pick out the z-component of fluid velocity this away: $\frac{dm}{dt} = \rho\vec{v} \cdot d\vec{A}$. Even more compact: $\frac{dm}{dt} = \vec{J} \cdot d\vec{A}$.

Current density as a vector: Conservation of mass

The current density is the amount of stuff per second per square meter that is flowing at points in our fluid bath. It increases with speed and with density and is the product of those two quantities:

$$\vec{J}(x, y, z) = \rho(x, y, z)\vec{v}(x, y, z)$$

Let's work up a differential equation that expresses the idea that mass is neither created nor destroyed in our fluid.

We investigate the inflow/outflow of mass in a small volume $dV = dx\,dy\,dz$ centered on the point (x, y, z). (I'm only going to draw it in two dimensions, to simplify the picture.)

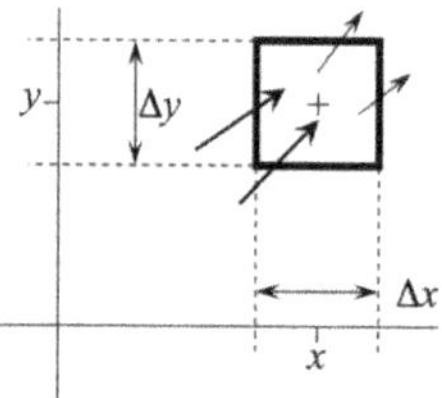

I've drawn the fluid flow so that the fluid's velocity (which is entirely in the xy plane) going *into* the box from the bottom and left sides is greater than the velocity *out* of the box through the top and right sides. As a result, we expect fluid to build up in the box, so the density should increase.

We can take as the average velocity for the fluid that enters the left side of the box the exact fluid velocity at the center of the left side:

$$\langle \vec{v}_{left} \rangle \approx \vec{v}\left(x - \frac{\Delta x}{2}, y, z\right)$$

The volume of fluid that flows into the box from the left side during time dt is shown as a shaded region in the following diagram. (Once again, everything in the figure lies in the xy plane.)

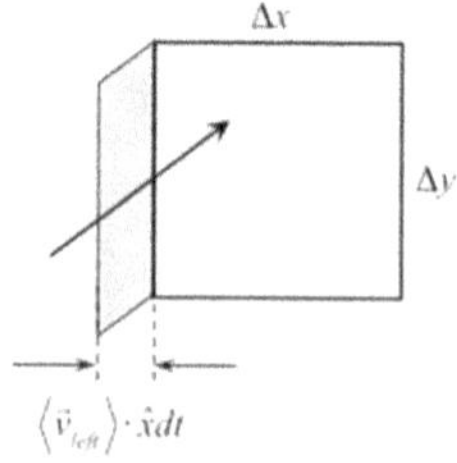

The volume of the region of fluid that flows into the left side of the box is

$$[\langle \vec{v}_{left} \rangle \cdot \hat{x}dt]\Delta y \Delta z \approx \vec{v}\left(x - \frac{\Delta x}{2}, y, z\right) \cdot \hat{x}dt\Delta y \Delta z$$

since the thickness of it in the x-direction is $\vec{v}dt \cdot \hat{x}$.

The volume that flows in from the bottom is $\vec{v}\left(x, y - \frac{\Delta y}{2}, z\right) \cdot \hat{y}dt\Delta x \Delta z$.

The volume that flows out of the top is $\vec{v}\left(x, y + \frac{\Delta y}{2}, z\right) \cdot \hat{y}dt\Delta x \Delta z$.

The volume that exits the right side is $\vec{v}\left(x + \frac{\Delta x}{2}, y, z\right) \cdot \hat{x}dt\Delta y \Delta z$.

There are additional terms for the front and back of the box, of course.

The *mass* that flows in from the left is the product of the local density and the in-flowing volume:

$$dm_{left} \approx \rho\left(x - \frac{\Delta x}{2}, y, z, t\right) \vec{v}\left(x - \frac{\Delta x}{2}, y, z, t\right) \cdot \hat{x}dt\Delta y \Delta z$$

$$= \vec{J}\left(x - \frac{\Delta x}{2}, y, z, t\right) \cdot \hat{x}dt\Delta y \Delta z$$

Note that I'm approximating density and velocity for the left side using the value of ρ and $\vec{v}$ at the center of the left face of the volume element.

Adding up the in-flowing and out-flowing mass for all six faces of the volume element (and omitting explicit time dependence, to save myself the effort of writing, "t" everywhere) gives

$$dm = \rho\left(x - \frac{\Delta x}{2}, y, z\right) v_x\left(x - \frac{\Delta x}{2}, y, z\right) dt\Delta y \Delta z$$

$$- \rho\left(x + \frac{\Delta x}{2}, y, z\right) v_x\left(x + \frac{\Delta x}{2}, y, z\right) dt\Delta y \Delta z$$

$$+ \rho\left(x, y - \frac{\Delta y}{2}, z\right) v_y\left(x, y - \frac{\Delta y}{2}, z\right) dt\Delta x \Delta z$$

$$- \rho\left(x, y + \frac{\Delta y}{2}, z\right) v_y\left(x, y + \frac{\Delta y}{2}, z\right) dt\Delta x \Delta z$$

$$+ \rho\left(x, y, z - \frac{\Delta z}{2}\right) v_z\left(x, y, z - \frac{\Delta z}{2}\right) dt\Delta x\Delta y$$

$$- \rho\left(x, y, z + \frac{\Delta z}{2}\right) v_z\left(x, y, z + \frac{\Delta z}{2}\right) dt\Delta x\Delta y$$

Of course, I could have written this in terms of the current, saving myself some typing:

$$dm = \left[J_x\left(x - \frac{\Delta x}{2}, y, z\right) - J_x\left(x + \frac{\Delta x}{2}, y, z\right)\right] dt\Delta y\Delta z$$

$$+ \left[J_y\left(x, y - \frac{\Delta y}{2}, z\right) - J_y\left(x, y + \frac{\Delta y}{2}, z\right)\right] dt\Delta x\Delta z$$

$$+ \left[J_z\left(x, y, z - \frac{\Delta z}{2}\right) - J_z\left(x, y, z + \frac{\Delta z}{2}\right)\right] dt\Delta x\Delta y$$

Let's assume that the tiny box is so small that we can rename Δx, etc. as dx, etc.

Exercise 12.6: In terms of partial derivatives

Rewrite the above expression for dm in terms of partial derivatives of the components of the current $\vec{J}$.

Solution

$$\lim_{\Delta x \to 0} \frac{J_x\left(x + \frac{\Delta x}{2}, y, z\right) - J_x\left(x - \frac{\Delta x}{2}, y, z\right)}{\Delta x} = \frac{\partial J_x(x, y, z)}{\partial x}$$

and similarly for y, z. As a result,

$$dm = -\frac{\partial J_x}{\partial x}\Delta x\Delta y\Delta z dt - \frac{\partial J_y}{\partial y}\Delta x\Delta y\Delta z dt - \frac{\partial J_z}{\partial z}\Delta x\Delta y\Delta z dt$$

$$= -\Delta x\Delta y\Delta z dt \left[\frac{\partial J_x}{\partial x} + \frac{\partial J_y}{\partial y} + \frac{\partial J_z}{\partial x}\right]$$

The divergence operator

The divergence operator $\vec{\nabla}$ is defined this way:

$$\vec{\nabla} \equiv \left[\hat{x}\frac{\partial}{\partial x} + \hat{y}\frac{\partial}{\partial y} + \hat{z}\frac{\partial}{\partial z}\right]$$

Exercise 12.7: In terms of partial derivatives

Show that your answer to the previous exercise is the same as

$$dm = -dt\,dx\,dy\,dz\,\vec{\nabla}\cdot\vec{J}$$

Solution

From the definition, $\vec{\nabla}\cdot\vec{J} = \frac{\partial J_x}{\partial x} + \frac{\partial J_y}{\partial y} + \frac{\partial J_z}{\partial x}$ so the earlier result is the same as $-\Delta x\Delta y\Delta z\,dt\,\vec{\nabla}\cdot\vec{J}$.

Recall that the mass inside the small volume $dV = dx\,dy\,dz$ is $m = \rho(x, y, z, t)dx\,dy\,dz$.

As a result, the time derivative of the mass inside the volume is

$$\frac{dm}{dt} = \frac{d[\rho(x, y, z, t)dx\,dy\,dz]}{dt} = \frac{\partial\rho(x, y, z, t)}{\partial t}dx\,dy\,dz$$

We are holding the volume dV fixed in space, so x, y, and z never change. That's why I turned the total time derivative into a partial derivative.

We also know that $dm = -dt\,dx\,dy\,dz\,\vec{\nabla}\cdot\vec{J}$, so it must be true that

$$\frac{dm}{dt} = -\vec{\nabla}\cdot\vec{J}dx\,dy\,dz$$

As a result,

$$\frac{\partial\rho(x, y, z, t)}{\partial t} = -\vec{\nabla}\cdot\vec{J}(x, y, z, t)$$

or

$$\frac{\partial\rho(x, y, z, t)}{\partial t} + \vec{\nabla}\cdot\vec{J}(x, y, z, t) = 0$$

The divergence of the current tells us about the net outflow (or inflow) of stuff at a point in space. An outflow causes the density to decrease.

This equation expresses the fact that mass is conserved in our fluid, even though the density and velocity of flow can change with time. Keep in mind that I worked this up using a volume element dV that was *fixed* in space: with x, y, z held constant, our time derivative is a partial derivative $\partial/\partial t$, not a total derivative d/dt.

The divergence theorem

Imagine that we stack up a large number of tiny boxes whose walls are transparent to the flowing fluid. The fluid flowing out of one face of a box will enter the adjacent face of its neighbor, so the loss of mass in one box will contribute to an increase in the masses of its neighbors, as long as all the faces of that box are in contact with the faces of neighboring boxes.

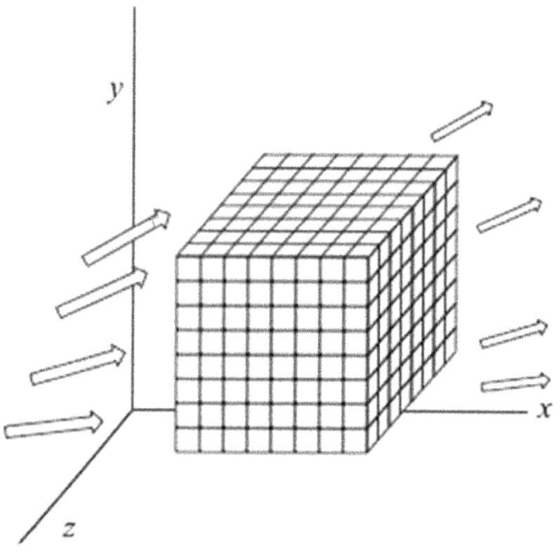

But the current through the box faces that lie on the surface of the pile of boxes **will** contribute to the net gain or loss of mass in the pile of boxes since the in/outflow from an exterior face is not canceled by the out/inflow from a neighbor.

Let's say the mass inside the ith box is m_i, while the mass of the entire assembly of boxes is M. The sum of the changes in the fluid mass inside each box will, of course, equal the net change in the fluid mass contained in the entire stack of boxes. We have

$$M = \sum_{i=1}^{i=N} m_i$$

$$\frac{dM}{dt} = \sum_{i=1}^{i=N} \frac{dm_i}{dt} = \sum_{i=1}^{i=N} -\vec{\nabla} \cdot \vec{J}(x_i, y_i, z_i) dx dy dz$$

Box faces that lie on the surface of the pile of boxes are the only ones that contribute to the net gain or loss of mass. In equations, this means

$$\frac{dM}{dt} = \sum_{\substack{\text{all} \\ \text{exterior} \\ \text{faces } k}} -\vec{J}(x_k, y_k, z_k) \cdot d\vec{A}_k$$

$$= \sum_{\substack{\text{all} \\ \text{blocks } i}} -\vec{\nabla} \cdot \vec{J}(x_i, y_i, z_i) dx dy dz$$

The calculus version of this, as we let the number of boxes go to infinity while shrinking their individual volumes, is the Divergence Theorem:

$$\oint_{\text{surface}} \vec{J}(x, y, z) \cdot d\vec{A} = \int_{\substack{\text{volume} \\ \text{bounded} \\ \text{by surface}}} \vec{\nabla} \cdot \vec{J}(x, y, z) dx dy dz$$

It is true for **all** vector fields, not just those describing fluid flows.

Exercise 12.8: Quickie Gauss's law

You learned in intro E&M that the flux of electric field lines leaving a surface equals the amount of charge contained inside the surface, scaled by a factor of ε_0. If we represent the charge density as $\rho(x, y, z)$, we can write Gauss's law this way:

$$\oint_{\text{surface}} \vec{E}(x, y, z) \cdot d\vec{A} = \frac{Q_{total}}{\varepsilon_0}$$

$$= \int_{\substack{\text{volume} \\ \text{bounded} \\ \text{by surface}}} \frac{1}{\varepsilon_0} \rho(x, y, z) dx dy dz$$

Prove that the divergence of the electric field at the point (x, y, z) is the same thing as the ratio of the charge density at that point and ε_0. In equations, show that Gauss's law is equivalent to the statement $\vec{\nabla} \cdot \vec{E} = \rho/\varepsilon_0$.

Solution

From the divergence theorem,

$$\oint \vec{E} \cdot \overrightarrow{dA} = \int_{\text{volume}} \vec{\nabla} \cdot \vec{E} dV$$

But it's also true that

$$\oint \vec{E} \cdot \overrightarrow{dA} = \int_{\text{volume}} \frac{\rho}{\varepsilon_0} dV$$

As a result,

$$\vec{\nabla} \cdot \vec{E} = \frac{\rho}{\varepsilon_0}$$

A reference for you: Information about two non-Cartesian coordinate systems

The unit vectors change which way they point as $\vec{x}$ moves around in space if we are working in a non-Cartesian coordinate system. This complicates the taking of derivatives.

Cylindrical coordinates: the unit vectors point in the directions of increasing r, θ, z

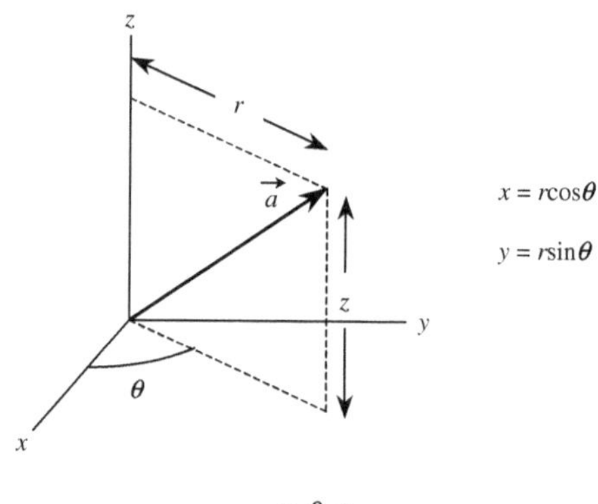

$$r, \theta, z$$

Spherical coordinates: the unit vectors point in the directions of increasing r, θ, ϕ

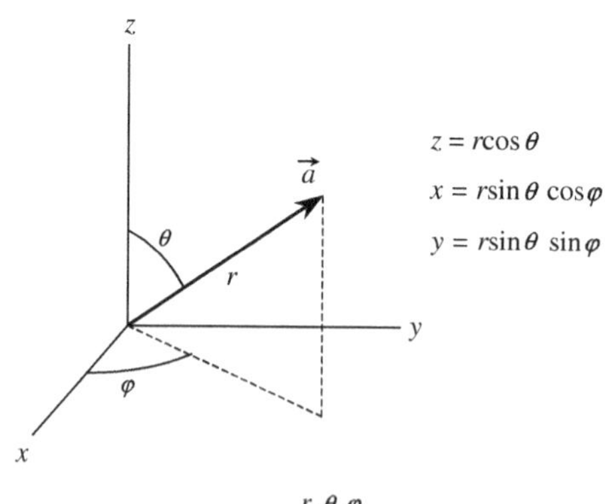

$$r, \theta, \varphi$$

Gradient in cylindrical coordinates

$$\vec{\nabla} f(r,\theta,z) \equiv \left[\hat{r}\frac{\partial}{\partial r} + \frac{\hat{\theta}}{r}\frac{\partial}{\partial \theta} + \hat{z}\frac{\partial}{\partial z} \right] f(r,\theta,z)$$

Gradient in spherical coordinates

$$\vec{\nabla} f(r,\theta,\varphi) \equiv \left[\hat{r}\frac{\partial}{\partial r} + \frac{\hat{\theta}}{r}\frac{\partial}{\partial \theta} + \frac{\hat{\varphi}}{r\sin\theta}\frac{\partial}{\partial \varphi} \right] f(r,\theta,\varphi)$$

Divergence in cylindrical coordinates

$$\vec{A}(r,\theta,z) \equiv A_r(r,\theta,z)\hat{r} + A_\theta(r,\theta,z)\hat{\theta} + A_z(r,\theta,z)\hat{z}$$

$$\vec{\nabla}\cdot\vec{A}(r,\theta,z) \equiv \frac{1}{r}\frac{\partial(rA_r)}{\partial r} + \frac{1}{r}\frac{\partial A_\theta}{\partial \theta} + \frac{\partial A_z}{\partial z}$$

Divergence in spherical coordinates

$$\vec{A}(r,\theta,\varphi) \equiv A_r(r,\theta,\varphi)\hat{r} + A_\theta(r,\theta,\varphi)\hat{\theta} + A_z(r,\theta,\varphi)\hat{\varphi}$$

$$\vec{\nabla}\cdot\vec{A}(r,\theta,\varphi) \equiv \frac{1}{r^2}\frac{\partial(r^2 A_r)}{\partial r} + \frac{1}{r\sin\theta}\frac{\partial(A_\theta\sin\theta)}{\partial \theta} + \frac{1}{r\sin\theta}\frac{\partial A_\varphi}{\partial \varphi}$$

Curl in cylindrical coordinates

$$\vec{\nabla}\times\vec{A}(r,\theta,z) \equiv \left[\frac{1}{r}\frac{\partial A_z}{\partial \theta} - \frac{\partial A_\theta}{\partial z} \right]\hat{r} + \left[\frac{\partial A_r}{\partial z} - \frac{\partial A_z}{\partial r} \right]\hat{\theta}$$

$$+ \frac{1}{r}\left[\frac{\partial(rA_\theta)}{\partial r} - \frac{\partial A_r}{\partial \theta} \right]\hat{z}$$

Curl in spherical coordinates

$$\vec{\nabla}\times\vec{A}(r,\theta,\phi) \equiv \frac{1}{r\sin\theta}\left[\frac{\partial(A_\phi\sin\theta)}{\partial \theta} - \frac{\partial A_\theta}{\partial \phi} \right]\hat{r}$$

$$+ \frac{1}{r}\left[\frac{1}{\sin\theta}\frac{\partial A_r}{\partial \phi} - \frac{\partial(rA_\phi)}{\partial r} \right]\hat{\theta} + \frac{1}{r}\left[\frac{\partial(rA_\theta)}{\partial r} - \frac{\partial A_r}{\partial \theta} \right]\hat{\phi}$$

Volume element for volume integration in cylindrical coordinates

$$dV = dr(rd\theta)dz = rdrd\theta dz$$

$$\int_{\text{volume}} f(r,\theta,z)dV = \int_{r=r_i}^{r=r_f} r \left[\int_{z=z_i}^{z=z_f} \left(\int_{\theta=0}^{\theta=2\pi} f(r,\theta,z)d\theta \right) dz \right] dr$$

Volume element for volume integration in spherical coordinates

$$dV = dr(rd\theta)(r\sin\theta d\phi) = r^2 \sin\theta dr d\theta d\phi$$

$$\int_{\text{volume}} f(r,\theta,\phi)dV$$

$$= \int_{r=r_i}^{r=r_f} r^2 \left[\int_{\theta=0}^{\theta=\pi} \sin\theta \left(\int_{\phi=0}^{\phi=2\pi} f(r,\theta,\phi)d\phi \right) d\theta \right] dr$$

Outward-facing area elements for surface integration in cylindrical coordinates

$$d\vec{A} = (rd\theta)dz\hat{r} = rd\theta dz\hat{r} \quad \text{(radial)}$$

$$d\vec{A} = \pm dr(rd\theta)\hat{z} = \pm rdrd\theta\hat{z} \quad \text{(end caps)}$$

Outward-facing area element for surface integration in spherical coordinates

$$d\vec{A} = (rd\theta)(r\sin\theta d\phi)\hat{r} = r^2 \sin\theta d\theta d\phi\hat{r}$$

Summary

We discussed

- gradients;
- force and potential energy;
- flux, flow, and divergence;
- current density and conservation of mass;
- the divergence operator and the divergence theorem;
- Gauss's law.

Problem 1: Working with non-Cartesian coordinates

It is common for us to confront problems in which the systems under consideration show certain geometrical symmetries. Probably spherical and cylindrical symmetries are the most common. Though it takes some getting used to, it is usually best to work with a non-Cartesian coordinate system in these analyses.

We've been discussing the differential operators associated with the gradient of a scalar function, the divergence of a vector function, and the curl of a vector function. Their physical meanings are unchanged when we switch coordinate systems—the curl tells us, for example, the amount of torque that might act on a submerged object, while the gradient tells us the direction (and magnitude) of most rapid change for a scalar function.

A complicating factor in working with cylindrical and spherical coordinate systems is that the directions of the unit vectors can change as we move around in space. See the expressions for various differential operators at the end of this unit's in-class material.

(a) The electric field for a point charge Q at the origin is

$$\vec{E} = \frac{Q}{4\pi\varepsilon_0 r^2}\hat{r}$$

(1) Show that (for $r \neq 0$) $\vec{\nabla} \cdot \vec{E} = 0$
(2) Show that (for $r \neq 0$) $\vec{\nabla} \times \vec{E} = 0$

Solution

Do these in spherical coordinates. See the expression a little ways back for the divergence in spherical coordinates. Since the electric field is radial and only depends on r, most derivatives will be zero. Plug in:

(a) 1: $r^2 E_r = \frac{Q}{4\pi\varepsilon_0}\hat{r}$, which is constant, so $\frac{1}{r^2}\frac{\partial r^2 E_r}{\partial r} = 0$ so that $\vec{\nabla}\cdot\vec{E} = 0$ as long as we're not at $r = 0$.

(a) 2: Same idea. Only the $\frac{\partial E_r}{\partial \varphi}$ and $\frac{\partial E_r}{\partial \theta}$ terms might contribute, but since E_r only depends on r, both of those derivatives are zero. Therefore, $\vec{\nabla} \times \vec{E} = 0$.

(b) Imagine that the electric field created by a spherically symmetric charge distribution is

$$\vec{E} = \beta r e^{-r/r_0}\hat{r}$$

where β and r_0 are positive constants. Calculate the charge density ρ that gives rise to this field.

Solution

$\vec{\nabla} \cdot \vec{E} = \frac{\rho}{\varepsilon_0}$ so take the divergence in spherical coordinates. Plug in:

$$\frac{\rho}{\varepsilon_0} = \frac{1}{r^2}\frac{\partial}{\partial r}(r^2 \beta r e^{-r/r_0}) = \frac{\beta}{r^2}\frac{\partial}{\partial r}(r^3 e^{-r/r_0})$$

$$= \beta e^{-r/r_0}\left[3 - \frac{r}{r_0}\right]$$

(c) In a static universe, in which all charges are at rest, it is always true that $\vec{\nabla} \times \vec{E} = 0$. Determine which of the following electric fields might be able to exist in a static universe. (ψ is a constant.)

$$(1)\ \vec{E}(x, y, z) = \psi[x\hat{x} + 2y\hat{y}]$$

$$(2)\ \vec{E}(x, y, z) = \psi[y\hat{x} + x\hat{y}]$$

$$(3)\ \vec{E}(x, y, z) = \psi[y\hat{x} - x\hat{y}]$$

$$(4)\ \vec{E}(r, \theta, z) = \psi[r\hat{r} + r\hat{\theta}]$$

Solution

Take the curls: this is a plug-and-chug exercise. The first two electric fields have zero curl, but the last two do not and cannot be produced by a static distribution of charge.

Problem 2: Vector calculus identities

Recall that you can safely change the order of differentiation when working with well-behaved functions

$$\frac{\partial}{\partial x}\left(\frac{\partial f(x, y)}{\partial y}\right) = \frac{\partial}{\partial y}\left(\frac{\partial f(x, y)}{\partial x}\right)$$

in two dimensions. This also holds true for functions of x, y, and z.

(a) Use this to prove that the curl of the gradient of a scalar function is always zero:

$$\vec{\nabla} \times (\vec{\nabla} U(x, y, z)) = 0$$

Solution

Just crank out all the derivatives in Cartesian coordinates:

$$\vec{\nabla} \times (\vec{\nabla} U(x, y, z)) = \hat{x} \left[\frac{\partial}{\partial y} \left(\frac{\partial U}{\partial z} \right) - \frac{\partial}{\partial z} \left(\frac{\partial U}{\partial y} \right) \right]$$

$$+ \text{ terms in the } \hat{y} \text{ and } \hat{x} \text{ directions.}$$

$$\frac{\partial}{\partial y} \left(\frac{\partial U}{\partial z} \right) = \frac{\partial}{\partial z} \left(\frac{\partial U}{\partial y} \right) \quad \text{so the } \hat{z} \text{ term is zero.}$$

The same is true for $\hat{x}, \hat{y}$. As a result,

$$\vec{\nabla} \times (\vec{\nabla} U(x, y, z)) = 0.$$

(b) In a similar fashion, prove that the divergence of the curl of a vector function is always zero: $\vec{\nabla} \cdot (\vec{\nabla} \times \vec{A}(x, y, z)) = 0$.

Solution

Same idea: just crank out all the derivatives in Cartesian coordinates. (Do you really need me to show this to you?)

Problem 3: Mrs. Rumpdock's leaky propane tank

A Department of Justice investigator, covertly surveilling the Rumpdocks' estate, discovers that the curved walls of their large propane tank are leaking. The tank is a cylinder of radius $0.5\,\text{m}$ and length $2\,\text{m}$; the investigator determines that the flow rate out of the now-porous walls of the tank varies with position. (The end caps are not leaking.) Placing the coordinate origin at the center of the tank, he finds that the outflow from the tank is well described as

$$\vec{J}(r = 0.5, \theta, z) = \alpha z^2 \hat{r} \,\text{kg/m}^2/\text{s}$$

The tank initially contains Γ kg of propane. How long will it take for the tank to become entirely empty of propane?

Solution

$\vec{J} = \alpha z^2 \hat{r}$; $\int \vec{J} \cdot \overrightarrow{dA}$ is the rate of loss of mass dM/dt.

$$\frac{dM}{dt} = \int_{\theta=0}^{\theta=2\pi} \int_{z=-1}^{z=1} \alpha z^2 \hat{r} \cdot r d\theta dz \hat{r}$$

$$= \int_{z=-1}^{z=1} \int_{\theta=0}^{\theta=2\pi} \alpha z^2 r d\theta dz = \left. \frac{2\pi \alpha z^3 r}{3} \right|_{z=-1}^{z=1} = \frac{2\pi \alpha}{3}$$

Time to empty is $\Gamma / \frac{2\pi \alpha}{3} = \frac{3\Gamma}{2\pi \alpha}$.

Problem 4: "Dr." Shazam's nutraceutical balls

An El Paso-based fraudster markets a product claimed to cure various forms of skin cancer with a single topical (on the surface of the skin) application. The inefficacious product is sold in spherical containers; the density of the mildly toxic product inside a container is

$$\rho(r, \theta, \phi) = \alpha(1 - \cos\theta)r^3$$

The origin of the coordinate system is at the center of the sphere, while the z-axis passes through the small filling port on the top of the container. The radius of a container is R. How many kilograms of bogus medication are loaded into each of Shazam's spheres?

Solution

$$M = \int_{\text{volume}} \rho dV = \int_{r,\theta,\varphi} \left[\alpha(1 - \cos\theta)r^3\right] r^2 \sin\theta dr d\theta d\varphi$$

$$= 2\pi\alpha \int_{r=0}^{r=R} r^5 \int \left[\begin{matrix}\theta = \pi \\ \theta = 0\end{matrix} (1 - \cos\theta)\sin\theta d\theta\right] dr$$

$$= 2\pi\alpha \int_{r=0}^{r=R} r^5 \left[\int_{\theta=0}^{\theta=\pi} -(1 - \cos\theta)d\cos\theta\right] dr$$

$$= -2\pi\alpha \int_{r=0}^{r=R} r^5[-1 - -1 - 1 + 1] = \frac{4\pi\alpha}{6}R^6$$

Rio Maggiore, Italy

Unit 13

The Maxwell Equations

What you saw in your introductory E&M course

Some intro E&M courses discuss integral forms of the master equations of electrodynamics—the Biot–Savart and Ampère equations, for example—since these are thought to be easier for beginning students to digest. But it is more useful to recast these as differential equations relating the fields and their derivatives at points in space–time

293

rather than global equations involving integrals over closed loops, surfaces, and volumes.

We can use some of what we've covered in the last two units to rewrite the equations in a more useful (and elegant) fashion.

Gauss's law in differential form

If you managed to make it all the way through the unit 12 exercises, you showed that the divergence of the electric field is proportional to the charge density and that this came about from the integral expression relating the flux of the electric field to enclosed charge.

In equations, the divergence theorem guarantees that

$$\oint_{\text{surface}} \vec{E}(x, y, z) \cdot d\vec{A} = \int_{\substack{\text{volume} \\ \text{bounded} \\ \text{by surface}}} \vec{\nabla} \cdot \vec{E}\, dx\, dy\, dz$$

since this is automatically true for **all** vector fields.

We also know that the definition of charge density guarantees that

$$\int_{\substack{\text{volume} \\ \text{bounded} \\ \text{by surface}}} \rho(x, y, z)\, dx\, dy\, dz = Q_{\text{enclosed}}$$

In its integral form, Gauss's law connects the flux of the electric field through a closed surface with contained charge this way:

$$\oint_{\text{surface}} \vec{E}(x, y, z) \cdot d\vec{A} = \frac{Q_{\text{enclosed}}}{\varepsilon_0}$$

Putting this all together,

$$\int_{\substack{\text{volume} \\ \text{bounded} \\ \text{by surface}}} \vec{\nabla} \cdot \vec{E}(x, y, z)\, dx\, dy\, dz = \oint_{\text{surface}} \vec{E}(x, y, z) \cdot d\vec{A} = \frac{Q_{\text{enclosed}}}{\varepsilon_0}$$

$$= \int_{\substack{\text{volume} \\ \text{bounded} \\ \text{by surface}}} \frac{\rho(x, y, z)}{\varepsilon_0}\, dx\, dy\, dz$$

so that, equating the integrands in the first and last integrals,

$$\vec{\nabla} \cdot \vec{E}(x, y, z) = \frac{\rho(x, y, z)}{\varepsilon_0}$$

This is the differential form of Gauss's law.

Exercise 13.1: Gauss's law for magnetic fields?

To date, there have been no observations of isolated magnetic charges: north poles are always paired with south poles, and the cleavage of a bar magnet midway between its ends only produces another pair of north and south poles, "neutralizing" the two halves.

Write the differential equation that is the magnetic equivalent of Gauss's law.

Solution

Since there's no such thing as magnetic charge, we have $\vec{\nabla} \cdot \vec{B} = 0$.

Faraday's law in differential form: Intro E&M and beyond

In your first contact with electromagnetism, you saw that Faraday's law describes how changes in magnetic flux can create electric fields. In integral form, Faraday's law is

$$\oint_{\substack{\text{closed} \\ \text{loop}}} \vec{E} \cdot d\vec{l} = -\frac{d\Phi_{\text{magnetic}}}{dt}$$

$$= -\frac{d}{dt} \int_{\substack{\text{surface} \\ \text{enclosed} \\ \text{by loop}}} \vec{B} \cdot d\vec{A}$$

Exercise 13.2: "My Dear Stokes"[1]

We discussed Stokes' theorem a couple of units ago. Use it to rewrite the left side of the equation

$$\oint_{\substack{\text{closed} \\ \text{loop}}} \vec{E} \cdot d\vec{l} = -\frac{d}{dt} \int \vec{B} \cdot d\vec{A}$$

as an integral over the area enclosed by the loop. Then, by equating the integrands of the left and right side integrals, write the differential form of Faraday's law.

---

[1]Salutation of a July 2, 1850, letter from William Thomson to George Stokes, cited in A.C.T. Wu and Chen Ning Yang, "Evolution of the Concept of the Vector Potential in the Description of Fundamental Interactions," *International Journal of Modern Physics* A Vol. 21, No. 16 (2006) 3235–3277.

Solution

$$\oint_{\substack{\text{closed} \\ \text{loop}}} \vec{E} \cdot d\vec{l} = \int_{\substack{\text{surface} \\ \text{enclosed}}} (\vec{\nabla} \times \vec{E}) \cdot d\vec{A} = -\frac{d}{dt} \int \vec{B} \cdot d\vec{A}$$

$$= -\int \left(\frac{d}{dt} \vec{B} \right) \cdot d\vec{A} \quad \text{so} \quad \vec{\nabla} \times \vec{E} = -\frac{d}{dt} \vec{B}.$$

Ampère's law: Intro physics and beyond

Ampère's law describes the relationship between currents and the magnetic fields they create. Its integral form is

$$\oint_{\substack{\text{closed} \\ \text{loop}}} \vec{B} \cdot d\vec{l} = \mu_0 I_{\text{enclosed}}$$

We can produce the differential version of this by rewriting the loop integral of the magnetic field using Stokes' theorem, then expressing the enclosed current as an integral over the current density flowing through the surface enclosed by the loop. The current density is a vector, of course, and it describes the number of Coulombs per second per square meter flowing in a current-carrying object.

Exercise 13.3: Differential Ampère

Rewrite the integral form of Ampère's law in terms of the current density $\vec{J}$ and the curl of the magnetic field. Then, by equating the integrands on both sides of the equation, write the differential form of Ampère's law.

Solution

$$\oint_{\substack{\text{closed} \\ \text{loop}}} \vec{B} \cdot d\vec{l} = \mu_0 I_{\text{enclosed}} = \mu_0 \int \vec{J} \cdot d\vec{A}$$

$$\oint_{\substack{\text{closed} \\ \text{loop}}} \vec{B} \cdot d\vec{l} = \int (\vec{\nabla} \times \vec{B}) \cdot d\vec{A} \quad \text{so} \quad \vec{\nabla} \times \vec{B} = \mu_0 \vec{J}$$

There's more to it than that

Ampère's law as originally formulated is incomplete: it fails to take into account the creation of magnetic fields by changing electric fields.

The integral form of the modified version of the law is this:

$$\oint_{\substack{\text{closed} \\ \text{loop}}} \vec{B} \cdot d\vec{l} = \mu_0 I_{\text{enclosed}} + \mu_0 \varepsilon_0 \frac{d}{dt} \int_{\substack{\text{surface} \\ \text{enclosed} \\ \text{by loop}}} \vec{E} \cdot d\vec{A}$$

Exercise 13.4: The whole thing

Work with the integral term on the right side of the above equation to modify your differential version of the original Ampère's law to include the effects of a time-varying electric field.

Solution

This one's easy! Combine the answers to the previous two exercises to write

$$\vec{\nabla} \times \vec{B} = \mu_0 \vec{J} + \mu_0 \varepsilon_0 \frac{d\vec{E}}{dt}$$

The Maxwell Equations

Assembling our four equations, we have what are collectively known as the Maxwell Equations. They are

$$\vec{\nabla} \cdot \vec{E} = \frac{\rho}{\varepsilon_0} \qquad \vec{\nabla} \cdot \vec{B} = 0$$

$$\vec{\nabla} \times \vec{E} = -\frac{\partial \vec{B}}{\partial t} \qquad \vec{\nabla} \times \vec{B} = \mu_0 \varepsilon_0 \frac{\partial \vec{E}}{\partial t} + \mu_0 \vec{J}$$

To my mind, the invention/discovery of the Maxwell Equations is one of the greatest intellectual achievements of all time by our species.

The asymmetry between the electric and magnetic fields arises from the mystifying absence of magnetic charge in our universe: there is no such thing as a non-zero density (or current) of magnetic charge anywhere to be found.

The constants μ_0 and ε_0 are determined from static measurements. The ratio of magnetic field strength to current in a long, straight wire carrying current I measured at a distance of 1 m is

$$\frac{|\vec{B}|}{I} = \frac{\mu_0}{2\pi}$$

while the ratio of electric field strength to charge at a distance of $1\,\mathrm{m}$ from a point charge Q is

$$\frac{|\vec{E}|}{Q} = \frac{1}{4\pi\varepsilon_0}$$

Wave equations

At points in space where the local charge and current density are zero, the curl equations reduce to

$$\vec{\nabla} \times \vec{E} = -\frac{\partial \vec{B}}{\partial t} \quad \vec{\nabla} \times \vec{B} = \mu_0\varepsilon_0 \frac{\partial \vec{E}}{\partial t}$$

We also know, of course, that the divergence of the fields in empty space is zero.

Let's come up with the wave equation. We start by taking the curl of the first curl equation and using the fact that we can commute the order of various differentiations. We then insert the second curl equation:

$$\vec{\nabla}\times(\vec{\nabla}\times\vec{E}) = -\vec{\nabla}\times\frac{\partial \vec{B}}{\partial t} = -\frac{\partial \vec{\nabla}\times\vec{B}}{\partial t} = -\frac{\partial\left(\mu_0\varepsilon_0\frac{\partial\vec{E}}{\partial t}\right)}{\partial t} = -\frac{\partial^2 \mu_0\varepsilon_0\vec{E}}{\partial t^2}$$

Crank out the left side:

$$\vec{\nabla}\times(\vec{\nabla}\times\vec{E}) = \vec{\nabla}\times\left[\left(\frac{\partial E_y}{\partial x} - \frac{\partial E_x}{\partial y}\right)\hat{z} + \left(\frac{\partial E_z}{\partial y} - \frac{\partial E_y}{\partial z}\right)\hat{x}\right.$$

$$\left. + \left(\frac{\partial E_x}{\partial z} - \frac{\partial E_z}{\partial x}\right)\hat{y}\right]$$

$$= \left[\frac{\partial\left(\frac{\partial E_x}{\partial z} - \frac{\partial E_z}{\partial x}\right)}{\partial x} - \frac{\partial\left(\frac{\partial E_z}{\partial y} - \frac{\partial E_y}{\partial z}\right)}{\partial y}\right]\hat{z}$$

$$+ \left[\frac{\partial\left(\frac{\partial E_y}{\partial x} - \frac{\partial E_x}{\partial y}\right)}{\partial y} - \frac{\partial\left(\frac{\partial E_x}{\partial z} - \frac{\partial E_z}{\partial x}\right)}{\partial z}\right]\hat{x}$$

$$+ \left[\frac{\partial\left(\frac{\partial E_z}{\partial y} - \frac{\partial E_y}{\partial z}\right)}{\partial z} - \frac{\partial\left(\frac{\partial E_y}{\partial x} - \frac{\partial E_x}{\partial y}\right)}{\partial x}\right]\hat{y}$$

Now, grind out the x-component:

$$\vec{\nabla}\times(\vec{\nabla}\times\vec{E})_x = \frac{\partial\left(\frac{\partial E_y}{\partial x}-\frac{\partial E_x}{\partial y}\right)}{\partial y} - \frac{\partial\left(\frac{\partial E_x}{\partial z}-\frac{\partial E_z}{\partial x}\right)}{\partial z}$$

$$= \frac{\partial^2 E_y}{\partial x\partial y} - \frac{\partial^2 E_x}{\partial y^2} - \frac{\partial^2 E_x}{\partial z^2} + \frac{\partial^2 E_z}{\partial x\partial z}$$

$$= \frac{\partial}{\partial x}\left(\frac{\partial E_y}{\partial y}+\frac{\partial E_z}{\partial z}\right) - \frac{\partial^2 E_x}{\partial y^2} - \frac{\partial^2 E_x}{\partial z^2}$$

Since the divergence of the electric field is zero in empty space,

$$\vec{\nabla}\cdot\vec{E} = \frac{\partial E_x}{\partial x}+\frac{\partial E_y}{\partial y}+\frac{\partial E_z}{\partial z} = 0 \;\Rightarrow\; \frac{\partial E_y}{\partial y}+\frac{\partial E_z}{\partial z} = -\frac{\partial E_x}{\partial x}$$

so that

$$\vec{\nabla}\times(\vec{\nabla}\times\vec{E})_x = \frac{\partial}{\partial x}\left(-\frac{\partial E_x}{\partial x}\right) - \frac{\partial^2 E_x}{\partial y^2} - \frac{\partial^2 E_x}{\partial z^2}$$

$$= -\left[\frac{\partial^2 E_x}{\partial x^2}+\frac{\partial^2 E_x}{\partial y^2}+\frac{\partial^2 E_x}{\partial z^2}\right]$$

$$= -\vec{\nabla}\cdot(\vec{\nabla}E_x)$$

That peculiar-looking operator at the end, the divergence of the gradient of E_x, is called the Laplacian and usually written as ∇^2:

$$\vec{\nabla}\times(\vec{\nabla}\times\vec{E})_x = -\nabla^2 E_x$$

The results for the y- and z-components follow in a similar fashion. We can now rewrite the equation $\vec{\nabla}\times(\vec{\nabla}\times\vec{E}) = -\frac{\partial^2\mu_0\varepsilon_0\vec{E}}{\partial t^2}$ as

$$\nabla^2\vec{E} = \mu_0\varepsilon_0\frac{\partial^2\vec{E}}{\partial t^2}$$

This is really three independent equations; one for each of the three components of the electric field. We'll discuss why we call this a "wave equation" shortly.

Exercise 13.5: And now for the magnetic field

To generate the wave equation for the electric field, I took the curl of both sides of the curl-of-E equation, then used the curl-of-B equation to eliminate the magnetic field from the right side of the resulting equation. The same procedure, but starting with the curl-of-B equation, will yield a wave equation for the magnetic field.

From the algebraic slog on the previous page, you can safely conclude that $\vec{\nabla} \times (\vec{\nabla} \times \vec{B}) = -\nabla^2 \vec{B}$.

Complete the process to show that the wave equation for the magnetic field is

$$\nabla^2 \vec{B} = \mu_0 \varepsilon_0 \frac{\partial^2 \vec{B}}{\partial t^2}$$

Solution

$$\vec{\nabla} \times \vec{B} = \mu_0 \varepsilon_0 \frac{\partial \vec{E}}{\partial t}$$

so

$$\vec{\nabla} \times (\vec{\nabla} \times \vec{B}) = \mu_0 \varepsilon_0 \frac{\partial \vec{\nabla} \times \vec{E}}{\partial t} \quad \text{or}$$

$$-\nabla^2 \vec{B} = \mu_0 \varepsilon_0 \frac{\partial}{\partial t} \left(-\frac{\partial \vec{B}}{\partial t} \right) = -\mu_0 \varepsilon_0 \frac{\partial^2 \vec{B}}{\partial t^2}$$

so that

$$\nabla^2 \vec{B} = \mu_0 \varepsilon_0 \frac{\partial^2 \vec{B}}{\partial t^2}$$

Plane wave solutions to the wave equation

Physically realizable electric and magnetic fields must always *always always* satisfy all four of the Maxwell Equations. It is not sufficient for a proposed solution to the wave equation to just satisfy the wave equation: I had assumed in my derivation that the divergence of the field is zero.

For example, all the second partial derivatives of the time-independent function $\vec{E}(x, y, z, t) = E_0\, x/x_0 \hat{x}$ are zero, satisfying the wave equation, but that field's divergence is non-zero. So, it can't be an actual solution to the full set of Maxwell Equations in empty space.

One especially interesting class of solutions are the plane waves: time-dependent electric and magnetic fields that depend on only one of the three spatial coordinates. (We call them "plane waves" because the electric field is everywhere the same in a plane defined by the axes of the missing coordinates.)

As an example, consider $\vec{E}(x, y, z, t) = \vec{E}_0 f(u)$, where f is a function that takes one argument (u), with $u \equiv az + bt$. The partial derivatives of f are easy to evaluate. Using the chain rule,

$$\frac{\partial f(u)}{\partial z} = \frac{df}{du}\frac{\partial u}{\partial z}$$

$$\frac{\partial^2 f(u)}{\partial z^2} = \left[\frac{d}{du}\frac{\partial f(u)}{\partial z}\right]\frac{\partial u}{\partial z} = \left[\frac{d}{du}\left(\frac{df}{du}\frac{\partial u}{\partial z}\right)\right]\frac{\partial u}{\partial z}$$

$$= \left[\frac{d^2 f}{du^2}\frac{\partial u}{\partial z} + \frac{df}{du}\left(\frac{d}{du}\frac{\partial u}{\partial z}\right)\right]\frac{\partial u}{\partial z}$$

Looks messy, right? But for $u = az + bt$ we have $\partial u/\partial z = a$, so $d(\partial u/\partial z)/du = 0$. As a result,

$$\frac{\partial^2 f(u)}{\partial z^2} = \frac{d^2 f}{du^2}\left(\frac{\partial u}{\partial z}\right)^2 = a^2\frac{d^2 f}{du^2}$$

Similarly,

$$\frac{\partial^2 f(u)}{\partial t^2} = b^2\frac{d^2 f}{du^2}$$

The partial derivatives of f with respect to x and y are zero, of course. As a result, we have

$$\frac{\partial^2 \vec{E}}{\partial z^2} = \frac{\partial^2 \vec{E}_0 f(az + bt)}{\partial z^2} = a^2\vec{E}_0\frac{d^2 f(u)}{du^2}$$

$$\frac{\partial^2 \vec{E}}{\partial t^2} = b^2\vec{E}_0\frac{d^2 f(u)}{du^2}$$

$$\frac{\partial^2 \vec{E}}{\partial z^2} = \frac{a^2}{b^2}\frac{\partial^2 \vec{E}}{\partial t^2}$$

This definition of the electric field will work in the wave equation as long as $a^2/b^2 = \mu_0\varepsilon_0$. (The same will be true for the magnetic field.)

In words, an electric field that is a function of $u \equiv az + bt$ will always satisfy the wave equation as long as a and b are related through $a^2/b^2 = \mu_0\varepsilon_0$.

Exercise 13.6: How fast can Pablo run?[2]

With the use of technology stolen from Area 51, Mrs. Rumpdock creates an electric field of the form $\vec{E}_0 f(u)$, with the function $f(u)$ shown in the following graph. Consistent with the demands of the wave equation, the argument of f is $u = z + t/\sqrt{\mu_0\varepsilon_0}$.

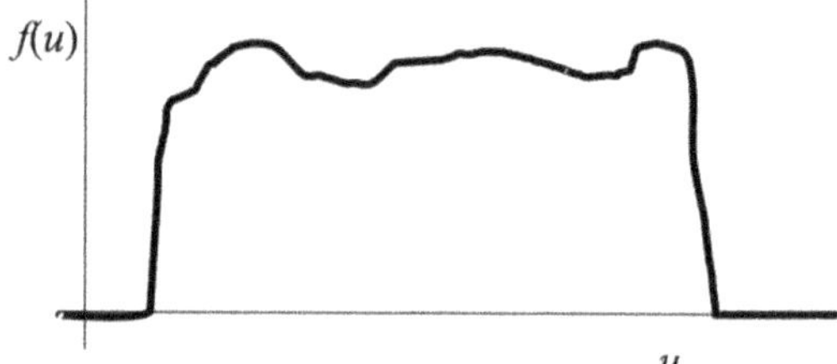

The locations of the various bumps and wiggles in the field change with time but always in a way that preserves the shape shown in the graph.

Determine how fast and in what direction the pattern is moving. (Express your answer in terms of μ_0 and ε_0.)

Solution

A bump at z_0 at time $t = 0$ will always be found (at a later time) at z, t satisfying

$$z + \frac{t}{\sqrt{\mu_0\varepsilon_0}} = z_0$$

So, the bump will be found at

$$z(t) = z_0 - \frac{t}{\sqrt{\mu_0\varepsilon_0}}$$

---

[2]Pablo is the name of my daughter's very charming Golden Doodle.

the pattern moves to the left with speed

$$c = \frac{1}{\sqrt{\mu_0 \varepsilon_0}}$$

because

$$v_{\text{bump}} = \frac{d\left(z_0 - \frac{t}{\sqrt{\mu_0 \varepsilon_0}}\right)}{dt} = -\frac{1}{\sqrt{\mu_0 \varepsilon_0}}$$

Exercise 13.7: Divergence

Consider the electric field from the previous exercise at the time $t = 0$. By drawing a Gaussian "pillbox" that spans a z-position at which the field changes dramatically, prove that if the electric field lies entirely in the x, y plane (so that $E_z = 0$), it will have zero divergence. (The same will be true for the magnetic field. Take note that a *constant* field in any direction will also yield zero divergence.)

Solution

Here's a diagram.

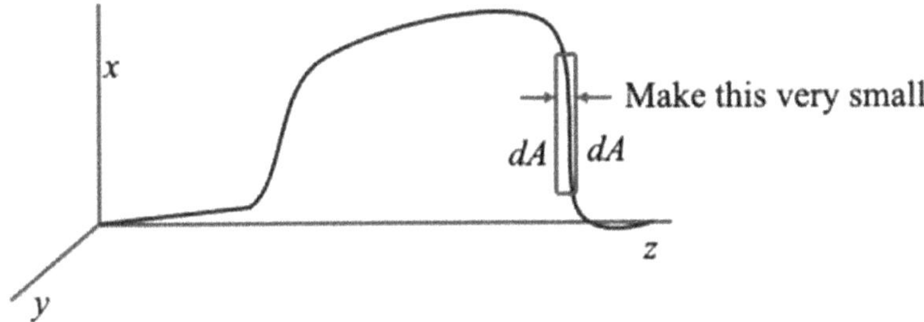

If the pillbox has area dA on the faces parallel to the xy plane, we'll have

$$\int \vec{E} \cdot d\vec{A} \approx E_z dA \ (\text{right}) - E_z dA \ (\text{left})$$

So to have zero divergence, we'll need $E_z = 0$.

Oh, in case you're curious, here's where that function $f(u)$ comes from.

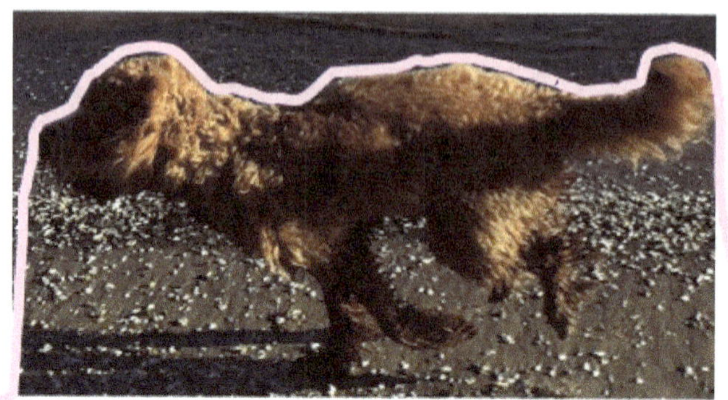

The connection between E and B

We haven't done anything yet to establish a connection between the electric and magnetic fields in vacuum. It should be obvious from two of the Maxwell Equations that there *is* a connection: wherever there is a time-varying magnetic field, there must also be an electric field with non-zero curl. And wherever there is a time-varying electric field (in vacuum), there must be a magnetic field with non-zero curl. In addition, the spatial patterns in the electric and magnetic fields travel at the same speed, namely $c = 1/\sqrt{\mu_0 \varepsilon_0}$.

Let's assume that

$$\vec{E} = f_1(z - ct)\hat{x} + f_2(z - ct)\hat{y} \quad \text{and} \quad \vec{B} = g_1(z - ct)\hat{x} + g_2(z - ct)\hat{y}$$

Plug these into the curl-of-E Maxwell equation. Only two of the terms on the left side are non-zero:

$$\vec{\nabla} \times [f_1(z - ct)\hat{x} + f_2(z - ct)\hat{y}] = -\frac{\partial[g_1(z - ct)\hat{x} + g_2(z - ct)\hat{y}]}{\partial t}$$

$$\frac{\partial f_1(z - ct)}{\partial z}\hat{y} - \frac{\partial f_2(z - ct)}{\partial z}\hat{x} = -\frac{\partial g_1(z - ct)}{\partial t}\hat{x} - \frac{\partial g_2(z - ct)}{\partial t}\hat{y}$$

Equating the terms in the x-direction (and also in the y direction) gives

$$\frac{\partial f_1(z - ct)}{\partial z} = -\frac{\partial g_2(z - ct)}{\partial t}$$

$$\frac{\partial f_2(z - ct)}{\partial z} = \frac{\partial g_1(z - ct)}{\partial t}$$

Defining $u = z - ct$, we can write the partial derivatives as follows:

$$\frac{df_1(u)}{du}\frac{\partial u}{\partial z} = -\frac{dg_2(u)}{du}\frac{\partial u}{\partial t}$$

$$\frac{df_2(u)}{du}\frac{\partial u}{\partial z} = \frac{dg_1(u)}{du}\frac{\partial u}{\partial t}.$$

Since $\partial u/\partial z = 1$ and $\partial u/\partial t = -c$, we have

$$\frac{df_1(u)}{du} = c\frac{dg_2(u)}{du}$$

$$\frac{df_2(u)}{du} = -c\frac{dg_1(u)}{du}$$

Except for the possibility of a spatially (and temporally) constant field, we see that $f_1 = cg_2$ and $f_2 = -cg_1$, so $E_x = cB_y$ and $E_y = -cB_x$.

Note that the electric and magnetic fields are perpendicular:

$$\vec{E} \cdot \vec{B} = E_x B_x + E_y B_y + E_z B_z$$

$$= cB_y B_x - cB_x B_y + 0$$

$$= 0$$

In addition,

$$|\vec{E}| = \sqrt{E_x^2 + E_y^2 + E_z^2}$$

$$= c\sqrt{B_y^2 + B_x^2 + 0}$$

$$= c|\vec{B}|$$

An electromagnetic wave with a maximum magnetic field of 1 Tesla will have a maximum electric field of about 300 megavolts per meter.

Exercise 13.8: Circular polarization

Consider an electromagnetic wave with an electric field $\vec{E} = f_1(z - ct)\hat{x} + f_2(z - ct)\hat{y}$.

Find a pair of functions f_1 and f_2 that will yield a circularly polarized electromagnetic wave whose electric field (for fixed time) rotates through one full turn per foot along the wave's direction of travel.

By that I mean that if we took a snapshot (so we're looking at the wave everywhere in space but at one instant of time), we'd see the electric field corkscrewing around as we moved along the z-axis.

Solution

Let's pick $t = 0$ and try a sine and a cosine for our two functions:

$$f_1(z - ct) = E_0 \cos\left(\frac{2\pi}{1\,\text{foot}}z - ct\right)\widehat{x}, \quad f_2(z - ct)$$

$$= E_0 \sin\left(\frac{2\pi}{1\,\text{foot}}z - ct\right)\widehat{y}$$

should do it.

Summary

We discussed:

- Gauss's law for electric and magnetic fields;
- Faraday's law;
- Ampère's law;
- the Maxwell Equations;
- the wave equation;
- plane waves and the speed of light;
- the connection between the electric and magnetic fields.

Problem 1: Ampère's outhouse

An unusual magnet is built from a pair of partially cylindrical conductors, as shown in cross-section in the figure on the right. The hollowed-out region between the conductors is just how it looks: the region in which the cylinders overlap has been emptied of material.

The left conductor carries a uniform current density of 35 million amperes per square meter, directed out of the page. The right conductor carries the same current density but into the page. The diameter of the cylinders is $a = 0.5\,\text{m}$. The centers of the cylinders are offset horizontally by $a/2$.

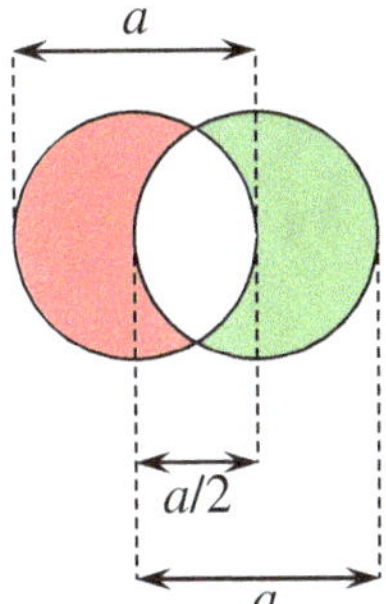

Prove that the magnetic field in the void between the conductors is constant, and determine its value.

I suggest you approach the problem this way: first, calculate the x, y components of the magnetic field inside a solid cylinder centered on the origin that carries a uniform current density. Then place a second cylinder, carrying the same current density but flowing in the opposite direction, that overlaps the first, as shown in the figure. (In the region of overlap, the current densities will cancel.)

Solution

Do it like so: calculate the magnetic field for a full cylinder of solid, constant current density. Then add the field for two overlapping cylinders with one centered on the origin.

From the Maxwell equations,

$$\vec{\nabla} \times \vec{B} = \mu_0 \vec{J} \quad \text{so} \quad \int_{\text{loop}} \vec{B} \cdot \vec{dl} = \mu_0 \vec{J} A$$

(A is the area of the enclosed loop)

Have the loop of radius r be centered on the middle of the cylinder so that

$$2\pi r B = \mu_0 J \pi r^2 \quad \text{or} \quad B = \frac{\mu_0 J r}{2}.$$

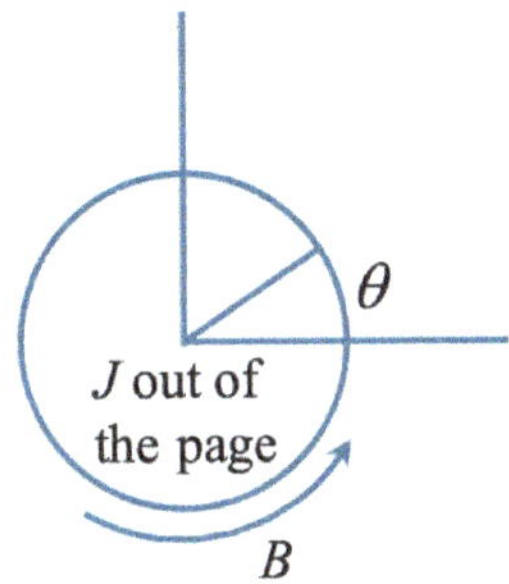

For the left cylinder (which is centered on the origin), a point $x = r \cos \theta$, $y = r \sin \theta$ will have

$$B_x = -r \sin \theta \mu_0 J, \quad By = r \cos \theta \mu_0 J$$

Now, overlap the two cylinders. The angles and distances from the centers of the cylinders are defined as shown in the following.

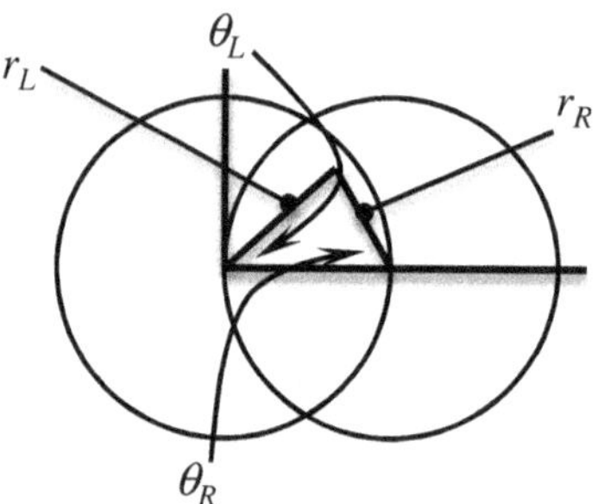

The magnetic field in the void between the cylinders is a super-position of the fields from the individual, void-free cylinders. Take note of the definitions of the distances and angles!

Note that $r_L \cos \theta_L + r_R \cos \theta_R = \frac{a}{2}$ so that

$$B_y = \frac{\mu_0 J}{2}[r_L \cos \theta_L + r_R \cos \theta_R] = \frac{\mu_0 J a}{4}$$

The y-component of the field is constant! Now, do x.

Note that $r_L \sin \theta_L = r_R \sin \theta_R$. As a result,

$$B_x = \frac{\mu_0 J}{2}[-r_L \sin \theta_L + r_R \sin \theta_R] = 0$$

For a current density of 3.5×10^7 amps per square meter, we get 5.5 Tesla.

Problem 2: Dumping the solenoid

Some years ago, the National High Magnetic Field Laboratory in Florida operated the world's strongest magnet,[3] a 45 Tesla hybrid

---

[3] https://nationalmaglab.org/about-the-maglab/around-the-lab/meet-the-magnets/
meet-the-45-tesla-hybrid-magnet/.

with both superconducting and conventional coils. Imagine that the field is solenoidal—a cylindrical volume holding a constant field—and that the diameter of the field-containing cylinder is 10 cm. (This is about three times wider than the actual NHMFL magnet bore.)

Imagine that a coil failure at time $t = 0$ shorts the magnet's current through a resistive load so that the current (and therefore the magnet's field) drops exponentially, with a time constant of 0.5 seconds. As a result, after the short develops, the magnetic field strength is $B(t) = 45e^{-2t}$. Calculate the electric field inside the magnet as a function of position and time.

Solution

Use a Maxwell equation and Stokes' law:

$$\vec{\nabla} \times \vec{E} = -\frac{\partial \vec{B}}{\partial t} = 2\vec{B}_0 e^{-2t}$$

so

$$\oint \vec{E} \cdot \vec{dl} = \int (\vec{\nabla} \times \vec{E}) \cdot \vec{dA}$$

$$2\pi r E = 2B_0 e^{-2t} \pi r^2$$

so that

$$E = B_0 e^{-2t} r$$

Problem 3: Mrs. Blepon's vat of toxic jello

On a stormy afternoon, Mrs. Blepon, unpleasant proprietress of *Blepon's Deli & Hardware Emporium*, loads the Deli's tank of lime jello with a toxic mix of green jello, pineapple juice, and electrical charge. Taking the center of the spherical tank of radius R as the origin, one of the Deli's trainees determines that the electrical field inside the tank is radial in orientation, with field strength $E(r) = E_0 \frac{r}{R} e^{-r/R}$. Calculate the charge density $\rho(r)$ inside the tank.

Solution

Use, in spherical coordinates, $\vec{\nabla} \cdot \vec{E} = \rho/\varepsilon_0$.

$$\vec{\nabla} \cdot \vec{E} = \frac{1}{r^2} \frac{\partial \frac{E_0 r^3}{R} e^{-r/R}}{\partial r} = \frac{E_0}{r^2 R} \frac{\partial r^3 e^{-r/R}}{\partial r}$$

$$\vec{\nabla} \cdot \vec{E} = \frac{E_0}{R} e^{-r/R} \left[3 - \frac{r}{R}\right] = \frac{\rho}{\varepsilon_0}.$$

so

$$\rho = \frac{\varepsilon_0 E_0}{R} e^{-r/R} \left[3 - \frac{r}{R}\right]$$

Problem 4: Science fiction

None of the following field configurations are physically realizable. For each of the following, explain clearly and explicitly why the particular electric and magnetic field combination is impossible. Note that I have not defined for you the charge or current densities, which might be different from zero. In addition, I have not specified the values of E_0 and B_0, though you should assume they are different from zero:

$$\text{(a)} \quad \begin{aligned} \vec{E}(x, y, z, t) &= E_0 \hat{x} \sin kx \\ \vec{B}(x, y, z, t) &= B_0 \hat{y} \sin ky \end{aligned}$$

$$\text{(b)} \quad \begin{aligned} \vec{E}(r, \theta, \phi, t) &= a E_0 \hat{\phi} r^2 \\ \vec{B}(r, \theta, \phi, t) &= b B_0 \hat{\phi} r \end{aligned}$$

with a and b positive constants bearing the correct units and

$$\text{(c)} \quad \begin{aligned} \vec{E}(x, y, z, t) &= E_0 \hat{x} \sin(kz - \omega t) \\ \vec{B}(x, y, z, t) &= B_0 \hat{y} \sin(kz + \omega t) \end{aligned}$$

with $E_0, B_0, k,$ and ω all positive.

Solution

(a) Take the divergence of the magnetic field to find that it is non-zero. That's impossible.

(b) Take the curl of the electric field to find that it is non-zero. But this is only possible if the magnetic field has a non-zero time derivative. But the magnetic field is time-independent so that doesn't work.

(c) Let's look at a curl equation. We need $\vec{\nabla} \times \vec{E} = \frac{\partial E_x}{\partial z} \hat{y} = kE_0 \cos(kz - \omega t)\hat{y}$. Now look at $-\frac{\partial \vec{B}}{\partial t} = -\omega B_0 \cos(kz + \omega t)\hat{y}$, which should equal the above for all z, t. *But* $kE_0 \cos(kz - \omega t) \neq -\omega B_0 \cos(kz + \omega t)$ so the proposed fields don't satisfy the Maxwell equations.

Assisi, Italy

Unit 14

A Covariant Formulation of Electrodynamics

Introduction: Things you already know

We're going to work up a version of the Maxwell Equations written entirely in terms of things that transform gracefully under Lorentz transformations. You already know more than you think about the ingredients for this reformulation.

313

The Maxwell Equations

We discussed the Maxwell Equations in differential form last unit:

$$\vec{\nabla} \cdot \vec{E} = \frac{\rho}{\varepsilon_0} \qquad \vec{\nabla} \cdot \vec{B} = 0$$

$$\vec{\nabla} \times \vec{E} = -\frac{\partial \vec{B}}{\partial t} \qquad \vec{\nabla} \times \vec{B} = \mu_0 \varepsilon_0 \frac{\partial \vec{E}}{\partial t} + \mu_0 \vec{J}$$

Since the divergence operator involves partial derivatives with respect to the space coordinates, it is safe to include the possibility of explicit time dependence in the list of arguments for the fields and charge density. The functions in the above equations are actually

$$\vec{E}(x, y, z, t) \quad \vec{B}(x, y, z, t) \quad \rho(x, y, z, t) \quad \vec{J}(x, y, z, t)$$

Since time is held fixed when taking a space derivative and since these equations describe the density and derivatives of the fields at points in space, there aren't any propagation-delay issues for us to tackle here.

The wave equation

You probably remember that we can take the curl of two of the Maxwell Equations to produce the wave equations

$$\nabla^2 \vec{E} = \mu_0 \varepsilon_0 \frac{\partial^2 \vec{E}}{\partial t^2} \qquad \nabla^2 \vec{B} = \mu_0 \varepsilon_0 \frac{\partial^2 \vec{B}}{\partial t^2}$$

They show that the speed of propagation in empty space of a wiggle in the fields is $c = 1/\sqrt{\mu_0 \varepsilon_0}$.

Note that the propagation speed of a pattern in the fields is entirely independent of any motion of the source relative to the observer. The laboratory staff measuring the space–time dependence of the fields will always find the same value for c.

So far, that's the only obvious place where relativity has insinuated itself into the master equations of electromagnetism.

The vector potential

To streamline the development of a covariant form of the Maxwell Equations, we will spend more time discussing the vector potential, then combine the scalar and vector potentials into the four potential A^μ.

We did a little with the vector potential in Unit 11. For a non-relativistic charge (so that we can neglect the time needed for information to propagate from the charges and currents to the observer), the vector potential is just the product of the scalar potential and the charge's velocity, divided by c^2. We have

$$\vec{A}(r) = \frac{V(r)}{c}\frac{\vec{v}}{c} \quad \text{or} \quad \frac{\varphi(r)}{c}\frac{\vec{v}}{c}.$$

We will discuss the more general case in this unit.

An important fact: the curl of the vector potential is exactly the same thing as the magnetic field. Since that's an essential attribute of the vector potential—we take its curl to calculate the magnetic field—we can add to the vector potential any other vector field we want, as long as the curl of that add-on field is zero. This will always be true, for example, for a vector field comprising the gradient of a scalar function since the curl of a gradient is zero. An adjustment of this sort, in which we add on something with zero curl, is called a *gauge transformation*.

The Laplacian of the scalar potential

Here's another thing you know, both from introductory physics and our discussions in this course. The connection between a conservative force F and the potential energy U of an object moving under its influence is $\vec{F} = -\vec{\nabla}U$. Since a stationary charge in an electric field experiences a force $\vec{F} = q\vec{E}$ and possesses potential energy $U = q\varphi$, it follows that $\vec{E} = -\vec{\nabla}\varphi$. And since the first Maxwell equation states that $\vec{\nabla} \cdot \vec{E} = \rho/\varepsilon_0$, we can conclude that $\vec{\nabla} \cdot \vec{E} = \vec{\nabla} \cdot (-\vec{\nabla}\varphi) = \rho/\varepsilon_0$ so that

$$\nabla^2 \varphi = -\rho/\varepsilon_0$$

Note the vagaries of notation: sometimes I use V for the potential (potential energy per Coulomb) and sometimes ϕ and sometimes φ.

There's a small technical issue here since the full-blown expression for the electric field also includes a term related to the time derivative of the vector potential. But we are always free to perform a gauge transformation on A so that its divergence is zero. In so doing, this term will disappear in the calculation of the divergence of the electric field.

Machinery for an eventual retooling of the Maxwell Equations

In developing a covariant version of the Maxwell Equations, we will want to build our equations from objects that behave well under Lorentz transformations, namely Lorentz scalars, four vectors, and Lorentz tensors.

We began discussing these sorts of objects in Unit 4, then spent more time on them in Unit 7. I think you are probably pretty comfortable with Lorentz scalars and four-vectors by now. The basic idea is that they behave in very specific, very sharply defined ways when we change frames of reference and reevaluate them according to observations made in the new frame. Here's a reminder, taken from Unit 7.

1. ***Lorentz scalars***

   - The sum, difference, product, or quotient of two Lorentz scalars is a Lorentz scalar.
   - The scalar product of any two four-vectors is a Lorentz scalar:

$$s = \sum_{\mu} A^{\mu} B_{\mu} = \sum_{\mu} A^{\mu} \sum_{\rho} B^{\rho} g_{\mu\rho} = \sum_{\mu,\rho} A^{\mu} B^{\rho} g_{\mu\rho}$$

   - The "contraction" of both indices of any two tensors is a Lorentz scalar:

$$b = \sum_{\mu,\nu} S^{\mu\nu} T_{\mu\nu} = \sum_{\mu,\nu} \left[S^{\mu\nu} \sum_{\rho,\sigma} g_{\mu\rho} g_{\nu\sigma} T^{\rho\sigma} \right]$$

$$= \sum_{\mu,\nu,\rho,\sigma} S^{\mu\nu} g_{\mu\rho} g_{\nu\sigma} T^{\rho\sigma}$$

2. ***Four-vectors***

   - The product of a Lorentz scalar and a four-vector is a four-vector.
   - The sum or difference of two four-vectors is a four-vector.
   - The contraction of a four-vector and one index of a tensor is a four-vector:

$$K^{\upsilon} = \sum_{\mu} S^{\mu\upsilon} A_{\mu} = \sum_{\mu} \left[S^{\mu\upsilon} \sum_{\rho} g_{\mu\rho} A^{\rho} \right] = \sum_{\mu} \sum_{\rho} S^{\mu\upsilon} g_{\mu\rho} A^{\rho}$$

3. **Tensors**

- The product of a Lorentz scalar and a tensor is a tensor.
- The sum or difference of two tensors is a tensor.
- The outer product of two four-vectors is a tensor: $T^{\mu\nu} = A^\mu B^\nu$.
- A product of two tensors done with the contraction of one index is a tensor: $A^{\mu\nu} = \sum_\rho [B^\mu{}_\rho C^{\rho\nu}] = \sum_\rho \{[\sum_\sigma g_{\rho\sigma} B^{\mu\sigma}] C^{\rho\nu}\}$.

4. **General covariance of "good" equations**

- An equation written as a relationship among Lorentz scalars, four-vectors, and Lorentz tensors that is known to be true in one inertial frame will be true in all inertial frames.

5. **Behavior under Lorentz transformations**

- Scalars: $s' = s$.
- Four vectors: $A'^\mu = \sum_v \Lambda^\mu{}_v A^v$.
- Tensors: $T'^{\mu\nu} = \sum_{\rho,\sigma} \Lambda^\mu{}_\rho \Lambda^\nu{}_\sigma T^{\rho\sigma}$.

Covariant derivatives

We will need to take derivatives in a way that follows the basic tenets of relativity, in which space and time are treated in similar ways. How to do this?

There is guidance in the equation expressing conservation of charge. Two units ago, we discussed conservation of mass in a fluid, and the same discussion works for electric charge. Defining the current density to be the number of Coulombs that move each second through $1\,\mathrm{m}^2$ aperture (with the current's direction taken to be the same as the local fluid velocity) gave us

$$\frac{\partial \rho(x,y,z,t)}{\partial t} = -\vec{\nabla} \cdot \vec{J}(x,y,z,t)$$

The meaning is obvious if you're comfortable with divergences: the density of electric charge at a point in space decreases (increases) according to the net outflow (inflow) of charge from (into) tha point.

I had worked this up for a nonrelativistic fluid, for which we were discussing current densities as kilograms per square meter per second.

It works the same way for current densities of electric charge: we define the current density as $\vec{J} \equiv \rho\vec{v}$.

Writing out all the partial derivatives, rearranging terms, and multiplying by a propitious[1] factor of 1 (that's the c/c in the time derivative), gives us, for the conservation equation:

$$\frac{\partial c\rho}{\partial ct} + \frac{\partial J_x}{\partial x} + \frac{\partial J_y}{\partial y} + \frac{\partial J_z}{\partial z} = \frac{\partial[c\rho]}{\partial[ct]} + \frac{\partial \rho v_x}{\partial x} + \frac{\partial \rho v_y}{\partial y} + \frac{\partial \rho v_z}{\partial z} = 0$$

This cries out for a relativistic generalization! We'll take the density $\rho = \gamma\rho_0$ to be the charge density, including any Lorentz contractions due to local motion inside the cloud of charge. ρ_0 is the distribution's proper density: the density measured by an observer moving with the swirling charge.

We define the four-current density as

$$J^\mu \equiv \rho_0 u^\mu = \gamma\rho_0(c, \vec{v})$$

while the (contravariant) four divergence is

$$\nabla^\mu \equiv \left(\frac{\partial}{\partial ct}, -\frac{\partial}{\partial x}, -\frac{\partial}{\partial y}, -\frac{\partial}{\partial z} \right) = \left(\frac{\partial}{\partial x^0}, -\frac{\partial}{\partial x^1}, -\frac{\partial}{\partial x^2}, -\frac{\partial}{\partial x^3} \right)$$

$$= \left(\frac{\partial}{\partial ct}, -\vec{\nabla} \right)$$

The covariant version of the four divergence is

$$\nabla_\mu \equiv \left(\frac{\partial}{\partial ct}, +\frac{\partial}{\partial x}, +\frac{\partial}{\partial y}, +\frac{\partial}{\partial z} \right) = \left(\frac{\partial}{\partial x_0}, -\frac{\partial}{\partial x_1}, -\frac{\partial}{\partial x_2}, -\frac{\partial}{\partial x_3} \right)$$

$$= \left(\frac{\partial}{\partial ct}, \vec{\nabla} \right)$$

With these, we can restate conservation of mass—or charge—this way: $\sum_{\mu=0}^{\mu=3} \nabla_\mu J^\mu = 0$.

---

[1] Mr. Wagner, my fabulous calculus teacher in high school, liked to use this phrase. "Propitious" means just what you'd think.

If I were using the Einstein summation convention, in which we sum over repeated indices, I could have written the charge conservation equation more compactly as $\nabla_\mu J^\mu = 0$.

The conservation equation is obviously correct for non-relativistic fluids or charge distributions for which γ is very close to 1 since it reduces to the non-relativistic version on the previous page. But it should be apparent to you that it ought to work fine for a relativistic system as long as we take into account that the local density has increased by a factor of γ due to Lorentz contraction.

A bit of refinement in our notation: it is more common for physicists to use the symbol ∂_μ than ∇_μ to represent these partial derivatives. We also use ∂^μ more than ∇^μ.

It may look strange that I've defined ∇^μ (and ∂^μ) with a negative sign in front of the space derivatives. That looks like a covariant rather than contravariant thing. But it is necessary so that ∂^μ will actually transform like a contravariant four vector while ∂_μ like a covariant four vector. (You can prove that they will transform this way if you are so inclined—it's not too difficult.)

Note the third bullet point under "3. Tensors" a few pages ago. It also applies to outer products that include ∂^μ as one of the four vectors. For example, $T^{\mu\nu} = \partial^\mu A^\nu$ is a tensor, as are the symmetrized and antisymmetrized versions of it: $T^{\mu\nu}_{\text{symmetrized}} = \partial^\mu A^\nu + \partial^\nu A^\mu$ and $T^{\mu\nu}_{\text{antisymmetrized}} = \partial^\mu A^\nu - \partial^\nu A^\mu$. By "symmetric" I mean that $T^{\mu\nu}_{\text{symmetrized}} = +T^{\nu\mu}_{\text{symmetrized}}$ while $T^{\mu\nu}_{\text{antisymmetrized}} = -T^{\nu\mu}_{\text{antisymmetrized}}$.

Exercise 14.1: Covariant derivatives and scalars

By evaluating the derivatives, prove that $\sum_\mu \partial_\mu x^\mu$ is a Lorentz scalar.

Solution

$$\sum_\mu \partial_\mu x^\mu = \frac{\partial ct}{\partial ct} + \frac{\partial x}{\partial x} + \frac{\partial y}{\partial y} + \frac{\partial z}{\partial z} = 1 + 1 + 1 + 1 = 4$$

which is a pure number and the same in all frames.

A theoretical underpinning: The Helmholtz theorem

The Helmholtz theorem lets us solve for the potentials in terms of (constant-in-time) charge and current distributions. This allows us to derive the connection between the electric and magnetic fields, the potentials, and the charge distributions and currents. Some of these conclusions will also apply to time-varying distributions of charge and current, but I don't want to do more than state the conclusions for those cases.

Suppose we are given an unknown vector function $\vec{F}$ but are told that its divergence and curl **are** known with $\vec{\nabla} \times \vec{F}(x, y, z) = \vec{C}(x, y, z)$ and $\vec{\nabla} \cdot \vec{F}(x, y, z) = D(x, y, z)$. Is that enough information for us to solve for $\vec{F}$?

According to the Helmholtz theorem, it is. The theorem states that we can always write any well-behaved vector function $\vec{F}$ in terms of its curl and gradient as long as the functions $\vec{C} = \vec{\nabla} \times \vec{F}$ and $D = \vec{\nabla} \cdot \vec{F}$ go to zero sufficiently rapidly at large distances.

The exact form of the solution is

$$\vec{F}(\vec{r}) = \vec{\nabla} \times \left(\frac{1}{4\pi} \int_{\text{all space}} \frac{\vec{\nabla}' \times \vec{F}(\vec{r}')}{|\vec{r} - \vec{r}'|} d^3 r' \right)$$

$$- \vec{\nabla} \left(\frac{1}{4\pi} \int_{\text{all space}} \frac{\vec{\nabla}' \cdot \vec{F}'(\vec{r}')}{|\vec{r} - \vec{r}'|} d^3 r' \right)$$

$$= \vec{\nabla} \times \left(\frac{1}{4\pi} \int_{\text{all space}} \frac{\vec{C}(\vec{r}')}{|\vec{r} - \vec{r}'|} d^3 r' \right)$$

$$- \vec{\nabla} \left(\frac{1}{4\pi} \int_{\text{all space}} \frac{D(\vec{r}')}{|\vec{r} - \vec{r}'|} d^3 r' \right)$$

An important point: the partial derivatives in the gradient and curl that are *outside the integrals* are taken with respect to x, y, z and *not* with respect to x', y', z'. And the primes on the curl and gradient operators *inside the integrals* indicate that those derivatives are taken with respect to x', y', z'.

When we apply the Helmholtz theorem to the electric field, we have

$$\vec{E}(\vec{r}) = \vec{\nabla} \times \left(\frac{1}{4\pi} \int_{\text{all space}} \frac{\vec{\nabla}' \times \vec{E}(\vec{r}')}{|\vec{r} - \vec{r}'|} d^3 r' \right)$$

$$- \vec{\nabla} \left(\frac{1}{4\pi} \int_{\text{all space}} \frac{\vec{\nabla}' \cdot \vec{E}(\vec{r}')}{|\vec{r} - \vec{r}'|} d^3 r' \right)$$

$$\vec{E}(\vec{r}) = \vec{\nabla} \times \left(\frac{1}{4\pi} \int_{\text{all space}} \frac{-\partial \vec{B}(\vec{r}')/\partial t'}{|\vec{r} - \vec{r}'|} d^3 r' \right)$$

$$- \vec{\nabla} \left(\frac{1}{4\pi} \int_{\text{all space}} \frac{\rho(\vec{r}')/\varepsilon_0}{|\vec{r} - \vec{r}'|} d^3 r' \right)$$

$$= -\vec{\nabla} \left(\frac{1}{4\pi} \int_{\text{all space}} \frac{\rho(\vec{r}')/\varepsilon_0}{|\vec{r} - \vec{r}'|} d^3 r' \right)$$

since I am not letting the currents change, forcing the magnetic field to be unchanging. I've made use of the Maxwell Equations to replace the curl and divergence of the electric field in the denominators inside the integrals.

For the magnetic field (whose divergence is always zero), we have

$$\vec{B}(\vec{r}) = \vec{\nabla} \times \left(\frac{1}{4\pi} \int_{\text{all space}} \frac{\vec{\nabla}' \times \vec{B}(\vec{r}')}{|\vec{r} - \vec{r}'|} d^3 r' \right)$$

$$- \vec{\nabla} \left(\frac{1}{4\pi} \int_{\text{all space}} \frac{\vec{\nabla}' \cdot \vec{B}(\vec{r}')}{|\vec{r} - \vec{r}'|} d^3 r' \right)$$

$$= \vec{\nabla} \times \left(\frac{1}{4\pi} \int_{\text{all space}} \frac{\mu_0 [\varepsilon_0 \partial \vec{E}(\vec{r}')/\partial t + \vec{J}(\vec{r}')]}{|\vec{r} - \vec{r}'|} d^3 r' \right)$$

$$= \vec{\nabla} \times \left(\frac{1}{4\pi} \int_{\text{all space}} \frac{\mu_0 \vec{J}(\vec{r}')}{|\vec{r} - \vec{r}'|} d^3 r' \right)$$

Once again, the time independence of the currents makes the time derivative zero. The term in the parentheses is the vector potential, since its curl is the magnetic field.

The vector potential

We have $\vec{B}(\vec{r}) = \vec{\nabla} \times \vec{A}(\vec{r}) = \vec{\nabla} \times \left(\frac{1}{4\pi} \int_{\text{all space}} \frac{\mu_0 \vec{J}(\vec{r}')}{|\vec{r}-\vec{r}'|} d^3 r' \right)$ so that $\vec{A}(\vec{r}) = \frac{1}{4\pi} \int_{\text{all space}} \frac{\mu_0 \vec{J}(\vec{r}')}{|\vec{r}-\vec{r}'|} d^3 r'$.

The expression for $\vec{A}$ is really three separate equations, one for each component of the vector potential:

$$A_x(\vec{r}) = \int_{\text{all space}} \frac{\mu_0}{4\pi} \frac{J_x(\vec{r}')}{|\vec{r} - \vec{r}'|} d^3 r', \quad A_y(\vec{r}) = \int_{\text{all space}} \frac{\mu_0}{4\pi} \frac{J_y(\vec{r}')}{|\vec{r} - \vec{r}'|} d^3 r',$$

$$A_z(\vec{r}) = \int_{\text{all space}} \frac{\mu_0}{4\pi} \frac{J_z(\vec{r}')}{|\vec{r} - \vec{r}'|} d^3 r'$$

We already know a lot about equations of this form since the scalar potential is

$$V(\vec{r}) = \int_{\text{all space}} \frac{1}{4\pi\varepsilon_0} \frac{\rho(\vec{r}')}{|\vec{r} - \vec{r}'|} d^3 r'$$

and it obeys the superposition principle as well as satisfying Poisson's equation

$$\nabla^2 V(\vec{r}) = -\frac{\rho(\vec{r})}{\varepsilon_0} = -\mu_0 c^2 \rho(\vec{r})$$

I've used the fact that $\mu_0 \varepsilon_0 = 1/c^2$ to generate the constants in the last term.

The vector potential superposes too, and (as was the case for the scalar potential) each component of it satisfies

$$\nabla^2 A_x(\vec{r}) = -\mu_0 J_x(\vec{r}'), \quad \nabla^2 A_y(\vec{r}) = -\mu_0 J_y(\vec{r}'),$$

$$\nabla^2 A_z(\vec{r}) = -\mu_0 J_z(\vec{r}')$$

or

$$\nabla^2 \vec{A}(\vec{r}) = -\mu_0 \vec{J}(\vec{r}')$$

We know this because the equations

$$V(\vec{r}) = \int_{\text{all space}} \frac{1}{4\pi\varepsilon_0} \frac{\rho(\vec{r}')}{|\vec{r} - \vec{r}'|} d^3 r'$$

and

$$A_x(\vec{r}) = \int_{\text{all space}} \frac{\mu_0}{4\pi} \frac{J_x(\vec{r}')}{|\vec{r} - \vec{r}'|} d^3 r'$$

are of the same form, with a function of position on the left side of the equation and a function of position in the integrand's numerator on the right side.

Remember that for the moment we're only dealing with charge and current densities that are unchanging: the number of Coulombs per cubic meter at a particular point never changes, no matter how long we wait.

We *can* have constant ρ when charges are moving: as long as the amount of charge $dq = \rho dV$ flowing into a small volume element equals the amount flowing out, we'll never have a change in the amount of charge contained by dV.

One more thing: in the above equations, the charge density ρ needs to include the increase in density associated with Lorentz contractions if our clouds of charge are moving rapidly. So, what I mean by ρ is really the same thing as $\gamma \rho_0$.

We'd like to describe the vector potential for a point charge in motion, but this sort of charge distribution does not satisfy the constraint that $\rho(x, y, z, t)$ is constant in time. (As the point charge moves, the charge density changes a great deal at each point the charge occupies during its travels.) However, as long as the charge is moving slowly compared to the speed of light, it's not a bad approximation to use

$$\vec{A}(\vec{r}, t) = \int_{\text{all space}} \frac{\mu_0}{4\pi} \frac{\rho(\vec{r}')\vec{v}(\vec{r}')}{|\vec{r} - \vec{r}'|} d^3 r' = \frac{\mu_0}{4\pi} \frac{q\vec{v}(t)}{|\vec{r} - \vec{r}_q(t)|}$$

where $\vec{r}_q(t)$ and $\vec{v}(t)$ represent the charge's position and velocity.

Exercise 14.2: Oscillating charge, ignoring the finite speed of light

Imagine that a charge q oscillates up and down at the origin so that its position is $\vec{r} = a \sin(\pi \times 10^9 t)\hat{z}$. Assume that the amplitude a is very small compared to one foot. Using the above expression for the vector potential (which does not take into account the propagation time associated with the finite speed of light), sketch the value of the vector potential along the positive x-axis for 1 foot $< x < 5$ feet at the times $t = 0$ and $t = 1$ ns.

Solution

At $t = 0$, $\vec{v}(t) = a\pi \times 10^9 \cos(\pi \times 10^9 t)\hat{z}$. When evaluating the vector potential along the x-axis, ignoring retardation, we get

$$\vec{A}(t) = \frac{\mu_0 q}{4\pi} \frac{a\pi \times 10^9}{x}\hat{z} = \frac{a \times 10^9 \mu_0 q}{4} \frac{1}{x}\hat{z}$$

so here it is, at 0 and 1:

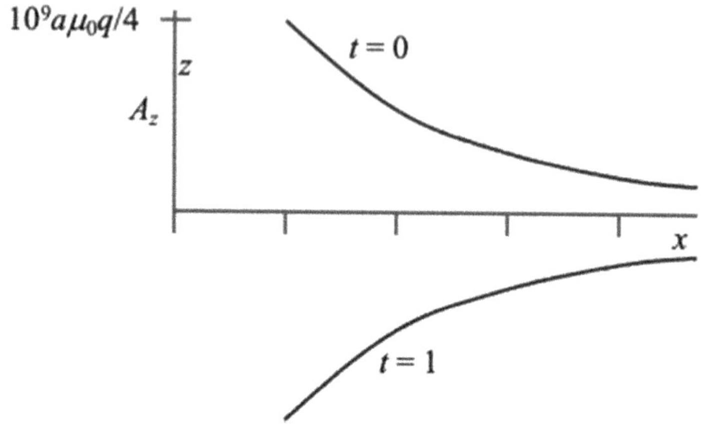

What if the charge and current distributions are time-dependent?

The more general solution is intuitively appealing. We know that nothing travels faster than the speed of light, so inside the superposition integral, we sum over where the charge was, not where it is right now. The time lag is greater for points in space farther from the observer; this retarded time is $t' = t - |\vec{r} - \vec{r}'|/c$.

What I mean is that the integral for the vector potential becomes

$$\vec{A}(\vec{r}, t) = \int_{\text{all space}} \frac{\mu_0}{4\pi} \frac{\rho(\vec{r}', t - |\vec{r} - \vec{r}'|/c)\vec{v}(\vec{r}', t - |\vec{r} - \vec{r}'|/c)}{|\vec{r} - \vec{r}'|} d^3 r'$$

$$= \int_{\text{all space}} \frac{\mu_0}{4\pi} \frac{\rho(\vec{r}', t')\vec{v}(\vec{r}', t')}{|\vec{r} - \vec{r}'|} d^3 r'$$

$$= \int_{\text{all space}} \frac{\mu_0}{4\pi} \frac{\vec{J}(\vec{r}', t')}{|\vec{r} - \vec{r}'|} d^3 r'$$

With time dependence, the current density is $\vec{J}(\vec{r}, t) \equiv \rho(\vec{r}, t)\vec{v}(\vec{r}, t)$ and the magnetic field is

$$\vec{B}(\vec{r}, t) = \vec{\nabla} \times \vec{A}(\vec{r}, t) = \frac{\mu_0}{4\pi} \vec{\nabla} \times \left[\int_{\text{all space}} \frac{\vec{J}(\vec{r}', t')}{|\vec{r} - \vec{r}'|} d^3 r' \right]$$

The time t' inside the integral is the retarded time: at each value of $\vec{r}'$ that the integral touches, t' represents the present time t set back by the amount of time a light beam would have required to arrive at the observer's position $\vec{r}$.

Sometimes the retarded time in the integrand can make the integral difficult to evaluate! Since the curl acts on unprimed coordinates, any time dependence in the current density will cause it to "react" to the curl's partial derivatives due to the presence of unprimed spatial coordinates in the retardation term.

We also use the retarded time in calculating the scalar potential when charges are in motion. To summarize, we have

$$t' \equiv t - |\vec{r} - \vec{r}'|/c$$

$$\vec{A}(\vec{r}, t) = \frac{\mu_0}{4\pi} \int_{\text{all space}} \frac{\vec{J}(\vec{r}', t')}{|\vec{r} - \vec{r}'|} d^3 r'$$

$$= \frac{\mu_0}{4\pi} \int_{\text{all space}} \frac{\rho(\vec{r}', t')\vec{v}(\vec{r}', t')}{|\vec{r} - \vec{r}'|} d^3 r'$$

$$V(\vec{r}, t) = \frac{\mu_0 c^2}{4\pi} \int_{\text{all space}} \frac{\rho(\vec{r}', t')}{|\vec{r} - \vec{r}'|} d^3 r' = \frac{\mu_0 c}{4\pi} \int_{\text{all space}} \frac{\rho(\vec{r}', t')c}{|\vec{r} - \vec{r}'|} d^3 r'$$

In spite of these complications, it is still true that the curl of the (time-dependent) vector potential is the magnetic field, though I'm not going to derive that.

Exercise 14.3: Revisiting the oscillating charge problem

Redraw your $t = 0$ graph from the previous exercise but taking into account the retardation caused by the finite speed of light, which you should take to be one foot per nanosecond.

Solution

Every foot we move away from the origin knocks the retarded time back by a nanosecond. (That's half a period.) So, we get something like this, for the vector potential *along the x-axis*:

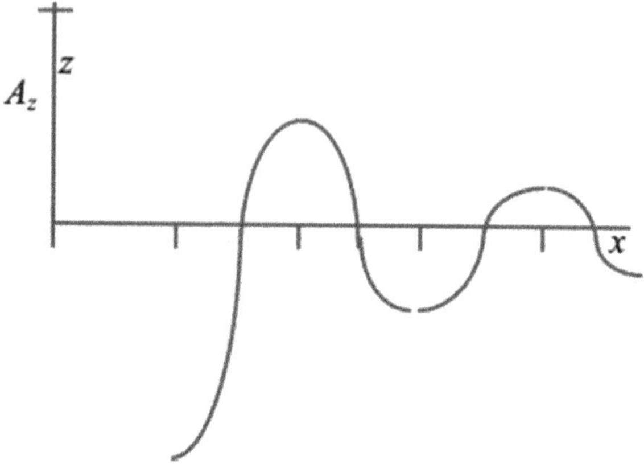

Exercise 14.4: Radiation magnetic field

Indicate on your graph from the previous exercise the direction of the curl of the vector potential. This corresponds, of course, to the magnetic field of the electromagnetic radiation emitted by the oscillating charge. Note that the y-axis is into the page.

Solution

Keep in mind that A points in the z-direction and depends only on x. The only non-zero derivative in the curl is this:

$$\vec{\nabla} \times \vec{A} = -\frac{\partial A_z}{\partial x}\,\widehat{y} \ \text{ so} \ldots$$

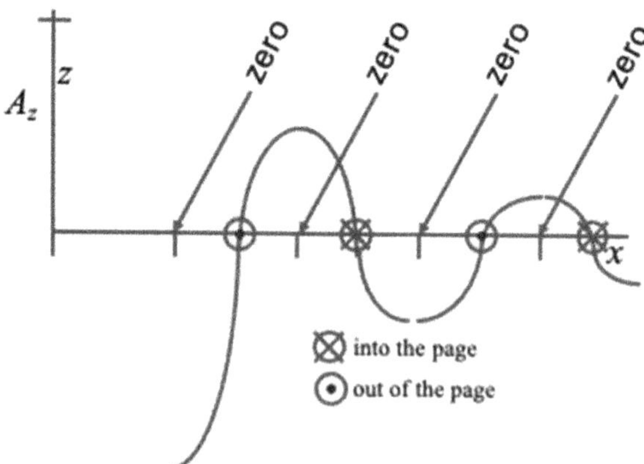

And now for the relativistic case

If the charge distributions are relativistic, their densities will increase due to Lorentz contraction. We define the four-vector potential this way:

$$A^\mu(\vec{r}, t) \equiv \left(\frac{V(\vec{r}, t)}{c}, \vec{A}(\vec{r}, t) \right)$$

If I write this for the *non-relativistic case*, using the integrals above, I have

$$A^\mu(\vec{r}, t) = \left(\frac{V(\vec{r}, t)}{c}, \vec{A}(\vec{r}, t) \right)$$

$$= \frac{\mu_0}{4\pi} \int_{\text{all space}} \frac{\rho(\vec{r}\,', t')(c, \vec{v}(\vec{r}\,', t'))}{|\vec{r} - \vec{r}\,'|} d^3 r' \quad \text{(non-relativistic case)}$$

It should be obvious to you that a good candidate for the relativistic case would be to use the four current density as the numerator in the integrand:

$$\int_{\text{all space}} \frac{\rho(\vec{r}\,', t')(c, \vec{v}(\vec{r}\,', t'))}{|\vec{r} - \vec{r}\,'|} d^3 r'$$

$$\Rightarrow \int_{\text{all space}} \frac{\gamma \rho_0(\vec{r}\,', t')(c, \vec{v}(\vec{r}\,', t'))}{|\vec{r} - \vec{r}\,'|} d^3 r' = \int_{\text{all space}} \frac{J^\mu(\vec{r}\,', t')}{|\vec{r} - \vec{r}\,'|} d^3 r'$$

That is, in fact, the correct expression for the four-vector potential:

$$A^\mu(\vec{r}, t) = \int_{\text{all space}} \frac{J^\mu(\vec{r}\,', t')}{|\vec{r} - \vec{r}\,'|} d^3 r'$$

There's a bit of subtlety for a point charge, though: a continuous distribution of charge increases in density when it is in motion, but a volume integral over a region of space that entirely contains the cloud of charge will yield the same answer whether or not the charge distribution is in motion. For a point charge, this amounts to the integral evaluating to something that is proportional to q, not γq.

Exercise 14.5: The fields are not the space components of four-vectors

It is tempting to think that the electric field ought to comprise the three space components of a four-vector of some sort, with the magnetic field filling out the space components of a different four-vector. But that's impossible. Why? When you have a good answer, write it down.

Solution

Since the electric and magnetic fields mix under boosts, you can't have something like $(E^0, \vec{E})$ behaving correctly under boosts.

Rewriting the Maxwell equations

We'd like to rewrite the Maxwell equations in terms of covariant derivatives and the four potential. Let's define the (antisymmetric) Lorentz tensor $F^{\mu\nu}$ as

$$F^{\mu\nu} \equiv \partial^\mu A^\nu - \partial^\nu A^\mu$$

This is going to turn out to be the same beast that we discussed in earlier units.

There's another tensor, the "dual" of $F^{\mu\nu}$, that we'll also want to use in a short while. We use the Levi-Civita[2] tensor density $\varepsilon_{\mu\nu\rho\sigma}$ to generate it from $F^{\rho\sigma}$ this way:

$$F^{*\mu\nu} = \sum_{\rho,\sigma} \varepsilon_{\mu\nu\rho\sigma} F^{\rho\sigma}$$

$\varepsilon_{\mu\nu\rho\sigma}$ is a rank-four object whose elements are $+1, -1$, or 0. When we contract two of its indices with $F^{\rho\sigma}$ to produce $F^{*\mu\nu}$, we end up with a tensor in which $\vec{E}/c \Rightarrow \vec{B}$ and $\vec{B} \Rightarrow -\vec{E}/c$ relative to their placements in $F^{\mu\nu}$.

Let's work out the elements of $F^{\mu\nu} \equiv \partial^\mu A^\nu - \partial^\nu A^\mu$ now.

---

[2]Tullio Levi-Civita (1873–1941), Italian mathematician. See material in *Encyclopedia Britannica* (https://www.britannica.com/biography/Tullio-Levi-Civita) or *Wikipedia* (Levi-Civita) about him.

Elements of the field strength tensor corresponding to the magnetic field

Recall that $\partial^\mu \equiv \left(\frac{\partial}{\partial ct}, -\frac{\partial}{\partial x}, -\frac{\partial}{\partial y}, -\frac{\partial}{\partial z} \right)$, and take note of the negative signs.

Since $\vec{B} = \vec{\nabla} \times \vec{A}$, we can write

$$B_x = -\partial^2 A^3 + \partial^3 A^2 = F^{32} = -F^{23}$$

$$B_y = -\partial^3 A^1 + \partial^1 A^3 = F^{13} = -F^{31}$$

$$B_z = -\partial^1 A^2 + \partial^2 A^1 = F^{21} = -F^{12}.$$

So far so good. We have six of the field strength tensor's elements in hand.

The four diagonal elements must be zero since F is antisymmetric. That makes ten.

Elements of the field strength tensor corresponding to the electric field

Start with the curl equation for the electric field, then rewrite the magnetic field in terms of the vector potential:

$$\vec{\nabla} \times \vec{E} = -\frac{\partial \vec{B}}{\partial t}$$

$$= -\frac{\partial \vec{\nabla} \times \vec{A}}{\partial t}$$

$$= -\vec{\nabla} \times \frac{\partial \vec{A}}{\partial t}$$

From this, we can conclude that the electric field is the same as something with zero curl added to $-\partial \vec{A}/\partial t$. We already know that the "something with zero curl" is $-\vec{\nabla}V$, the negative of the gradient of the scalar potential. From this, we have

$$\vec{E}(\vec{r}, t) = -\vec{\nabla}V(\vec{r}, t) - \frac{\partial \vec{A}(\vec{r}, t)}{\partial t}$$

We can rewrite this using our new notation as follows:

$$E_x = +\partial^1 V - c\partial^0 A_x, \quad E_y = +\partial^2 V - c\partial^0 A_y, \quad E_z = +\partial^3 V - c\partial^0 A_z$$

or, in terms of the four vector potential,

$$E_x = +c\partial^1 A^0 - c\partial^0 A^1, \quad E_y = +c\partial^2 A^0 - c\partial^0 A^2,$$

$$E_z = +c\partial^3 A^0 - c\partial^0 A^3$$

Dividing both sides of this equation by c and recalling the above definition of $F^{\mu\nu}$ gives

$$E_x/c = \partial^1 A^0 - \partial^0 A^1 = F^{10} = -F^{01}$$

$$E_y/c = \partial^2 A^0 - \partial^0 A^2 = F^{20} = -F^{02}$$

$$E_z/c = \partial^3 A^0 - \partial^0 A^3 = F^{30} = -F^{03}$$

We are done:

$$F^{\mu\nu} = \begin{bmatrix} 0 & -E_x/c & -E_y/c & -E_z/c \\ E_x/c & 0 & -B_z & B_y \\ E_y/c & B_z & 0 & -B_x \\ E_z/c & -B_y & B_x & 0 \end{bmatrix}$$

If you look back at earlier units (3 and 7, if I recall correctly), you'll see that this agrees with the definition of $F^{\mu\nu}$ that I asserted back then without proof. And since F is constructed from the outer product of two objects that transform like four-vectors, it automatically transforms like a tensor under Lorentz transformations.

The dual of F is

$$F^{*\mu\nu} = \begin{bmatrix} 0 & -B_x & -B_y & -B_z \\ B_x & 0 & E_z/c & -E_y/c \\ B_y & -E_z/c & 0 & E_x/c \\ B_z & E_y/c & -E_x/c & 0 \end{bmatrix}$$

It too is a tensor.

Exercise 14.6: Zip

The trace of $F^{\mu\nu}$ is $\sum_\mu F^\mu{}_\mu = \sum_\mu [\sum_\nu g_{\mu\nu} F^{\mu\nu}] = \sum_\mu [\sum_\nu F^{\mu\nu} g_{\nu\mu}]$. Note that the sum over ν is just the matrix multiplication of F and g. Prove that the trace of F is zero. The trace of a tensor is a Lorentz scalar. Why?

Solution

The easy way: $F^{\mu\nu}$ has zeroes on the diagonal, so its trace is zero. This fact is preserved when it is multiplied by $g_{\mu\nu}$ so that $F^{\mu}{}_{\nu}$ will also have zeroes on the diagonal. As a result, its trace will also be zero.

Derivatives of the field strength tensors

Recall the third bullet in the list of four-vector properties: the contraction of a four-vector and one index of a tensor is a four-vector. That works fine when the four vector is the four-divergence.

Let's see what we get if we do the following contractions: $\sum_\mu \partial_\mu F^{\mu\nu}$ and $\sum_\mu \partial_\mu F^{*\mu\nu}$.

Differentiating F

Crank out the differentiations, one component at a time:

$$\sum_\mu \partial_\mu F^{\mu 0} = 0 + \frac{1}{c}\left(\frac{\partial E_x}{\partial x} + \frac{\partial E_y}{\partial y} + \frac{\partial E_z}{\partial z}\right) = \frac{1}{c}\vec{\nabla}\cdot\vec{E}$$

$$\sum_\mu \partial_\mu F^{\mu 1} = -\frac{1}{c^2}\frac{\partial E_x}{\partial t} + 0 + \frac{\partial B_z}{\partial y} - \frac{\partial B_y}{\partial z} = -\frac{1}{c^2}\frac{\partial E_x}{\partial t} + (\vec{\nabla}\times\vec{B})_x$$

$$\sum_\mu \partial_\mu F^{\mu 2} = -\frac{1}{c^2}\frac{\partial E_y}{\partial t} - \frac{\partial B_z}{\partial x} + 0 + \frac{\partial B_x}{\partial z} = -\frac{1}{c^2}\frac{\partial E_y}{\partial t} + (\vec{\nabla}\times\vec{B})_y$$

$$\sum_\mu \partial_\mu F^{\mu 3} = -\frac{1}{c^2}\frac{\partial E_z}{\partial t} + \frac{\partial B_y}{\partial x} - \frac{\partial B_x}{\partial y} + 0 = -\frac{1}{c^2}\frac{\partial E_z}{\partial t} + (\vec{\nabla}\times\vec{B})_z$$

Putting these together, we obtain

$$\sum_\mu \partial_\mu F^{\mu\nu} = \left(\frac{1}{c}\vec{\nabla}\cdot\vec{E}, -\frac{1}{c^2}\frac{\partial\vec{E}}{\partial t} + \vec{\nabla}\times\vec{B}\right)$$

We know from the Maxwell Equations that $\vec{\nabla}\times\vec{B} = \mu_0\varepsilon_0\frac{\partial\vec{E}}{\partial t} + \mu_0\vec{J} = \frac{1}{c^2}\frac{\partial\vec{E}}{\partial t} + \mu_0\vec{J}$, so we can replace $-\frac{1}{c^2}\frac{\partial\vec{E}}{\partial t} + \vec{\nabla}\times\vec{B}$ with $\mu_0\vec{J}$. We can also use the Maxwell equation $\vec{\nabla}\cdot\vec{E} = \rho/\varepsilon_0$ to replace the divergence term.

With both these "improvements," we can rewrite the four-divergence of F as

$$\sum_\mu \partial_\mu F^{\mu v} = \left(\frac{1}{c\varepsilon_0}\rho, \mu_0 \vec{J}\right)$$

$$= (\mu_0 c\rho, \mu_0 \vec{J}) = \mu_0(c\gamma\rho_0, \vec{v}\gamma\rho_0)$$

$$= \mu_0\rho_0\gamma(c, \vec{v}) = \mu_0\rho_0 u^v$$

$$= \mu_0 J^v$$

To summarize,

$$\sum_\mu \partial_\mu F^{\mu v} = \mu_0 J^v$$

is equivalent to the Maxwell Equations

$$\vec{\nabla} \cdot \vec{E} = \frac{\rho}{\varepsilon_0}$$

$$\vec{\nabla} \times \vec{B} - \frac{1}{c^2}\frac{\partial \vec{E}}{\partial t} = \mu_0 \vec{J}$$

Differentiating F^*

Now, take the four-divergence of F^*:

$$\sum_\mu \partial_\mu F^{*\mu 0} = 0 + \left(\frac{\partial B_x}{\partial x} + \frac{\partial B_y}{\partial y} + \frac{\partial B_z}{\partial z}\right) = \vec{\nabla} \cdot \vec{B}$$

$$\sum_\mu \partial_\mu F^{*\mu 1} = -\frac{1}{c}\frac{\partial B_x}{\partial t} + 0 - \frac{1}{c}\frac{\partial E_z}{\partial y} + \frac{1}{c}\frac{\partial E_y}{\partial z} = -\frac{1}{c}\frac{\partial B_x}{\partial t} - \frac{1}{c}(\vec{\nabla} \times \vec{E})_x$$

$$\sum_\mu \partial_\mu F^{*\mu 2} = -\frac{1}{c}\frac{\partial B_y}{\partial t} + \frac{1}{c}\frac{\partial E_z}{\partial x} + 0 - \frac{1}{c}\frac{\partial E_x}{\partial z} = -\frac{1}{c}\frac{\partial B_y}{\partial t} - \frac{1}{c}(\vec{\nabla} \times \vec{E})_y$$

$$\sum_\mu \partial_\mu F^{*\mu 3} = -\frac{1}{c}\frac{\partial B_z}{\partial t} - \frac{1}{c}\frac{\partial E_y}{\partial x} + \frac{1}{c}\frac{\partial E_x}{\partial y} + 0 = -\frac{1}{c}\frac{\partial B_z}{\partial t} - \frac{1}{c}(\vec{\nabla} \times \vec{E})_z$$

Putting these together, we obtain

$$\sum_\mu \partial_\mu F^{*\mu v} = \left(\vec{\nabla} \cdot \vec{B}, \quad \frac{1}{c}\left[-\frac{\partial \vec{B}}{\partial t} - \vec{\nabla} \times \vec{E}\right]\right)$$

But we know from the Maxwell Equations that

$$\vec{\nabla} \cdot \vec{B} = 0 \quad \text{and} \quad \vec{\nabla} \times \vec{E} = -\frac{\partial \vec{B}}{\partial t} \quad \text{so} \quad \sum_{\mu} \partial_{\mu} F^{*\mu\nu} = 0.$$

Summarizing once more,

$$\sum_{\mu} \partial_{\mu} F^{*\mu\nu} = 0$$

is equivalent to the Maxwell Equations

$$\vec{\nabla} \cdot \vec{B} = 0$$

$$\vec{\nabla} \times \vec{E} = \frac{\partial \vec{B}}{\partial t}$$

That was our goal, to write a system of equations equivalent to the Maxwell Equations that uses nothing but objects with good Lorentz transformation properties.

A summary

1. The Maxwell Equations are local equations, describing the relationships between the space and time derivatives of the electric and magnetic fields and the sources of those fields, namely the electric charge and electric current at points in space–time. These coupled equations are commonly written this way:

$$\vec{\nabla} \cdot \vec{E} = \frac{\rho}{\varepsilon_0} \qquad \vec{\nabla} \cdot \vec{B} = 0$$

$$\vec{\nabla} \times \vec{E} = -\frac{\partial \vec{B}}{\partial t} \qquad \vec{\nabla} \times \vec{B} = \mu_0 \varepsilon_0 \frac{\partial \vec{E}}{\partial t} + \mu_0 \vec{J}$$

The constants μ_0 and ε_0 are obtained from static measurements of the magnetic field induced by a constant current and the Coulomb force between two stationary charges.

2. We obtain the wave equations by taking the curl of the two Maxwell Equations that contain the curl of the fields. The wave equations relate the second derivatives in space and time of a field. They are

$$\nabla^2 \vec{E} = \mu_0 \varepsilon_0 \frac{\partial^2 \vec{E}}{\partial t^2} \qquad \nabla^2 \vec{B} = \mu_0 \varepsilon_0 \frac{\partial^2 \vec{B}}{\partial t^2}$$

and reveal that the speed of propagation in empty space of a disturbance in the fields is $c = 1/\sqrt{\mu_0 \varepsilon_0}$. The propagation speed of a pattern in the fields is entirely independent of any motion of the source relative to the observer.

3. We can express the electric and magnetic fields in terms of the scalar and vector potentials. They are calculated as integrals over all of space of the charge and current densities, with the complication that the time used in the expressions inside the integrals includes a retardation term which varies with distance from the observer. We have

$$t' \equiv t - |\vec{r} - \vec{r}\,'|/c$$

$$V(\vec{r}, t) = \frac{\mu_0 c^2}{4\pi} \int_{\text{all space}} \frac{\rho(\vec{r}\,', t')}{|\vec{r} - \vec{r}\,'|} d^3 r' = \frac{\mu_0 c}{4\pi} \int_{\text{all space}} \frac{\rho(\vec{r}\,', t')c}{|\vec{r} - \vec{r}\,'|} d^3 r'$$

$$\vec{A}(\vec{r}, t) = \frac{\mu_0}{4\pi} \int_{\text{all space}} \frac{\vec{J}(\vec{r}\,', t')}{|\vec{r} - \vec{r}\,'|} d^3 r'$$

$$= \frac{\mu_0}{4\pi} \int_{\text{all space}} \frac{\rho(\vec{r}\,', t')\vec{v}(\vec{r}\,', t')}{|\vec{r} - \vec{r}\,'|} d^3 r'$$

The charge density ρ includes (position and time-dependent) increases in local density caused by Lorentz contraction. A better (but equivalent) way to define the potentials makes use of the four-current density.

For a point charge q at retarded position $\vec{r}\,'$, the integrals evaluate to

$$V(\vec{r}, t) = \frac{\mu_0 c^2}{4\pi} \frac{q}{|\vec{r} - \vec{r}\,'|} \qquad \vec{A}(\vec{r}, t) = \frac{\mu_0}{4\pi} \frac{q\vec{v}(t')}{|\vec{r} - \vec{r}\,'|}$$

4. The four-current density is $J^\mu \equiv \rho_0 u^\mu = \gamma \rho_0 (c, \vec{v})$ where ρ_0 is the charge density measured by an observer at rest with respect to the local cloud of charge.

5. We define the four-vector potential this way:

$$A^\mu(\vec{r}, t) \equiv \left(\frac{V(\vec{r}, t)}{c}, \vec{A}(\vec{r}, t) \right)$$

It can be calculated in terms of the four-current density this way:

$$A^\mu(\vec{r}, t) = \int_{\text{all space}} \frac{J^\mu(\vec{r}\,', t')}{|\vec{r} - \vec{r}\,'|} d^3 r'$$

Note the use of the retarded time inside the integral.

6. We define the contravariant and covariant gradient operators this way:

$$\partial^\mu \equiv \left(\frac{\partial}{\partial ct}, -\frac{\partial}{\partial x}, -\frac{\partial}{\partial y}, -\frac{\partial}{\partial z}\right) = \left(\frac{\partial}{\partial x^0}, -\frac{\partial}{\partial x^1}, -\frac{\partial}{\partial x^2}, -\frac{\partial}{\partial x^3}\right)$$

$$= \left(\frac{\partial}{\partial ct}, -\vec{\nabla}\right)$$

$$\partial_\mu \equiv \left(\frac{\partial}{\partial ct}, +\frac{\partial}{\partial x}, +\frac{\partial}{\partial y}, +\frac{\partial}{\partial z}\right) = \left(\frac{\partial}{\partial x_0}, -\frac{\partial}{\partial x_1}, -\frac{\partial}{\partial x_2}, -\frac{\partial}{\partial x_3}\right)$$

$$= \left(\frac{\partial}{\partial ct}, \vec{\nabla}\right)$$

The contraction of these with an appropriate four-vector (covariant for ∂^μ and contravariant for ∂_μ) always yields a Lorentz scalar. The contraction of one of these with a rank-two tensor yields a four-vector. The outer product of one of these with a four vector yields a rank-two tensor.

7. We define the electromagnetic field strength tensor as $F^{\mu\nu} \equiv \partial^\mu A^\nu - \partial^\nu A^\mu$.

   Its dual is $F^{*\mu\nu} \equiv \sum_{\rho,\sigma} \varepsilon_{\mu\nu\rho\sigma} F^{\rho\sigma}$.

   The elements of these tensors are as follows:

$$F^{\mu\nu} = \begin{bmatrix} 0 & -E_x/c & -E_y/c & -E_z/c \\ E_x/c & 0 & -B_z & B_y \\ E_y/c & B_z & 0 & -B_x \\ E_z/c & -B_y & B_x & 0 \end{bmatrix}$$

$$F^{*\mu\nu} = \begin{bmatrix} 0 & -B_x & -B_y & -B_z \\ B_x & 0 & E_z/c & -E_y/c \\ B_y & -E_z/c & 0 & E_x/c \\ B_z & E_y/c & -E_x/c & 0 \end{bmatrix}$$

8. By differentiating F and F^*, we obtain the Maxwell Equations in covariant form:

$$\sum_\mu \partial_\mu F^{\mu\nu} = \mu_0 J^\nu \quad \Leftrightarrow \quad \left[\begin{array}{l} \vec{\nabla} \cdot \vec{E} = \dfrac{\rho}{\varepsilon_0} \\[2ex] \vec{\nabla} \times \vec{B} - \dfrac{1}{c^2}\dfrac{\partial \vec{E}}{\partial t} = \mu_0 \vec{J} \end{array} \right.$$

$$\sum_{\mu} \partial_{\mu} F^{*\mu\nu} = 0 \iff \begin{bmatrix} \vec{\nabla} \cdot \vec{B} = 0 \\ \\ \vec{\nabla} \times \vec{E} = \dfrac{\partial \vec{B}}{\partial t} \end{bmatrix}$$

9. We can express the fact that charge is conserved in terms of the four-current density this way:

$$\sum_{\mu=0}^{\mu=3} \partial_{\mu} J^{\mu} = 0$$

Summary

We discussed:

- gauge transformations;
- covariant derivatives;
- the Helmholtz theorem;
- vector potential;
- retarded time and causality;
- radiation magnetic field;
- the Maxwell Equations in terms of covariant derivatives of the field strength tensor.

And that's the end of the story!

Problem 1: In a different context, this would be inappropriate

In this problem, assume that the speed of light is $0.3\,\text{m}$ per nanosecond.

A charge q sails along in the x, y plane, following a trajectory that is parallel to the x-axis and passes through the point $(x, y) = (0, 30\,\text{m})$ at time $t = 0$. An instrument package is at rest at the origin. A large number of stationary synchronized clocks, at rest with respect to the instrument package, are placed throughout the reference frame.

At $t = 0$, a camera in the origin-placed instrument package opens its shutter briefly and takes a photograph of the region defined by $y > 0$. The moving charge happens to be close to one of the clocks in the photograph.

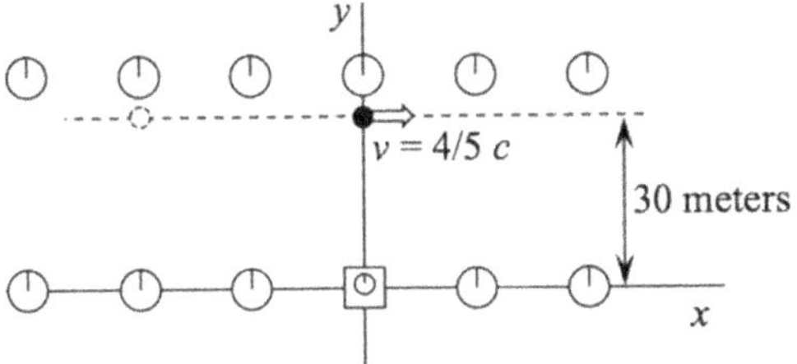

(a) In the photograph, where does the moving charge appear to be, and what is the reading on the stationary clock it is passing?

Solution

The charge's true position at time t is $(0.8ct, 30)$, using units of meters and nanoseconds. The observer is at the origin, so (neglecting retardation) the observer-charge distance is $\sqrt{(0.8ct)^2 + 30^2} = \sqrt{(0.24t)^2 + 30^2}$. At retarded time t', the separation is

$$r' = \sqrt{30^2 + (0.24t')^2}.$$

Since $t' = t - r'/c$ we have

$$r' = \sqrt{30^2 + (0.24(t - r'/c))^2}.$$

Square both sides and solve for r':

$$(r')^2 = 30^2 + (0.24(t - r'/c))^2$$

Let's switch back to using v, and do some algebra.

$$r'^2 = 30^2 + v^2t^2 - 2r'v^2t/c + \frac{v^2 r'^2}{c^2}$$

$$r' = -\gamma^2 v^2 t/c \pm \gamma^2\sqrt{(v^2t/c)^2 + (30^2 + v^2t^2)/\gamma^2}$$

For $t = 0$, we get $r' = 30\gamma = 50$.

As a result, the observer at the origin sees the moving charge when it was 50 m away. Recall the Pythagorean theorem! This separation happens when the charge was at $x = -40$ since $40^2 + 30^2 = 50^2$. And since the clock is moving at $0.8c$, it was at $x = -40$ at time

$-40\,\text{m}/0.24$ m per nanosecond $= -500/3$ nanoseconds according to the stationary clocks.

Recall that the definitions of the scalar and vector potentials for a moving charge include retardation effects:

$$V(\vec{r}, t) = \frac{\mu_0 c^2}{4\pi} \frac{q}{|\vec{r} - \vec{r}\,'|} \qquad \vec{A}(\vec{r}, t) = \frac{\mu_0}{4\pi} \frac{q\vec{v}(t')}{|\vec{r} - \vec{r}\,'|}$$

where the retarded time is $t' \equiv t - |\vec{r} - \vec{r}\,'|/c$ and the retarded position is the charge's position at the retarded time, not the current time t. Your answers to part (a) are the retarded position and time corresponding to time $t = 0$.

There is a curious thing here: if you were to grind through the differentiations to calculate the gradient of the scalar potential, you'd find that it pointed toward the retarded position of the charge, indicated by the dashed circle in the above figure. But in doing this, you'd be leaving out the time derivative of the vector potential since the electric field comprises both the gradient of the scalar potential and the time derivative of the vector potential.

The vector potential, calculated at the retarded time, points in the x-direction and grows stronger as the retarded position of the moving charge approaches the point $x' = 0$. As a result, its time derivative is in the positive x-direction. When you include this time derivative, it causes the electric field to point toward the instantaneous position of the charge rather than where the charge appeared in the photograph.

(b) In a later photograph, the charge is observed to be at the point $(x, y) = (0, 30)$. What is the true position of the charge when this photograph is taken?

Solution

The light emitted as the charge passes through that point required $30/0.3 = 100$ nanoseconds to travel to the origin. So now, in reality, the moving charge has traveled for another 100 nsec since passing through $x = 0$. That means it is now at $x = 0.24$ m per nanosecond $\times$ 100 nanoseconds $= 24$ m.

The behavior of the direction of the electric field may sound like it violates that most basic tenet of relativity, that *nothing* (including information) can travel faster than the speed of light. So, what happens if the charge stops the instant it reaches $(x, y) = (0, 30)$ rather than continuing its motion in the positive x-direction? How do the fields look?

What happens is this. Conservation of charge guarantees that field lines can only begin and end on positive and negative charges. Far from the charge, observers see the field lines pointing toward the position the charge would assume if it had not stopped. Closer to the charge, observers see the field pointing toward the now-stopped charge. There is a transition region in which the field lines show dramatic kinks, which propagate outwards at the speed of light. That's electromagnetic radiation, associated with the deceleration of the charge. In this particular case, it's referred to as *bremsstrahlung* radiation; if you've taken German, you might recognize the root verb *bremsen*—to brake. It's how a dentist's x-ray machine makes x-rays. Since "strahlung" means radiation, I suppose the English naming is redundant, like asking for an ATM machine.

Here's a picture of some of the field lines near a charge that had been moving at $0.5c$, then stopped suddenly. (I haven't included the bunching of lines in the plane perpendicular to the charge's initial velocity.)

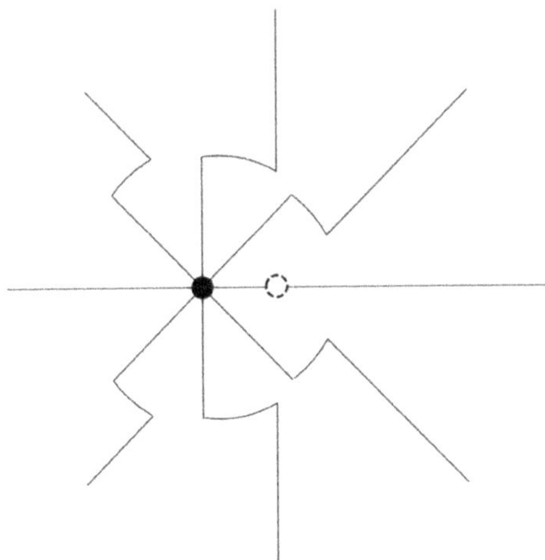

Problem 2: You'll want to refer to the Maxwell Equations

Prove that $\sum_{v=0}^{v=3} \partial_v \left[\sum_{\mu=0}^{\mu=3} \partial_\mu F^{\mu v} \right] = 0$.

Solution

This one's really easy! From Maxwell,

$$\sum_\mu \partial_\mu F^{\mu v} = \mu_0 J^v$$

$$\sum_v \partial_v \left[\sum_\mu \partial_\mu F^{\mu v} \right] = \mu_0 \sum_v \partial_v J^v$$

and, from charge conservation

$$\sum_v \partial_v J^v = 0$$

so that

$$\sum_v \partial_v \left[\sum_\mu \partial_\mu F^{\mu v} \right] = 0$$

Problem 3: Indexology

Recall that we change contravariant objects into covariant objects by multiplying them by the metric tensor, once for each index to be lowered. Recall that the (flat spacetime) metric is

$$g_{\mu v} = g^{\mu v} = \begin{bmatrix} 1 & 0 & 0 & 0 \\ 0 & -1 & 0 & 0 \\ 0 & 0 & -1 & 0 \\ 0 & 0 & 0 & -1 \end{bmatrix}$$

As a result, $F_{\mu v} = \sum_\beta g_{\beta v} \left[\sum_\alpha g_{\mu\alpha} F^{\alpha\beta} \right] = \sum_\beta g_{\beta v} F_\mu{}^\beta$. And since the metric tensor is symmetric, I can rewrite this as $F_{\mu v} = \sum_\beta g_{v\beta} \left[\sum_\alpha g_{\mu\alpha} F^{\alpha\beta} \right] = \sum_{\alpha,\beta} g_{\mu\alpha} g_{v\beta} F^{\alpha\beta}$.

(a) Prove that the elements of $F_{\mu\nu}$ holding the *electric* field components are opposite in sign to those in $F^{\mu\nu}$, while those holding the *magnetic* field components are identical in sign.

Solution

This is just a matter of grinding out the matrix multiplications. First, do the product, after writing the 4×4 matrices corresponding to

$$F_\mu^\beta = \sum_\alpha g_{\mu\alpha} F^{\alpha\beta}$$

You'll find that some elements have their signs flipped. Then do the multiplication

$$F_{\mu\nu} = \sum_\beta g_{\nu\beta} F_\mu^\beta$$

to finish the calculation.

(b) Prove that the elements of $F^*_{\mu\nu}$ holding the *magnetic* field components are opposite in sign to those in $F^{*\mu\nu}$, while those holding the *electric* field components are identical in sign.

Solution

Once again, it's just a matter of grinding out the matrix multiplications. You can find the explicit 4×4 representations of the matrices you need earlier in the book. Just plug and chug!

As I mentioned earlier, "a product of two tensors done with contraction of one index is a tensor." That is just a fancy way of describing matrix multiplication of a covariant and contravariant tensor. The fact that it is a tensor means that it will transform properly under changes of reference frame. And note that the identity matrix and a matrix filled entirely with zeroes will not change under a Lorentz transformation.

The product of the matrices A and B is done this way: $C_\mu^{\ \nu} = \sum_{\alpha=0}^{\alpha=3} A_{\mu\alpha} B^{\alpha\nu}$. That's an example of what I mean by "a product of two tensors with contraction of one index."

(c) Imagine that in one frame of reference, it is found that the electric and magnetic fields are orthogonal (perpendicular) at all points in space: $\vec{E} \cdot \vec{B} = 0$.

By calculating $K^\mu{}_\nu = \sum_{\alpha=0}^{\alpha=3} F^{\mu\alpha} F^*{}_{\alpha\nu}$, prove that the electric and magnetic fields are orthogonal in all frames.

Solution

Again, just grind out the matrix multiplications. What you should find is that

$$\sum_{\alpha=0}^{\alpha=3} F^{\mu\alpha} F^*_{\alpha\nu}$$

$$= \frac{1}{c} \begin{bmatrix} (E_xB_x + E_yB_y + E_zB_z) & 0 & 0 & 0 \\ 0 & (E_xB_x + E_yB_y + E_zB_z) & 0 & 0 \\ 0 & 0 & (E_xB_x + E_yB_y + E_zB_z) & 0 \\ 0 & 0 & 0 & (E_xB_x + E_yB_y + E_zB_z) \end{bmatrix}$$

$$= \frac{\vec{E} \cdot \vec{B}}{c} \begin{bmatrix} 1 & 0 & 0 & 0 \\ 0 & 1 & 0 & 0 \\ 0 & 0 & 1 & 0 \\ 0 & 0 & 0 & 1 \end{bmatrix}$$

As a result, if the electric and magnetic fields are perpendicular in one frame (so that their dot product is zero), they'll be perpendicular in all frames since that product of the two Lorentz tensors transforms well under changes of reference frame.

Problem 4: Theorems to the rescue

In Mrs. Rumpdock's neighborhood near Spokane, weird things involving electromagnets are an ordinary part of everyday life.

(a) One day, Rumpdock's daughter Hikalope discovers that there is an electric field in their back yard that is well described as $\vec{E} = A[x\hat{y} - y\hat{x}]$. (The origin of the coordinate system is centered above the Rumpdocks' septic tank. The ground defines the x, y plane; z is up.)

Hikalope places a circular loop of radius b on the ground, centered at the point $(x, y) = (3b, 5b)$. She proceeds to measure the line integral $\oint_{\text{loop}} \vec{E} \cdot d\vec{l}$, traveling around the loop so that her path is counterclockwise when seen from above.

A high school dropout of low morals and low intelligence, Hikalope is unaware of the theorems of vector calculus and performs her measurement the hard way. But you know better. Determine that line integral in as graceful and concise a manner as you can.

Solution

Use Stokes' theorem:

$$\oint \vec{E} \cdot d\vec{l} = \int (\vec{\nabla} \times \vec{E}) \cdot d\vec{A}.$$

$$\vec{\nabla} \times \vec{E} = \left(\frac{\partial E_y}{\partial x} - \frac{\partial E_x}{\partial y} \right) \hat{z} = A(1 - (-1)) = 2A$$

This is constant, so

$$\oint \vec{E} \cdot d\vec{l} = 2A \times \pi b^2.$$

(b) Convinced that the source of the electric field must be a large blob of positive electric charge inside the septic tank, Mr. Rumpdock threatens to sue everyone with in-depth knowledge of the Maxwell Equations.

Acting as a consultant to the plaintiffs, you are asked to prove that Mr. Rumpdock cannot possibly be right. With a little thought, you should be able to suggest a measurement proving them wrong. Write the proof in three or four lines. Do so.

Solution

Surround the septic tank with a large cube and measure the electric field on its surfaces. Since

$$\vec{\nabla} \cdot \vec{E} = \frac{\rho}{\varepsilon_0}, \quad \int_{\text{cube}} \vec{\nabla} \cdot \vec{E}\, dV = \int_{\text{cube}} \frac{\rho}{\varepsilon_0} dV = Q_{\text{enclosed}}$$

But

$$\vec{\nabla} \cdot (x\hat{y} - y\hat{x}) = 0$$

so

$$Q_{\text{enclosed}} = 0$$

Sagittarius/Rose

Afterword

But why?

I began this book by explaining why some of us chose to become physicists. One of my professors would tell us that "Physics is supposed to make your head feel funny,"[1] and he was right. Early on, my friends and I learned that there are more interesting things to think about than masses sliding down inclined planes.

---

[1] Sidney Coleman (1937–2007), "Sidney Coleman dies at 70: Luminary of theoretical physics was a member of the Harvard faculty for 43 years," *The Harvard Gazette*, November 29, 2007, https://news.harvard.edu/gazette/ story/2007/11/sidney-coleman-dies-at-70/.

We know that a Newtonian universe cannot support life and that a Newtonian space–time would render the Earth as hot as the surface of a star.[2] Carbon-based life could not exist, even in a quantum mechanical world, without the purely relativistic Pauli Exclusion Principle.[3]

I could have found deep satisfaction in prosecuting white collar criminals or building medical devices for impoverished communities. But the fever-dream strangeness of the physical world still thrills me, five decades into my career, so this has been a good direction.

There is loveliness in the crystalline perfection of classical mechanics and electrodynamics, but it is the hallucinatory aspects of relativity and quantum mechanics that excite me the most. This is why I so love to teach: I want to share this with other people.

Teaching relativity

As I mentioned in the first unit, I teach special relativity in ways that are similar to how I first learned it, relying on the absolute constancy of the speed of light to force on us a whole raft of surprising conclusions.

General relativity is a different matter, though: as a graduate student, my teachers would invariably adopt a more mathematical approach than was appropriate for me. I am an experimentalist, not a theorist, and the blizzard of Christoffel symbols and differential forms that my professors blasted out at us left many of us in the (pedagogical) dust. So, I come at the subject in a more geometrical fashion, loosely following some of the development in Taylor and Wheeler's *Exploring Black Holes: Introduction to General Relativity*.[4]

---

[2] "Olber's Paradox: Why Is The Sky Dark at Night?," American Museum of Natural History, https://www.amnh.org/exhibitions/journey-to-the-stars/educator-resources/stars/olbers-paradox, visited September 5, 2024.

[3] "January 1925: Wolfgang Pauli announces the exclusion principle," American Physical Society Archives, https://www.aps.org/archives/publications/apsnews/200701/history.cfm, visited September 5, 2024.

[4] Edwin F. Taylor and John Archibald Wheeler, *Exploring Black Holes: Introduction to General Relativity First Edition*, Addison Wesley Longman (2000).

I have found that this works nicely! My students, mostly physics majors in the second or third year, often find that the three units of general relativity feel more natural and sensible to them than the other topics we cover.

It is curious that general relativity isn't usually taught to undergraduates. It really should be included in the physics canon, along with classical mechanics, electrodynamics, quantum mechanics, and statistical mechanics. Perhaps this is a self-perpetuating omission since so few of my professorial colleagues learned enough about the subject to feel confident in teaching it. What a shame! And with the ever-increasing flow of remarkable observations—of coalescing binary black holes, binary neutron stars, and so forth—the subject's role in our ever-deepening understanding of the universe will increase steadily.

I have taught Physics 225, the University of Illinois's relativity course, six times. Since our semesters are $14\frac{1}{2}$ weeks long, I've divided the material into 14 units, each beginning on a Monday with an hour-long lecture sketching the topics for the week. Most of the actual learning takes place in a two-hour problem session, in which small groups of two or three students work a sequence of problems on the blackboards (or, in these modern times, whiteboards!) of the classroom. I would always teach the first session of the week, prowling around as my students worked, then meet with the other instructors to apprise them of what I had seen. The problem sessions were required, and the informal, relaxed setting gave me a chance to get to know some of my charges. We spent almost the entire time on our feet.

The problems in this book identified as "exercises" comprise the full set of in-class investigations I would have my students tackle. I did not provide them with the solutions, although I have included them for you. The slightly more complex puzzlers identified as "problems" were assigned as homework, to be submitted for grading a week later. Again, students were not shown the problems' solutions until after the due date. I expected a student to need about four hours to work each week's homework assignment.

I've always found that homework (with significant penalties for late submission) and examinations help students keep up with the course. This way, they are less likely to discover a week before the end of the term that they have yet to master the subject material!

I hope to include in an appendix, to be published online, the two midterm exams, as well as the final for my six offerings of Physics 225. I do not plan to include solutions, however!

Whenever possible, I let my students take tests in the evening so they could have as much time as they wanted to work an exam. I wrote the midterms hoping that most students would finish in an hour and a half; in actuality, nearly everyone was done in two hours.

I did not allow students to use calculators, smartphones, or computers. Each exam included a sheet of useful equations. Since we would occasionally have problems with cheating, I constructed several versions of each exam, placing all the figures at the very end of the exam booklet and assigning different titles to otherwise identical problems.

The first midterm exam covered Units 1–5, the second Units 6–10. The final exam was comprehensive, covering material from the entire course.

So who are the Rumpdocks, anyway?

Two things: I don't like to be bored, and I hold grudges.

I mentioned earlier in this book that I did not have a burning desire to analyze objects sliding down inclined planes. The same goes for writing homework problems about inclined planes. So, I craft little story problems in which, for example, "Mrs. Blepon" is flying along in a relativistic spaceship with a light clock and a bomb.

About holding a grudge: there was a hardware store in the town where I lived as a graduate student owned by the very nasty "Mrs. Blepon." She would follow me around when I would go into her store to buy nuts and bolts because she thought I would steal from her. A few years later, when I was back as an assistant professor, she continued to be as nasty as when I had been a poorly paid student. So, for my amusement, I visit (entirely harmless) vengeance on Mrs. Blepon by blowing her up from time to time.

Now, let me tell you about the "Rumpdocks."

They were American criminals in eastern Washington State who began bribing Liberian officials before the end of that country's disastrous civil war. They sold fake college diplomas and transcripts to anyone with an internet connection and a credit card.[5] Their intent was to obtain documents attesting to the legitimacy of a string of fake universities that they would claim were situated in Monrovia.

At a time when life expectancy in the country was 39 years and a seventh of its children died in infancy,[6] they ultimately controlled significant portions of Liberia's diplomatic corps, the Ministry of Justice, and the Ministry of Education.[7]

I learned about them and published a report outlining their activities. Pretending to be Liberian diplomats, they wrote to my university and threatened a lawsuit. An associate provost suggested he'd fire me, so we had a punch-up in the *Chronicle of Higher Education*, and my administration backed down.[8]

The Rumpdocks and their allies began bombarding my family with obscene messages. Five years later, sitting with fellow investigators from eight law enforcement agencies, I watched a Federal judge sentence the "Rumpdocks" to prison.

It's a complicated story that runs through a dozen states, Washington D.C., Paris, India, Ghana, and Liberia. I wrote a book about it with support from the John Simon Guggenheim Foundation.

So, of course, I'll drop the foul Mrs. Rumpdock and members of her family into black holes.

The images at the beginning of each unit

Most are just for fun, to provide visual relief from the arid mix of text, diagrams, and equations. Many of them are from France, where my family and I have lived from time to time.

But the Rose/Sagittarius graphic is different.

Sagittarius, the archer, is often associated with Chiron, the noblest of centaurs in Greek mythology. Chiron was a healer and a

---

[5]See, for example, *United States of America vs. Dixie Ellen Randock*, "Declaration of Brian Breen in Support of Objections to Presentence Report," June 5, 2008. [[00-004]].

[6]Index Mundi: http://www.indexmundi.com. Data are compiled from *CIA-The World Factbook*. In 2005, Liberia's life expectancy at birth was 38.89 years and its infant mortality rate was 16.2%. [[00-006]] [[00-005]].

[7]*Op. cit.*, *United States of America vs. Richard John Novak*, "Plea Agreement," March 20, 2006. [[00-001]].

[8]"Illinois Professor Shuts Down His Web Site on 'Diploma Mills'," *Chronicle of Higher Education*, Andrea Foster, October 24, 2003.

teacher, and his life was threaded with tragedy. John Updike's third novel, *The Centaur*, is a lovely, sad book that won a National Book Award. It weaves the myth's storyline into a modern setting. So, Chiron, especially in his role as teacher, has always been an important personal symbol for me.

The connection to general relativity is accidental but delightful: the massive black hole at the center of our galaxy is in the constellation Sagittarius.

Now, for the rose. My marvelous daughter is named for two of her great-grandmothers, Cordelia and Rose. If you remember Shakespeare's tragedy *King Lear*, Cordelia is Lear's strong-willed youngest daughter, and the only one of his children with integrity. Late in the play, she raises an army in an ultimately unsuccessful attempt to protect her father from his malevolent oldest daughters.

My grandmother Rose earned a law degree in 1914 (so early!) from the Brooklyn College of Law and, though she never practiced, was an advocate for separate courts for women in order to provide them with fair trials.

So, the connection to Cordelia, the strong woman who protects her father, through the image of a rose, has long provided me with a protective talisman. For years and years, every printed circuit board I've designed for my experiments has included a rose graphic in its silkscreen layers.

In these days of rising artificial intelligence tools, I thought it would be fun to ask ChatGPT to design an image that included a reference to both Sagittarius and a rose. And after only a few tries, it offered me the image that is on the cover of this book.

Index

Supplementary Material

Several examinations from Physics 225 are available online as supplementary material. For more detail, please visit https://www.worldscientific.com/worldscibooks/10.1142/14266#t=suppl

9 789898 198123 94